Careers
in Power Engineering and Boiler Operation

Your Guide to a Secure Future

by
Stanley Douglas Guiling

RENNER LEARNING RESOURCE CENTER
ELGIN COMMUNITY COLLEGE
ELGIN, ILLINOIS 60123

First Edition

PowerPlant Press
P.O. Box 431219
Pontiac, MI. 48343

**Renner Learning Resource Center
Elgin Community College
Elgin, IL 60123**

DISCLAIMER

This book is intended as an informational guide only. Every effort has been made to ensure that the information presented in this book is accurate, but no guarantee of accuracy is stated or implied. The author and the publisher have no liability and cannot be held responsible by any person or entity for any misunderstanding, misapplication or misuse of the information presented in this book. The author and the publisher cannot be held responsible for any personal injury suffered by an individual as a result of using this book. No guarantees of employment are implied by this book. The author and publisher are supplying information but are not supplying career counseling or other professional services. Assistance of appropriate qualified professionals should be obtained if needed. Anyone not wishing to abide by this written agreement can return this book for a complete refund.

Careers in Power Engineering and Boiler Operation

Your Guide to a Secure Future

© 1993 by Stanley Douglas Guiling

All Rights Reserved

No Portion of this publication may be reproduced or distributed in any form, written or electronic, nor stored in a database or retrieval system without the express written consent of the author

Guiling, Stanley Douglas
Careers in Power Engineering and Boiler Operation
1st Edition
Includes Index
ISBN 1-881861-20-1
1. Steam engineering I. Guiling, Stanley Douglas, 1952-
II. Title
TJ275.G   1993
621.8'93

SAN: 297-9306
Library of Congress Card Number 93-092661

331.7621
G956c
101743

Dedicated to my wife Sandi, without her help this project would not have been possible

Stephanie

&

Scott

Table of Contents

Chapter 1..1
Your Career as a Power Engineer

Chapter 2..34
Mounting the Job Campaign

Chapter 3..40
Wage Rates Around the U.S.

Chapter 4..46
N.I.U.L.P.E.

Chapter 5..68
The International Union of Operating Engineers

Chapter 6..72
License Requirements for States and Cities

Chapter 7..520
Regulations for Canadian Provinces and Territories

Chapter 8..790
Schools Teaching Power Engineering

Chapter 1

Your Career as a Power Engineer

This is a book designed for you. A book designed for people trying to better themselves. This book will tell you how to acquire the skill level necessary to get a stable, productive job as a Boiler Operator. Once you have achieved the goal of becoming a Boiler Operator, this book tells how to pursue the path to becoming a highly paid Power Engineer.

Having never met anyone who said "When I was a child I dreamed of being a Boiler Operator", you might wonder why this is a profession you would be interested in. If you are interested in a career where you are among the first to be hired, the last to be laid off and in the upper third pay scale of the company, Boiler Operation and Power Engineering can be your ticket to financial security.

What is Boiler Operation and Power Engineering? A Power Engineer is defined as one skilled in the management of energy conversion.[1] Simply stated, it is the operation and maintenance of steam and hot water boilers, steam turbines, internal combustion engines, motors, pumps, water treatment apparatus, large refrigeration equipment, and all their related auxiliaries. This equipment is found in small to large factories and large office buildings. Power Engineering is a more descriptive term than Stationary Engineering, although they mean the same thing. The term Stationary Engineering is used to de-

---

1   Peter Burno, President N.I.U.L.P.E.

Your Career as a Power Engineer	Chapter 1

scribe Power Engineers who operate steam boilers in factories, and to separate Stationary Engineers from Marine and Locomotive Engineers. Marine Engineering and Locomotive Engineering are terms used to describe the operation of steam boilers and steam auxiliaries on ships and steam locomotives.

This book will answer the following questions:

- Who can become a Boiler Operator?
- What is a boiler and what does a Boiler Operator do?
- Why are Boiler Operators necessary?
- How can I get the training necessary to become a Boiler Operator?
- Why are Boiler Operators licensed?
- Where do I go to take the license examination?
- When are the boiler examinations given in different cities?
- How much money can I make as a Boiler Operator?

Who can become a Boiler Operator? Anyone who has good sight, hearing, and is in good physical condition. You need good eyesight to judge the water level in the boiler at a distance. You will become familiar with the sounds in the boiler plant and a difference in sound will be your first warning of something amiss. Because of the climbing and physical tasks that must be performed in the boiler room, a person in good physical condition is necessary. You should have an interest in how things work. A certain amount of mechanical aptitude is necessary. No one should enter a trade they are not interested in, otherwise they will become bored with it and not put forth an honest effort to learn about the trade. There are no gender or

Chapter 1   Your Career as a Power Engineer

age barriers to prevent anyone from becoming a Boiler Operator. While this trade has traditionally been male, many women are now pursuing careers as Boiler Operators and Power Engineers. Many people in their 50's have entered the trade of boiler operation.

What is a boiler? The gas hot water heater in a house is a small type of boiler. It is an enclosed pressure vessel in which water is heated to provide a useful service, namely hot water for bathing or doing dishes. Fuel is burned on one side of the pressure vessel, and the heat is transferred through the steel to the water on the other side of the pressure vessel. All boilers, regardless of size, operate on the same principle. Many houses have "hydronic" heat. That is a fancy way of saying that water is heated in a boiler and flows through pipes that form a loop from the boiler, around the house and back into the boiler to be heated again.

If the boiler is designed to create steam, then the steam is used to do the heating. After it has heated the house through steam radiators, the steam turns back into water to be reheated in the boiler.

What is a Boiler Operator? A Boiler Operator is a skilled professional who ensures the steam or hot water needed to heat large buildings is available when needed. They manage the steam output in factories so the plant is ready for full production as the work day begins. Boiler Operators take care of the equipment and auxiliaries needed to keep the steam plant in top running order.

Why is a Boiler Operator necessary? After all, you may have a small boiler in your house and you do not need a Boiler Operator.

The boilers that require Operators are much larger in size. They heat large office complexes or factories. These are boilers large enough to power steam presses or generate electricity. Their controls are more sophisticated than the home hydronic boiler so they can burn fuel more efficiently. They usually run at higher pressures than the home boiler.

<u>Why are Boiler Operators licensed?</u> Most professions are licensed to assure the public that the person is qualified to perform his craft. Similarly, Boiler Operators and Power Engineers are licensed to ensure competency of the individuals licensed to operate boilers and related equipment. Steam boilers contain tremendous power and hold the potential for disaster if they are operated improperly. Licensing laws help insure the public safety and the continuation of power for the business that employs licensed Boiler Operators and Power Engineers. The insurance company that underwrites the policy on the boiler will give a discount if the boilers are cared for by licensed personnel.

<u>Where and when are the license examinations given?</u> The Code section of this book details all the necessary information about exam locations and times.

Steam is the transfer medium by which fuel such as wood, gas, coal, oil, or nuclear power can create electricity. The electricity that powers the electric lights and other appliances in the home comes from large electrical generating stations. Boilers ten stories high or higher provide steam to large turbine generators that create the electricity.

Some of the people who are Boiler Operators today started as Boiler Tenders or Machinists Mates in the Navy. If you are looking for a career after you get out of the military, this could be a good place to start. Most states or cities having boiler licenses will allow your Navy experience to account for the experience needed to write for a Third Class Power Engineer's License.

Let's take a minute and describe the licenses. Different states and cities have various licensing requirements for different grades of Boiler Operators and Power or Stationary Engineers. The licenses required in your area are described in this book.

These licenses vary widely from state to state. If some generalizations were to be made, the license structure and pay scales vary similar to this:

- Low Pressure Boiler Operator or Fireman .. $9 to $13 per hour
- High Pressure Boiler Operator or Fireman ...$10 to $14 per hour
- Third Class Power Engineer ....$12 to $15 per hour
- Second Class Power Engineer..................$15 to $20 per hour
- First Class Power Engineer ....$17 to $23 per hour

History of Power Engineering

Steam is basic power. Without steam the industrial revolution would never have occurred. Before steam all work was done manually or by beasts of burden. The vast majority of people lived on the farm and spent most of their time growing the food and tending to the livestock necessary to survive.

Steam boilers drove steam engines in England that pumped water from coal mines so that deeper shafts could be dug into the earth to get to the coal. The coal in turn was used in the great iron smelters of Bedlam to turn iron ore into the steel needed to build bigger and better steam engines and boilers.

These steam engines were used to run line shafts in factories. Line shafts were long rods that ran overhead the length of the factory. Pulleys were located along the rod above the mechanical equipment. Leather belts were attached to the line shaft pulleys to run mechanical looms that spun thread into cloth. Before steam engines, a hand loom had taken a person days to weave enough cloth to make one shirt. The mechanical loom could turn out enough cloth in minutes to make a hundred shirts. Low cost cloth created a demand for the goods and the textile industry was born.

Steam power allowed factories to be built at locations convenient for transporting the finished goods. When factories were run by water power, they were few in number and capacity due to the small number of sites that allowed dams to be built to generate power. The steam-powered factories were built on sites around the country in close proximity to each other. Cities grew up around these factories as people left the uncertainties of farming for a factory job. Factory jobs were not pleasant, but jobs were available and as demand for manufactured goods rose, so did the demand for people to manufacture them. Life on the farm was as equally harsh as life in the factory because the farm was subject to weather and pestilence that could wipe out a year's work in one day.

As factories were built, the first hired to operate the boilers and steam engines were the Boiler Operator and Power Engineer. The entire factory and by extension the industrial revolution could not have existed without them. They were entrusted to start the boiler, raise sufficient steam to operate the engine, and have the line shafts turning when the steam whistle blew to announce the beginning of another work day. If the steam plant broke down, the entire plant personnel would have to be sent home because there was no way to operate the machinery manually.

We have quickly gone from a dependence upon the muscles of man and beast to a dependence upon the power producing machinery of mans' imagination and determination. We have passed the point at which brute force is the measure of a human being into an era where the genius of the mind can finally be turned into tangible products that can be mass produced and made available to the common man.

It was steam engines that drove the metal lathes and shapers that created the first electric dynamos in 1882 and then ran those dynamos to create electricity to power Edison's light bulb. The electric motor eventually took the place of the steam engine in factories, but the power that drives them still comes from large steam turbines at over 3,000 generating plants in the United States alone.

So great was the change in America due to Edison and the production of electricity that when he died in 1931 President Hoover refused to shut down the electrical plants for one minute to honor Edison for the difference that he had made in American life. Electricity was considered too vital to shut off.[2]

---

[2] Ronald W. Clark, " Edison, The Man Who Made the Future," G. P. Putnam's Sons, New York, 1977

The science of boilers and boiler operation gained importance in the early 1900's. Poor boiler design and operator error caused numerous deaths and injuries. Mechanical Engineers of that era were concerned mainly with developing safer boilers and operating practices. This movement led to the formation of Codes for the construction of boilers by the American Society of Mechanical Engineers. These Codes are still the basis for the construction of boilers today.

The Job Market Today

Everyone has seen the effects of the greed of corporate America in the 1980's and 1990's. Unscrupulous individuals used borrowed money to acquire corporations, sell off the assets, lay off the workers, and shut down the company. With the decline of the industrial jobs in America, the high-paying low-skill jobs are being quickly eliminated. This is the reality of the situation. Just as a company and the head of that company look out for their own best self-interests, you have to look out for your own best self-interests. Start thinking of yourself as your own company. You have liabilities (your car payment, house payment, groceries, etc.) and you have assets (your skills that you are willing to sell to the highest bidder). The better your skills, the higher the bidding for your services. Once your assets outweigh your liabilities you will start gaining the financial security you deserve.

God has given you a certain amount of time to spend on this Earth. Spend your time wisely. Use your time working for positive change. You can change your future. You can stop having events overwhelm you and instead make positive decisions that will enable you to overcome your circumstances.

Consider two imaginary individuals, one called Joe and the other called Jack. Both were told that the factory they worked

Chapter 1   Your Career as a Power Engineer

in was about to close. Joe blamed the company and the union for the loss of his job. Joe took all the unemployment the State had to offer and after he was out of benefits he started looking for work. To his dismay Joe could find no work for the hourly rate he earned working for the factory. Not only was the pay for the jobs he could find in the $7.00 to $8.00 an hour range; there were no medical benefits. Joe continued to complain about his life's misfortunes and the world situation in general and lived unhappily ever after.

Jack also got his layoff notice but he immediately went to the schools in his area that offered training in Boiler Operation that qualified him to write for a High Pressure Boiler Operator's license. After he passed the course, he took the license examination and passed. He got a job in a laundry doing repair work and operating the boiler for $10.50 per hour. He continues to study and will soon be eligible to take his Third Class Power Engineer's exam. He is looking forward to getting higher licenses and higher paying jobs as a Power Engineer. He is anticipating a bright future where he is in demand by employers needing skilled people to run and maintain their boiler rooms.

While the particular individuals in this example may be fictitious, their response to circumstances is not. You have met people such as Joe and Jack.

A skilled trade beats a college degree any day of the week for employment security. Today white collar as well as blue collar workers are getting layoff notices. Middle to high level managers are getting cut from their jobs. They are trying to find similar positions with their college degree and their track record of managing people and finding that no one is interested. The companies they are applying at are laying off peo-

Your Career as a Power Engineer   Chapter 1

ple in these same positions. Knowing what Shakespeare said may help you pass a college exam, but it does not do much for a company looking to survive in tough times. Boiler Operators and Power Engineers are better career choices than most other skilled trades for the following reasons:

- Many of the skilled trades such as carpenter, sheet metal worker, electrician, pipefitter, or painter are construction oriented. If you work construction you fall victim to the economy quicker than anyone. When companies start losing money the first thing they do is stop any new construction plans they were contemplating.

- In the northern climates, most construction work slows down in the winter, leading to layoffs.

- The high paying union jobs in the electrical and pipefitting trades are controlled by the union. It is hard to get into the union, and if you succeed, your lack of seniority will guarantee you a quick layoff if the economy takes a downturn.

- The electrical union frowns on its workers doing outside work, making you dependent upon the union for steady employment.

- In the construction trade the benefit package you are offered may not be as good as one offered to a Power Engineer that works for a company.

- The construction trades are out in the elements until the shell of the building is completed and the windows are installed.

Chapter 1   Your Career as a Power Engineer

Boiler Operation and Power Engineering encompass many different types of jobs. You could be a Boiler Operator whose job is in the boiler room. They watch the boilers and make sure all the auxiliary equipment needed to keep the boilers running is maintained. They take hourly readings of the equipment to make sure there are no unusual changes in equipment operation since the last set of readings were taken. Boilers require water of a higher purity than the water you drink. The Boiler Operator takes water tests to ensure the quality of the water supplied to the boilers. The operator adjusts the water purifying equipment to make sure that the water quality is maintained, adding chemicals as needed.

In a large utility plant generating electricity many Boiler Operators and Power Engineers share the responsibilities. A turbine crew is assigned to make sure the turbine is running normally. A pump crew handles the operation and preventative maintenance of the many pumps that a large powerplant requires. A fuel handling crew in a coal fired facility unloads the coal from the barges or rail cars and then compacts the coal so that air does not infiltrate to ignite the coal from spontaneous combustion. They load the coal onto conveyors that transfer the coal into the boiler room for burning in the boilers.

Air compressors for factories are also located in the boiler room and become the responsibility of the Boiler Operator.

Large office buildings and modern factories usually have large centrifugal chillers for air conditioning that cool the building in the summer. The Boiler Operator is required to operate the refrigeration equipment and the boilers. Depending upon the city or state, this may require an additional license to operate the cooling equipment.

Your Career as a Power Engineer Chapter 1

Building Engineers are Boiler Operators and Power Engineers that maintain and operate the heating and cooling equipment in these buildings. In modern buildings this equipment is controlled by a computer, so some computer knowledge is necessary. In addition, they repair air conditioning, electrical, and plumbing equipment. This is the largest growing job market for Boiler Operators and Power Engineers. Manufacturing plants are closing while new high-tech office buildings are being built.

The generation of steam and electricity for a factory or hospital is known as cogeneration. Many new facilities are turning to cogeneration as a way of saving energy costs. This will mean more jobs are available in Power Engineering as these facilities are built.

No job is without its bad points. Shift work is the rule rather than the exception in boiler operation. Rotating shifts are common. You may work alone in some boiler rooms depending upon the shift. Because there may only be four Boiler Operators to share the work load in a plant operated 24 hours a day 365 days a year, overtime and Holiday work is necessary. Depending upon your point of view, this can be good or bad. The money is great but time off is important too. Boiler operation still remains one of the best professions to enter.

By pursuing the goal of becoming a Boiler Operator and Power Engineer, you can achieve a stable career with a rewarding financial future.

GETTING INTO THE PROFESSION OF BOILER OPERATION

Chapter 1   Your Career as a Power Engineer

You may be asking; how do you get one of these jobs? Start by telephoning all the different plants, companies or buildings where you think you would like to work. Ask to speak to the Chief Engineer or Facilities Manager. Get the correct spelling and pronunciation of their name and direct line phone number. Tell the Chief Engineer that you are interested in becoming a Boiler Operator and ask his advice on the best schools to attend. You may also inquire on how he entered the profession and what duties are required for becoming a Boiler Operator. Ask how many Boiler Operators he has working under him and what duties they perform. After you get the information, ask for his mailing address and send him a thank you letter.

You want to find the best schools that teach what you need to know to pass the boiler license examination. Ask about employee turnover and what the Chief Engineer looks for in people that he hires. Look in this book and find the closest city that has a license law. Talk to the License Examiner in that city and ask him the schooling requirements that he will accept before he will allow you to write for the exam. Be very pleasant to the Examiner as he holds the key to your future in his hands. If his city requires experience before they will allow you to write for an examination, ask him if he could suggest any companies that might hire someone without experience and whose boilers are big enough to qualify to write for the examination. If he does not know any companies that hire inexperienced people, ask where most of the people coming in to be examined are working.

After conversations with many Chief Engineers and Boiler Operators you should have a clear idea of the best schools to attend. Find out their schedule and when to apply. Register and get into the school. If they offer a one year course that allows you to write for a High Pressure Boiler Operator's license, talk

13

to the administrator of the school and see if there is anything you can do to help secure your position in a class.

If there are no classes in your immediate area, ask a community college close to you to start one. If enough people call, the college is sure to start a class.

Janitorial and maintenance jobs working around boilers are another way of getting the experience necessary to write for the boiler license examination. Taking a low paying janitorial position is not the quickest or best way to get into the boiler operating trade, but the jobs are easy to get and you are not planning on staying in that position for long anyway. Make sure before you take this position that the person in charge will write you a letter verifying your experience working around the boilers. Maintenance around boilers means cleaning the boiler once a year for inspection or more often if required. The fire side or the side of the boiler that has the products of combustion must be brushed and vacuumed. The water side must be cleaned with mechanical scrapers if the water treatment has been inadequate.

Maintenance jobs are a step up the ladder toward becoming a Boiler Operator. For Building Engineer positions, it is almost a necessity. You take care of the mechanical equipment of buildings. This includes lighting, electrical, plumbing, carpentry, locks, and any other repairs. The wider the variety of jobs you can handle, the more valuable you become to a company.

The best method to becoming a Boiler Operator or Power Engineer is through an apprenticeship program. An apprenticeship is a four year program that allows the student to write for a First Class Power Engineer's License and a First Class Refrigeration Operator's License after satisfactory completion. You get paid for the hours you spend in school as well as the

Chapter 1   Your Career as a Power Engineer

hours that you work for the company that is a participating sponsor of the apprenticeship. You pay for your tuition which is usually very nominal. These are excellent programs. The only trouble with apprenticeships is that they are hard to find and even harder to get into. The International Union of Operating Engineers in certain cities may have an apprenticeship program, and their numbers are listed in this book.

National Association of Power Engineers

Join the National Association of Power Engineers in your area. A license is not required. The local chapter can be found by calling 413/592-6273. You will meet people who are working as Boiler Operators and Power Engineers who will help you pursue your career. The N.A.P.E. is an educational association dedicated to improving the skills of Boiler Operators and Power Engineers. They conduct classes in all phases of boiler operation. You may have heard the term "Networking" as a way of finding a job. Networking is nothing more than meeting people who may be in a position to help you get a job. Cultivate their friendship and ask to tour the plant they work at. Get involved in your local chapter of the N.A.P.E. and hold an office if possible. The N.A.P.E. is a quality organization. The officers of the N.A.P.E. are unpaid volunteers whose only interest is in the furtherance of the Power Engineering craft.

History of the National Association of Power Engineers.

The National Association of Power Engineers is the oldest engineering society in America. The National Association of Power Engineers was started in 1879 by Henry Cozens. Henry Cozens was Chief Engineer of the Providence, Rhode Island Courthouse. This was a time when powerplants were being installed at a faster rate than qualified personnel could be found to run them. Direct current dynamos and alternating current generators were new inventions that were not fully understood by the people hired to operate them. Henry Cozens saw the need for an educational association that would allow power engineers to share information about the powerplants they operated. His engineering association in 1882 placed announcements in trade journals calling for a National Association of Stationary Engineers. The first convention was held on October 25, 1882 in Pythagoras Hall in New York City. Engineers from Rhode Island, New York, Massachusetts, Delaware, Pennsylvania, Michigan and Illinois attended. The preamble of the constitution set forth the goals of the association. "This Association shall not at any time be used for the furtherance of strikes, or for the purpose of interfering in any way between its members and their employers in regard to wages, recognizing the identity of interests between employer and employee, and will not support any project or enterprise that will interfere with harmony between them. Neither shall it be used for political or religious purposes. Its meetings shall be devoted to the business of the Association, and at all times preference shall be given to the education of Engineers and securing the enactment of Engineers' license laws to prevent destruction of life and property in the generation and transmission of steam as a motive power."[3]

National Association of Power Engineers members included Herbert Hoover, Dean Potter of Purdue University, Dean Woolwrich of the University of Texas, Senator and former Astronaut John Glenn, Charles Kettering of General Motors, and Henry Ford, founder of Ford Motor Company.

3   NAPE History

List of Active Chapters of the National Association of Power Engineers:

Connecticut

    Chapter #2 Stamford

    Chapter #10 New Haven

    Chapter #11 Meridan

District of Columbia

    Chapter #1 Washington, D.C.

Florida

    Chapter #4 Miami

Georgia

    Chapter #2 Atlanta

Illinois

    Chapter #1 Chicago

    Chapter #6 Joliet

    Chapter #49 Elgin

    Chapter U.S. Navy #1 Great Lakes

Indiana

    Chapter #1 Muncie

    Chapter #3 Terre Haute

    Chapter #4 Indianapolis

    Chapter #7 Evansville

    Chapter #8 Kokomo

Chapter #16 Fort Wayne

Iowa

Chapter #2 Des Moines

Maryland

Chapter # 5 Wheaton

Chapter #13 Pasedena

Chapter #14 Prince Georges

Chapter #20 Pax River

Massachusetts

Chapter #4 Springfield

Chapter #6 Fall River

Chapter #17 Lowell

Chapter #32 Worcester

Michigan

Chapter #1 Detroit

Chapter #12 Kalamazoo

Chapter #23 Flint

Minnesota

Chapter #2 Minneapolis

Chapter #4 Winoma

Chapter #12 Queen City (Rochester)

Chapter #18 St. Cloud

Missouri

Chapter 1 — Your Career as a Power Engineer

Chapter #2 St. Louis

Chapter #3 Kansas City

Chapter #4 Columbia

Nebraska

Chapter #1 Omaha

Chapter #2 Lincoln

Chapter #5 Norfolk

Nevada

Chapter #1 Las Vegas

New Hampshire

Chapter #1 Manchester

Chapter #3 Keene

New Jersey

Chapter #1 Central New Jersey

Chapter #6 Bergen

New York

Chapter #1 New York City

Chapter #3 Flower City (Rochester)

Chapter #6 Nassau County

Chapter #7 Westchester

Chapter #24 Phoenix

Chapter #37 Niagara Falls

Chapter #38 Southern Tier

Chapter #44 New York City
Ohio
Chapter #36 Cincinnati

Chapter #38 Columbus

Chapter #519 Cleveland
Pennsylvania
Chapter #18 Pittsburgh
Rhode Island
Chapter #1 Providence
South Carolina
Chapter #3 Florence

Chapter #7 Upstate
Virginia
Chapter #1 North Virginia

Chapter #2 Richmond

Chapter #92 Central
Wisconsin
Chapter #2 Sheboygan/Manitowac

Chapter #14 Madison

Chapter #15 Green Bay

Chapter #17 Union Grove[4]

---

[4] Sarah Larkin, President of NAPE supplied the listing of current NAPE Chapters

Chapter 1   Your Career as a Power Engineer

BOILER EXAMINATIONS

Let's describe how a Boiler Operator's examination is conducted for the City of Detroit. First, call and ask the Examiner to mail you an application for the High or Low Pressure Boiler Operator's Exam or go downtown to the City-County Building and ask for both items in person. After you have filled out the application, you must have letters from all the companies you worked for or a completion certificate from a recognized school that will qualify you to write for the license. The letters from each company must be current and signed by the Chief Engineer, the Director of Personnel, and be notarized. It cannot be emphasized enough the importance of getting letters that describe your duties so you can use them for your license examinations. If you wait, the company that you worked for may have gone out of business and there will be no way to prove you worked around the boilers there. Ask for the letters while you are on good terms with the company. If you quit and come back asking for a letter, they may not provide it for you.

After you have filled out the application you must have your signature notarized. This means having a Notary of the Public verify the signature on the application belongs to you. The Notary of the Public closest to you can be found in the yellow pages of your phone book. The cost should be from $2.00 to $5.00. Many banks offer this service for free if you have an account with their bank.

After you have filled out the application and received your letters of work experience and certificates of completion for schooling, take them to the examiner in person. He will check to make sure everything is in order and set an examination date. The Examiner will keep the originals of your work experience. Make sure you have photocopies of the letters of work experience for your records before you apply for the license. When the examination date arrives, give yourself plenty of time so that you are not late for the appointment. Bring any other documentation that the Examiner has asked you to bring. After paying the clerk for the examination, the Examiner will give you sheets of blank paper with questions about boilers and auxiliaries that you will have to draw and explain. After you have drawn the boilers and auxiliaries the examiner has required, you will be given an oral examination that will last from one to three hours depending upon how readily you can answer his questions.

The examiner realizes that some people are nervous taking exams and will try to ask a question in several different ways if you cannot answer the question correctly. Do not try to bluff the Examiner with nonsense. If you do not know the answer, just say so. Making up an answer is a sure ticket to failing the exam. The Examiner's job is to ensure that incompetent people are not given a license to operate a boiler unsafely that could endanger the lives of others.

If you pass the exam you will pay the clerk for the license and be on your way. If the Examiner states that you have not answered the questions properly, you will have to wait 90 days before you qualify for re-examination. Failing an exam is nothing to be ashamed of. Many people have not passed every test in school so do not become discouraged if you do not pass the boiler examination on the first try. Some people get their egos hurt by failing the exam and never go back. The only person they hurt is themselves so do not fall into that trap.

Do not become impudent with the Examiner. An Examiner was giving an exam to an individual when another fellow

Chapter 1   **Your Career as a Power Engineer**

barged into the examining room. He threw down his application on the Examiner's desk and told the Examiner "You might as well sign this now. I can answer any question you've got." The Examiner took the application, calmly tore off the name of the applicant and placed it under the glass on his desk. He handed the application back to the intruder and said "You come back in 90 days. I'll remember you." The stunned applicant then sheepishly turned and left the examination room.

Passing the license examinations are a requirement in obtaining your goals of job and financial security. The proper attitude and study habits are essential requirements in passing the license examinations.

Setting Up the Career Path

You must have an organized plan of action to achieve any goal. What follows is an organized schedule that keeps you on track to become a First Class Power Engineer.

(1) Set up a career path for yourself.

(2) Define your goals and set deadlines for yourself.

(3) Break each step into a series of smaller steps.

(4) Visualize yourself as having passed each step that is approaching.

(5) Do not let setbacks keep you from achieving the next step. If you fall down, pick yourself back up and keep moving toward the goal. If you watch a young child learning to walk they fall down many times. They continue to get back up and keep trying and this is what you must do.

(6) You can achieve your goals if you keep a positive attitude and do not let anyone discourage you.

(7) Higher licenses are important. Do not let anyone discourage you from getting higher licenses.

(8) Lastly, be grateful for all that you do have. Remember that in the past 6,000 years of recorded history, only in the last 100 years have people had automobiles, electric lights, air conditioning, and indoor plumbing. Some people tend to concentrate on what they do not have, making themselves and everyone around them miserable.

Career Path

1st Year

1st Step

Call Chief Engineers and License Examiners and ask them what they think are the best schools to learn Boiler Operation. Ask them if they know any places that hire part time workers around boilers.

2nd Step

Apply at all locations that have boilers and have jobs that will allow you to write for licenses. Apply at hospitals, laundries, private schools, public school systems, rest homes, shopping malls, and large apartment complexes. Subscribe to Power Magazine and Power Engineering. If you ask for the magazine as a Power Engineer, the subscription may be free.

Chapter 1 — Your Career as a Power Engineer

Power Magazine
Subscription Department
P.O. Box 521
Hightstown, NJ 08520-0521
609/426-5667

Power Engineering
P.O. Box 1440
Tulsa, OK 74101
708/382-2450

When you start getting Power and Power Engineering magazines fill out the reader service cards to get the free brochures from different manufacturers. Get a small two drawer filing cabinet with hanging files to file your brochures as they arrive.

Call the technical schools in your area and ask if they have a one year boiler course that is acceptable to write for a license. Some of those schools are listed in the back of this book.

3rd Step

Sign up for school. Buy "Practical Boiler Operation, The Guide to License Examinations." An Order Blank is on the last page of this book. Write down the names and phone numbers of other students in class. Talk to the other students at break periods and find out where they are working and if they have boilers there. If they do, ask to visit their plant when the Chief Engineer is there. Find out if they allow you to take pictures. Have a tour sheet similar to the one in figure #1. Write all the pertinent information about the equipment in the plant on the tour sheet. This will help you in school and will also help you if you apply for a job at that plant later. You will im-

Your Career as a Power Engineer   Chapter 1

press the Chief Engineer with how much you remembered about his plant.

Plant Tour Sheet

	Boiler 1	Boiler 2	Boiler 3	Boiler 4
Manufacturer				
Boiler Type				
Date Built				
Maximum Allowable Working Pressure				
Capacity in Pounds of Steam Per Hour				
Fuel(s)				
Fuel Delivery System				
Safety Valve Manufacturer				
Safety Valve(s) Capacity				
Pump Manufacturer and Type				
Number of Pumps Total				
Pump Motor Manufacturer and Type				
Feedwater Turbine Manufacturer and Type				
Injector Manufacturer and Type				
Burner Manufacturer and Type				
Burner Control Manufacturer and Type				
Type of Flame Sensing Device				
Forced Draft Fan Manufacturer and Type				
Induced Draft Fan Manufacturer and Type				

Figure 1

Chapter 1 — Your Career as a Power Engineer

4th Step

Find out who are passing the tests in school and cultivate their friendship. Start a study group and invite them to join. There should be six people maximum in this group. The rules of the group should be established from the onset. If any member fails to study and be an active member of the group then their place should be opened up for another. Give each individual a topic to research and questions to answer. Set a deadline of two weeks. Practice asking and answering questions with your group. This will help you when you talk to the License Examiner.

Take the examination and get the license.

Second Year

Start applying at all facilities that hire High Pressure Boiler Operators with boilers large enough that qualify you for higher licenses. Some of these are: paper mills, hospitals, factories, shopping centers, breweries, and public utilities. Drive around and look for tall chimneys; these may be facilities you can apply at. Call and ask for the name of the Chief Engineer. Call the Chief Engineer and ask for a tour. During the tour, ask his advice about getting higher licenses. After the tour write a follow-up thank you letter. Within two weeks send a resume to the head of personnel and one directly to the Chief Engineer. Be sure to spell their names correctly. Sign up at the unemployment office as a High Pressure Boiler Operator. Many of the better jobs are never placed in the want-ads but will be posted at the unemployment office through their job placement service. Keep copies of any letters of recommendation from previous employers and the certificates of completion from schools. They will be one of your greatest assets as you apply for jobs. Get a briefcase or a leather folder and put all your documents and updated resume together. When a good job is advertised in the paper you can quickly send them a resume. A sample resume is shown in figure #2.

Send your resume to employment agencies that have their fee paid by the employer. Do not waste time and money at employment agencies where you are expected to pay the fee.

Keep taking classes and meeting students that may be able to help you in the future. Get all the Sunday newspapers from the major cities in your area and look for job ads. Take your Third Class Power Engineer's exam.

Chapter 1   Your Career as a Power Engineer

Resume

John B. Goode

11000 Evergreen Lane
New Orleans, LA 70112
1-555-123-4567

Job Objective:

To advance my career in Boiler Operation, ultimately becoming a Power Engineer

Education:

1-1991 to present

Praline Parish Community College

200 Justin Lane

New Orleans, LA 70112

Currently completing courses in Power Engineering

9-1985 to 6-1989

New Orleans High School

300 Wilson Avenue

New Orleans, LA 70112

High School Diploma

Successfully completed High School with a 3.2 Grade Point Average

Experience:

1-1991 to present

Mechanical Maintenance at Serendipity Rest Home. Responsible for all heating, plumbing, and light electrical duties.

7-1989 to 1-1991

Assembly line worker at Global Motors, riveting truck frames at Commanchero truck assembly plant

Hobbies:

Playing guitar

References:

Gladly provided upon request

Figure 2

Third Year

Continue studying and taking classes. Start going on all tours of utility power plant boilers that you can. Your study group should have enough people to arrange a tour. If not, ask people in the school. Tape what the tour guide says. Take your Second Class License exam. Start taking more responsibility and find out what the Chief Engineer does. Learn how to write specifications for bidding and how to find vendors when parts are needed. Learn how a computerized preventative maintenance system works. Tackle any plumbing, pipefitting, welding, and electrical jobs that you can. Make yourself indispensable.

In Detroit, Michigan there is a two year wait between the Second and First Class examination. A lot of people lose interest in that two year waiting period and get out of the study habit. Now is a good time to get your refrigeration licenses. In Detroit the Third Class Refrigeration Operator's license takes three months experience. The First Class Refrigeration Operator's license requires either two years experience or one year of school and one years experience. Take one year of refrigeration school after you pass your Second Class Steam License. After you receive your certificate from going to school, take the Refrigeration Operator's exam. After passing the Refrigeration Operator's exam, go back to studying for your First Class Steam License.

Attending school for five years is not easy, but it is not as hard as you think. Once you set your mind on your goal, do not let circumstances sidetrack you. Five years from now you are going to be five years older whether you went through this program or not. A minister from the Urban Alternative of Dallas, Texas tells the story that before he left to go into the minis-

Chapter 1   Your Career as a Power Engineer

try he and his friends would hang around the street corner wasting time. Now he is the head of one of the largest churches in Dallas and has a nationwide radio program. When he goes back to his hometown to visit his parents, the same friends are still hanging around the same street corner he left years ago. It is easy to get into a routine of unhappiness. You think of things that might have been, but you do not pursue your dreams because you are afraid you might fail. Because you are afraid of failure you have already failed. You have a greater potential for success than you realize. You have the skills and talent that it takes to study and pass these examinations and get higher licenses. Do not let low self-esteem deprive you of the financial security that you deserve in these perilous economic times. Many people have problems with low self-esteem. It is not easy to hold your head up high when other people treat you unfairly. Just realize that they are trying to make themselves feel better at your expense. If they can put you down it makes them feel superior. Do not let them play these mind games on you. You have the same right to be respected as anyone else. Keep your eye on the goal of achieving financial security by being the best you can be at what you do.

Why place such emphasis on getting licenses? Because licenses are the yardstick that people are measured by when the high paying jobs are given to people. You may be the best Boiler Operator in the world but without a license you will have a hard time convincing anyone of that. Experience shows that people who operate boilers in states or cities without license laws have an elementary idea of what Boiler Operation and Power Engineering is all about. They may be fine in the plant they operate, following the same steps day after day, but if they had to get a job in a new plant they would be lost. By studying and passing license examinations you will have a skill that is not employer specific. You are no longer dependent upon one employer for your financial security.

Searching the Want Ads

Jobs for Power Engineers are in all the major metropolitan newspapers. Because of the different names that the jobs are called, there is no specific place in the want ads to turn to. Some of the different names for Power Engineers are:

- Air Conditioning and Heating Operators
- Assistant Chief Engineer
- Assistant Powerplant Engineer
- Boiler Operator
- Boiler Tender
- Building Engineer
- Chief Engineer
- Director of Plant Operations
- Engineer
- Facilities Engineer
- Fireman
- First Class Engineer
- Fourth Class Engineer
- General Maintenance
- Heating Plant Operator
- High Pressure Boiler Operator
- Low Pressure Boiler Operator

- Maintenance
- Nuclear Powerplant Operator
- Oiler
- Operating Engineer
- Physical Plant Operator
- Powerplant Operator
- Powerhouse Engineer
- Refrigeration Operator
- Second Class Engineer
- Stationary Engineer
- Steam Engineer
- Third Class Engineer
- Turbine Operator
- Watch Engineer
- Water Tender

Some of the ads for hospitals are in the medical section rather than the general or technical section. The better jobs are usually in display ads that can be placed anywhere throughout the want ad section.

Chapter 2
Mounting the Job Campaign

Seeking a job can be similar to mounting a military offensive. You have to marshall your resources together and overwhelm your opponents (other job seekers). You will do this by being better organized and being the best candidate for the position.

To assist in your campaign, an IBM compatible personal computer, word processing program, and a letter quality printer are recommended. Personal computers are now available for less than $1,000 and some of them come with word processing software already installed. Some printers are in the $300 range.

Why buy a computer? So you can print a resume that is directed toward the job being advertised. Each cover letter can be modified slightly to correspond closely to the job being advertised. Time is essential. With the basic resume and cover letter already installed, you do not have to type each one over and over again. Never send a photocopy of your resume. It looks cheap and sends that signal to the personnel department and the Chief Engineer. Buy expensive paper. A heavy paper with a color other than white, (cream, gray, natural) makes a good impression. There can be no misspellings in either your resume or cover letter. The spell checker on your word processor is a good first step, but also have someone proficient in English proofread them. Resumes should be one page. The purpose of the resume is to get a job interview. In the interview, you will fill in the gaps of education and work experience that you were not able to fit into your resume. There are

Chapter 2 — Mounting the Job Campaign

other reasons for using a personal computer rather than a typewriter. With a computer you can store your cover letters and resumes on disk. When a company calls you for an interview four weeks after you sent them a resume, you can look up the cover letter you sent them to refresh yourself on what you wrote. The world is moving toward personal computers more each day and proficiency with computers is becoming a requirement for employment.

It is a wise idea to make a folder with all the job ads stapled to a copy of the cover letter and resume for each job. Then you can refer to them as needed.

Have a leather folder or briefcase to keep your high school diploma, completion certificates from technical schools and licenses. If you have any letters of recommendation, these should also be placed into the folder. Then when you are called for an interview, you do not have to search to find everything.

Once you have mailed your resume, it is a good idea to call three days later to make sure the resume was received. Do not pin your financial future to the reliability of the postal service. If the personnel department says that they have your resume, tell them that you are very interested in the position and are looking forward to being interviewed. Do not call the personnel department again or else they may start thinking you are a pest. If you have not heard from the personnel department seven days from the time you mailed the resume, call and speak to the Chief Engineer. Ask if he has received your resume. If he has, tell him how interested you are in the position and that you are looking forward to discussing the position with him. Try to sound upbeat and enthusiastic. Do not call him again.

Before the interview spend some time at the library researching the company. What do they manufacture? Are they the second largest manufacturer of widgets in the U.S.A.? Finding out everything you can about the company can help you in the interview. By incorporating some of the facts you have learned about the company into the interview, you show that you are indeed interested in working for their company. Call ahead and know how to pronounce the interviewer's name before the interview. You may be interviewed by more than one person. Be sure to memorize their name. If it is hard to pronounce, immediately repeat it to the person to avoid embarrassment later.

Arrive at the interview fifteen minutes early. Make sure you go alone. For Boiler Operator and Power Engineering positions, a dress shirt, dress pants, and shined dress shoes are appropriate. A suit may make the Chief Engineer think that you are afraid to get your hands dirty. Your hair should be neatly trimmed. When you shake hands, look the person in the eye and give a firm but not bone-crushing handshake. Smile but do not laugh. Relax. Look interested. Lean forward in the chair as the interviewer in personnel or the Chief Engineer asks you questions. Use proper grammar and speak clearly. Answer the questions completely but do not volunteer answers without being asked. You have to be a salesman. You are selling yourself as the best candidate for the position.

Emphasize what you can do for the company as opposed to what you want the company to do for you.

Do not deal in negatives. No matter how bad your last job was, relate the positive aspects of the job and do not mention any ill will that may have occurred. Do not ask about salary or benefits at this time. If the company is unionized, your pay is set and cannot be negotiated. If the company is not unionized, your salary is normally determined by what you were making at your previous employment.

After the interview, make sure you have the mail code and address for the interviewer in personnel and the Chief Engineer. Send them a thank you letter similar to figure #3 in appreciation for the time they spent with you and relate how interested you are in working for their fine organization. This is very important. They may have interviewed twenty people for the position and faces and names become blurred after awhile. Few people send follow-up thank you letters and it will help make a positive impression in the mind of the interviewers.

If they call to offer you the job; ask about salary and benefits. They will normally offer you the same or slightly more than you are currently making. If you are currently unemployed, you are not in a position to bargain. If the boiler equipment is large enough to allow you to write for higher licenses, it is best to take the position as a starting point to build upon. If you have a job as a Boiler Operator or Power Engineer currently, it does no harm to bargain with the personnel department to try and raise the salary. Keep a credit card calculator in your wallet. Memorize the fact that there are 2,080 hours in a year at 40 work hours per week. If you are used to hourly wages, an offer of $27,000 per year can be quickly divided by 2,080 to show that they are offering $12.98 per hour. A biweekly offer of $1,200 means that they are offering $15.00 per hour; 1,200 divided by 80.

It cannot be emphasized enough the importance of going on tours. You increase your knowledge of boiler operation, you meet the Chief Engineers and are able to make a positive impression on them, and you find out if any jobs are going to be-

Mounting the Job Campaign   **Chapter 2**

come available in the near future. Look on the license board where the all licenses are posted and note how many issues each Engineer has on his license. More than thirty issues may mean that an Engineer working there is close to retirement.

    A job search is just like any other job. You need to spend forty hours a week sending resumes and cover letters, researching companies, going for interviews, and following up with thank you letters. If you follow this formula and you have the training and licenses, multiple job offers can be your reward.

Chapter 2   Mounting the Job Campaign

April 1, 1993

Mr. Jack Steamhammer
Chief Engineer
South General Hospital
400 Bayou Street
New Orleans, LA 70112

Dear Mr. Steamhammer:

Thank you for the interview for the Boiler Operator's position at your fine Hospital on March 31st. I was impressed by the pride taken by your staff in the upkeep of the powerplant. Knowing that South General is a leader in health care only serves to increase my enthusiasm for the position you have available.

I look forward to talking with you further about the job and what I can offer your facility.

Sincerely yours,

John B. Goode

John B. Goode

Figure 3

Chapter 3
Wage Rates Around The U.S.

How much money can you make? What follows are the wage rates from around the United States for 1990 and 1991; the latest available as of this writing.

Average Hourly Earnings from the U.S. Department of Labor Wage Survey for Stationary Engineers in Selected Cities in the United States for 1990 and 1991

The * means 1991 rates.

Arkansas

Little Rock, Arkansas....................................$12.91*

Memphis, TN-AR-MS................................$14.44*

California

Los Angeles-Long Beach, California............$19.40

Oakland, California.......................................$19.77*

San Francisco, California.............................$21.32*

Stockton, California......................................$14.47*

Colorado

Colorado Springs, Colorado........................$10.35

Denver, Colorado...$15.96

Chapter 3 — Wage Rates Around The U.S.

Connecticut
Connecticut..$15.73

District of Columbia
Washington, D.C, MD-VA...........................$16.73*

Florida
Jacksonville, Florida......................................$17.00

Georgia
Atlanta, Georgia...$16.15*
Macon-Warner-Robins, Georgia...................$17.95
Chatanooga, TN-GA......................................$13.59, $14.17*

Illinois
Decatur, Illinois...$16.72*

Indiana
Indianapolis, Indiana.....................................$17.77*
Cincinnati, OH-KY-IN..................................$14.74*

Iowa
Des Moines, Iowa..$14.10*
Omaha, NE-IA...$11.68

Kansas
Kansas City, MO-KS.....................................$15.51*

Kentucky

Cincinnati, OH-KY-IN $14.74*

Louisiana

Lake Charles, LA-TX $17.04*

Maine

Maine $15.69

Maryland

Baltimore, Maryland $15.27*
Washington, D.C., MD-VA $16.73*

Massachusetts

Massachusetts (Southeastern) $13.25*

Michigan

Ann Arbor, Michigan $18.24
Detroit, Michigan $18.45

Minnesota

Minneapolis, St.Paul, MN-WI $15.64*

Mississippi

Memphis, TN-AR-MS $14.44*

Missouri

Kansas City, MO-KS $15.51*
St. Louis, Missouri $16.70*

Nebraska

Chapter 3 — Wage Rates Around The U.S.

Omaha, NE-IA.....................$11.68

Nevada

Las Vegas-Tonopah, Nevada.....................$18.95

New Jersey

Middlesex-Somerset-Hunterdon,NJ...........$15.49, 16.53*

Monmouth-Ocean, New Jersey.....................$17.37*

Newark, New Jersey.....................$19.08*

Trenton, New Jersey.....................$16.01, $16.83*

Philadelphia, PA-NJ.....................$15.05

New Mexico

Almogordo, New Mexico.....................$12.80*

New York

Albany-Schenectady-Troy, New York.........$15.02, $15.73*

Buffalo, New York.....................$16.83, $16.88*

New York City, New York.....................$18.41*

North Carolina

Durham-Raleigh, North Carolina.................$15.27*

North Dakota

North Dakota.....................$11.44, $13.11*

Ohio

Cincinnati, OH-KY-IN.....................$14.74*

Cleveland, OH.....................$15.37*

43

Lima, OH..$16.17*
Portsmouth-Chillicothe-Gallipolis, OH.........$15.39
Sandusky, OH...$16.18*
Toledo, OH...$16.72

Oklahoma

Altus-Lawton, Oklahoma............................$15.74

Pennsylvania

Philadelphia, PA-NJ....................................$15.05
Pittsburgh, Pennsylvania..............................$15.79*
Scranton, Wilkes Barre, Pennsylvania..........$13.76*
York, Pennsylvania.....................................$14.77*

Tennessee

Chattanooga, TN-GA...................................$13.59, $14.17*
Memphis, TN-AR-MS.................................$14.44*

Texas

Northwest Texas..$13.51
Beaumont, Port Arthur, Texas......................$17.04
Lake Charles, Louisiana-Texas....................$17.04*
Dallas, Texas...$13.11
El Paso-Las Cruces, Texas..........................$12.80*
Alamogordo, New Mexico-Texas................$12.80*
Fort Worth-Arlington, Texas.......................$16.86*

Houston, Texas..$14.73*

San Antonio, Texas.......................................$12.07*

Wichita Falls, Texas.....................................$15.74*

Vermont

Vermont...$11.37

Virginia

Newport News-Norfolk, Virginia Beach,
Virginia...$13.24*

Petersburg-Richmond, Virginia...................$16.12*

Southwest Virginia.......................................$14.52

Washington

Seattle, Washington.....................................$16.85

Wisconsin

Madison, Wisconsin....................................$14.72

Milwaukee, Wisconsin...............................$16.01*

Minneapolis-St. Paul, MN-WI...................$15.64*

* 1991 rates

Chapter 4
N.I.U.L.P.E
The National Institute for the Uniform Licensing of Power Engineers

Certain parts of the United States do not have license laws for Power Engineers. As you read the codes included in this book, you will notice the vast differences in qualifications required and the different grades of licenses throughout the United States for those cities and states that license their Power Engineers. Because of the lack of licensing in certain parts of the United States, and the lack of a uniform method of judging the competence of engineers on a nationwide basis, members of the National Association of Power Engineers started an agency in 1972 which set a national standard. That agency is the National Institute for the Uniform Licensing of Power Engineers (N.I.U.L.P.E.). It is the only license that applies on a national level for Power Engineers.

Classification of N.I.U.L.P.E. Licenses

There are five levels of licenses, listed from lowest to highest:

1. Fourth Class Engineer

2. Third Class Engineer

3. Second Class Engineer

4. First Class Engineer

5. Chief Engineer

Chapter 4 — Classification of N.I.U.L.P.E. Licenses

the N.I.U.L.P.E. program has Examiners in the following states: California, Connecticut, District of Columbia, Georgia, Illinois, Iowa, Maryland, Massachusetts, Michigan, Minnesota, Missouri, Kansas, Nebraska, Nevada, New Hampshire, New York, North Carolina, North Dakota, Ohio, Pennsylvania, Rhode Island, South Carolina, Virginia, Washington, Wisconsin and USAHSC (U.S. Army).

States in which the N.I.U.L.P.E. program is active, but there are also state license laws are: the District of Columbia, Maryland, Massachusetts, Minnesota, and Ohio.

States not having their own licensing examination system will recognize the N.I.U.L.P.E. license.

N.I.U.L.P.E. LICENSE REQUIREMENTS

4th Class Engineer

- Minimum Age: 18
- Education Required: 2 years high school, 2 years approved apprenticeship, or approved on the job training
- Experience Required: 2 Years
- Examination Required: Oral, Written, Practical
- Maximum Engine (Prime Mover) Horsepower Requiring Licensed Power Engineer Unsupervised: 500 HP
- Maximum Boiler Horsepower Requiring Licensed Power Engineer Unsupervised: LP 150; HP 25 BHP
- Refrigeration License: Unsupervised: 100 tons
- Fees: Examination: $10; Renewal: $4

3rd Class Engineer

- Minimum Age: 20
- Education Required: High School Grad. or G.E.D., 3 years approved apprenticeship, or approved on the job training
- Experience Required: 3 years
- Examination Required: Oral, Written, Practical
- Minimum Time in Previous Grade: 1 Year
- Maximum Engine (Prime Mover) Horsepower Requiring Licensed Power Engineer Unsupervised: 1,000 HP
- Maximum Boiler Horsepower Requiring Licensed Power Engineer Unsupervised: LP Unlimited; HP 200 BHP
- Refrigeration License: Unsupervised: 500 Tons
- Fees: Examination: $15; Renewal: $5

2nd Class Engineer

- Minimum Age: 20
- Education Required: High School; 4 Years approved apprenticeship or approved on the job training
- Experience Required: 4 years
- Examination Required: Oral, Written, Practical
- Minimum Time in Previous Grade: 1 year
- Maximum Engine (Prime Mover) Horsepower Requiring Licenses Power Engineer Unsupervised: 2,500 BHP

| Chapter 4 | Classification of N.I.U.L.P.E. Licenses |

- Maximum Boiler Horsepower Requiring Licensed Power Engineer Unsupervised: 500 BHP
- Refrigeration License: Unsupervised: 1,000 tons
- Fees; Examination: $18.50; Renewal: $6

1st Class Engineer

- Education Required: High School; Technical School; Job Experience
- Experience Required: 6 years
- Examination Required: Oral, Written, Practical
- Minimum Time in Previous Grade: 2 Years
- Maximum Engine (Prime Mover) Horsepower Requiring Licensed Power Engineer Unsupervised: 7,500 HP
- Maximum Boiler Horsepower Requiring Licensed Power Engineer Unsupervised: 1,500 HP
- Refrigeration License: Unsupervised: 5,000 tons
- Fees: Examination: $20; Renewal: $7

Chief Engineer

- Education Required: High School; Technical School; Plant and Management
- Experience Required: 10 Years
- Examination Required: Oral, Written, Practical
- Minimum Time in Previous Grade: 4 Years
- Maximum Engine (Prime Mover) Horsepower Requiring Licensed Power Engineer Unsupervised: Unlimited

- Maximum Boiler Horsepower Requiring Licensed Power Engineer Unsupervised: Unlimited
- Refrigeration License: Unsupervised: Unlimited
- Fees: Examination: $25; Renewal: $8[5]

N.I.U.L.P.E. Examiners

The following examiners responded to letters of permission to print their names, addresses and phone numbers. These are unpaid volunteers who graciously donate their time and energy to further the Power Engineering craft.

California

Fagan, William L., C.P.E.
177F Riverside Dr.
Newport Beach, CA 92663
714/638-9808

Hughes, Robert J.
1436 Granada St.
Vallejo, CA 95491-7633
704/644-3747

Kelly, William, C.E.
2856 W. Lincoln Ave., L4
Anaheim, CA 92801
714/761-2814

---

[5] the N.I.U.L.P.E. Green Book and list of Examiners, courtesy James Westergard, Executive Secretary of N.I.U.L.P.E.

Ward, Jack D.
8711 Chester St.
Paramount, CA 90723
310/633-0409
Chr-N.I.U.L.P.E.-Exam. CA.
Pres: Chap #3 CA. N.A.P.E.

Connecticut

Cleroux, James B.
8 West Drive
East Hampden, CT 06424
203/267/2658

Gosselin, Ralph R.
94 Chapman Rd.
Marlborough, CT 06447
203/295-0674

District of Columbia

Brooks, Ronald G.
7P Research Rd.
Greenbelt, MD. 20770

Mitchell, Robert P.
7410 Garrison Rd.
Hyattsville, MD 20784
301/577-3083

Peevy, Orion J.
4071 Kassel Circle
Dale City, VA 22193
703/590-3593 (Home)
202/371-9342 (Work)

Sohns, Helmut
7706 Bristow Dr.
Annandale, VA 22003
703/941-1106 (Home)
301/495-4440 (Work)
Past National President, NAPE

Wagner, Robert J.
9700 Galsworth Court
Fairfax, VA 22032-2802
202/783-0601
Pres. NAPE DL-1
Bd. Memb. NIULPE

Yochum, John P.
4141 N. Henderson Rd.
Arlington, VA 22203
703/522-2284

Georgia

Arnold, Jack W.
11100 Bowen Road
Roswell, GA 30075-2241
404/998-1002 (Home)
404/552-7046 (Work)
404-641-0842 (Fax)
Sec. GA. Bd. of Exam.-NIULPE

Barr, Walter L.
P.O. Box 715
Social Circle, GA 30279
404/242-2188

Floyd, Larry M.
1526 Brookstone Dr.
Graniteville, SC 29829
803/593-5765

Chapter 4   N.I.U.L.P.E. Examiners

Hatchett, Monroe T.
80 Osner Drive NE
Atlanta, GA 30342
404/256-2705

Myrick, Jerry E.
45 Old Alcovy Road
Covington, GA 30209-9370
404/786-4955
GA V.P.

Staulcup, James O.
728 Deerwood Place
Evans, GA 30809
706/863-2187 (Work)
706-791-5927 (Home)
Pres. GA.-State Exam Board-NIULPE

Illinois

Bordeau, Rich
2072 Lawson Blvd.
Gurnee, IL 60031
708/249-0223

Keaty, Tom
7625 W. 157th Place
Orland Park, IL 60462
312/532-3941

Klima, Dr. Karel M.,P.E.,CEM
Program Manager
US DOE Chicago
9800 South Cass Avenue, Bldg. 201-1
Argonne, IL 60439-4899
Licensed Chief Examiner
Chairman, Energy/Ecology Committee/NAPE/NIULPE

Smiley, Roscoe R.
44 East 102nd St.
Chicago, IL 60628
312/264-2077 (Home)
312/744-3514 (Work)

Stephens, Jr. Ezekiel
1450 E. 55th Place
Apt. 618
Chicago, IL 60637
312/667-0428

Wileman, Wallace K.
840 W. Kinzie
Chicago, IL 60622
312/829-1060
Nat'l Sec. N.I.U.L.P.E

Iowa

Hall, W. James
2494 X. Ave.
Grimes, IA 50111
515/986-3380
IA Secretary

Love, William
3261 320th street
Waukee, Iowa 50263
515/987-1728
IA Chairman
Past National President, NAPE

Maryland

Doyle, Eugene C.
15722 Pointer Ridge Dr.
Bowie, MD 20716
301/249-6148

Larkin, Henry M.
103 Herbert Ct.
Glen Burnie, MD 30176-3269
301/768-3269
Past National President, NAPE

Sines, Stanley L.
11302 Stephens Ln.
Beltsville, MD 20705
301/937-7466

Massachusetts

Carlotto, John
464 Ventura St.
Ludlow, MA 01056
413/583-4842
Past National President, NAPE

Chickering, Winford
P.O. Box 12
W. Chesterfield, NH 03466
603/256-6351
MA Chairman

Gould, Alfred W.
42 Johnston Ave.
Whitinsville, MA 01588
508/234-7933

N.I.U.L.P.E. Examiners   Chapter 4

Harvey, James A.
P.O. Box 64
Richmond, VT 05477
802/434-2230
MA Secretary

Judd, William F.
5-7 Springfield St.
Chicopee, MA 01013
413/592-6273
Past National President, NAPE

Keelty, Leif
26 Slim Brown Rd.
Slim Brown Rd.
Milton, VT 05468
802/893-1524

Magiera, Edward J.
5 Sable Ave.
No. Dartmouth, MA 02714
508/990-0809

Michigan

Bhimani, Nitin K.
7241 Bridgetown
Caledonia, MI 49316-9194
616/698-2440

Erickson, Arthur L.
3199 Pine Hill Place
Flushing, MI 48433-2451
313/732-1033

Hamilton, Kenneth L.
P.O. Box 192
Cloverdale, MI 49035
616/623-5396 MI Secretary

Knapp, Michael Lee
P.O. Box 224
Comstock, MI 49041
616-342-1189 (Home)
616-384-6507 (Work)
Past National President, NAPE

Knapp, Wesley L.
5781 E. Comstock Ave.
Kalamazoo, MI 49001-3438
616/342-5911
MI Chairman

Mentzer, Donald J.
10051 North 16th St.
Plainwell, MI 49080
616/685-5529

Ouimette, William L.
201 North Birdsall Dr.
Battle Creek, MI 49017
616/968-0874
Region Director, 4, NAPE

Minnesota

Berger, William C.
4215 Halifax No.
Robbinsdale, MN. 55422

Bistodeau, Richard
101 57th Place N.E.
Minneapolis, MN 55432
612/574-0039
MN Chairman

Weltzin, Howard C.
12490 Genesse Way
Apple Valley, MN 55124
612/431-1578
MN Secretary

Missouri

Deal, John A.
14322 Jesse James Farm Road
Kearney, MO 64060-9624
816/635-5878 (Home)
816/556-2810 (Work)
MO-KS Chr.

McDaniel, Jr., Kirk
P.O. Box 410122
Kansas City, MO 64141-0122
816/225-0700
MO-KS Office
Sec. of Bd.
Examiner #312

Nash, P.E., Lorin B.
10204 E. 87th St.
Raytown, MO 64138
816/358-5846

Nebraska

Burley, Gil
5331 High St.
Lincoln, NE 68506
402/489-2670

Moore, Richard E.
308 Mohawk St.
Syracuse, NE 68446
402/269-3093

Chapter 4 — N.I.U.L.P.E. Examiners

Nevada

Barbier, P.J.
3188 Brazos St.
Las Vegas, NV 89109
702/732-4932

Cox, Gary E.
Flamingo Hilton Hotel
3555 Las Vegas Blvd. S
Las Vegas, NV 89109
702/733/3320
NV Secretary

Daugherty, John E.
3066 E. St. Louis Ave.
Las Vegas, NV 89104
702/457-4503
NV Chairman

Williams, Franklin
4139 Monthill Ave.
Las Vegas, NV 89121
702/454-5964

New Hampshire

Duby, Kevin
16 Frost Place
Brattleboro, VT 05302
802/254-2917

Harvey, James A.
P.O. Box 64
Richmond, VT 05477
802/434-2230
NH Secretary

Isakson, Richard
HCR 10, Box 628
Wellington Drive
Keene, NH 03431
NH Chairman

Pecorelli, Ralph
HCR-10 Box 170
Westmoreland Road
Keene, NH 03431
603/363-4402

Werbinski, Frank J.
29 Ridge Rd.
N. Swanzey, NH 03431
603/352-4111 ext. 2007 (Work)
603/357-3822 (Home)

New York

Bussell, Edwin T.
46 Colonial Road
Emerson, NJ 07630
201/261-8692

Karre, Richard J.
P.O. Box 23
Limerick, NY 13567
315/639-3170
NY Secretary

Lambert, Samuel F.
4071 Edison Ave.
Bronx, NY 10466
212/324-4448
Past National President, NAPE

Chapter 4 — N.I.U.L.P.E. Examiners

Schmid, James F.
63 Mark Drive
Manahawkin, NJ 08050
609/597-8128
NY Chairman
Past National President, NAPE

North Carolina

DeHart, Rita M.
Training Manager
Cogentrix Inc.
9405 Arrowpoint Blvd.
Charlotte, NC 28273-8110
704/525-3800
NC Secretary

Ohio

Beltramo, Peter A.
1750 Muskegon Drive
Cincinnati, OH 45255
513/474-4152
Dist. 2

Bullard, Richard L.
2296 Angel Ave.
Toledo, OH 43611
419/729-4528
Bd. Chairman-Dist. 5

Burkhard, David L.
7082 Wil-Lou Lane
North Ridgeville, OH 44039
216/327-7082
Board V.P.-Dist. 4

Eldridge, Gerold W.
3729 Nine Mile Rd.
Cincinnati, OH 45255
513/752-0388
OH Job Service Dir. - Dist. 3

Ferrel, Robert J.
2825 Castleton
Toledo, OH 43613
Dist. 5

Fisher, John K.
4167 Bennet Drive
Hamilton, OH 45011-9351
513/860-0388
Dist. 1 - Examiner
Commission #489

Lainhart, Charles L.
1019 Gail Ave.
Fairfield, OH 45014
513/863-1749
OH Secy/Treas.

McCabe, John R.
P.O. Box 392
Moundsville, WV 26041
304/845-8872

Nejbauer, Richard H.
853 N. Fairfield Rd.
Beaver Creek, OH 45014-1803
513/429-2196
Dist. 2

Oldiges, Robert W.
240 Berry Ave.
Bellevue, KY 41073
606/491-4092

Chapter 4 — N.I.U.L.P.E. Examiners

Reed, Mark L.
4847 Robinson Rd.
Sylvania, OH 43560
419/882-5889
Dist. 5 and MI.

Snider, James D.
911 Woodlawn Avenue
Napoleon, OH 43545
419/592-4230 (Home)
419/445-8015 ext. 223 (Work)
NIULPE Examiners Commission #478
District 5 Examiner

Pennsylvania

Dishart, Urban E.
15 Falkirk Dr.
Pittsburgh, PA 15235-4605
412/795-1513

Henning, E. James
272 Northview Estates
Indiana, PA 15017-3810
412/349-4927

Perlinger, Arthur
7861 Thon Dr. Penn Hills
Verona, PA. 15147-1528
412/241-8217
PA Secretary

Rhode Island

Arthur, Carlyle G.
36 Baker St.
Warren, RI 02885
401/245-9138
RI Secretary

63

Mogey, Robert
15 King Phillip Rd.
North Kingstown, RI 02852
401/884-6892
RI Chairman

Pendlebury, Frederick E.
144 Blackrock Rd.
Coventry, RI 02816-8004
401/821-2719

South Carolina

McCormick, Alvin R.
122 S. Charleston Road
Darlington, S.C. 29532
803/393-0674

Scott, W. Francis
117B St. James St.
Georgetown, SC 29440
803/546-9502 (Home)
803/527-7453 (Work)

Virginia

Fullmer, Robert L.
Rt. 3, Box 5535
Berryville, VA 22611
703/750-5656
Nat'l Education Director, NAPE

Kimmel, Franklin D.
7529 Albemarle Dr.
Manassas, VA 22111
703/361-6416

Chapter 4 — N.I.U.L.P.E. Examiners

Washington

Bogar, Darrell A.
W-2403 Dalton Ave.
Spokane, WA 99205
509/327-6796
WA Chairman

Brierley, Gilbert C.
E. 8616 Boone
Spokane, WA 99212
509/926-9406

Morgan, Richard A.
1314 Gary
Cheney, WA 99004
509/235-8048
WA Secretary

Wisconsin

Budde, Dean
1285 N. Clayton Ave.
Neenah, WI 54956
414/725-4097

Burno, Peter H.
1343 Veek Road
Stoughton, WI 53589-3766
608/873-8656
N.I.U.L.P.E. President
Past National President, NAPE

Carle, Larry C.
2311 Hayes Ave.
Racine, WI 53405
414/637-8741

N.I.U.L.P.E. Examiners

Demerath, Stephen F.
1124 S. Roosevelt St.
Green Bay, WI 54301
414/435-5777

Ewald, Donald E.
509 Terrill St.
Chippewa Falls, WI 54729
715/723-3419

Hansen, Jerome R.
753 River Heights
Menomonie, WI 54751
715/235-5486

Schulze, Edwin J.
8125 W. Cleveland Ave.
West Allis, WI 53219
414/541-2769

Smith, Don E.
4605 Glendale Ave.
Green Bay, WI 54313
414/865-7294

Westergard, James P.
1436 Fritz Road
Verona, WI 53593
608/845-7559
WI. Secretary
Executive Secretary NIULPE
Past National President, NAPE

USAEHSC (U.S. Army)

Bruschi, Robert A.
CEHSC-M-HM
Ft. Belvoir, VA 22060
703/806-3002

Mercer, Michael J.
Prime Power School
CEHSC-M-T-D
Ft. Belvoir, VA 22060-5516
USAEHSC President
703/704-1499

Stewart, Alan M.
Prime Power School
HHC, USAEHSC Bldg. 371
Fort Belvoir, VA 22060-5516
703/704-1510
USAEHSC Secretary

Chapter 5
International Union of Operating Engineers

IUOE Stationary Engineers' Training Programs

History of the International Union of Operating Engineers

The National Union of Steam Engineers was founded in 1896 in Chicago, Illinois. This was the forerunner of the International Union of Operating Engineers. Engineers of that era were subject to long working hours and poor working conditions. The founders of the International Union of Operating Engineers felt their interests would be best served by forming a union that addressed the specific problems of Engineers who worked on steam operated hoisting and construction equipment and in the powerhouses of factories in that day.

According to the seventy-five year history of the I.U.O.E., when trade unions were forming in the late 1800's and early 1900's,

"it was inevitable that free men in a prosperous, growing country should rebel against these intolerable conditions. Skilled Engineers were becoming increasingly aware of the vital importance of their craft in the development of the nation. Without their knowledge and ability, there would have been no railroads built, canals dug, nor vast projects such as dams and waterworks, as well as the sewage disposal plants so essential to the life of fast growing cities."

Chapter 5 — International Union of Operating Engineers

The National Union of Steam Engineers was accepted into the American Federation of Labor in May of 1897 and the first charted Union was in Denver, Colorado in June of 1897.

Organized labor fought for and won many of the benefits that the working individual takes for granted today, namely:

- Increased pay
- Child labor laws
- Two weeks vacation
- The 8 hour work day
- Overtime for work over 8 hours in a day
- The 40 hour work week
- Workmen's Compensation Laws

The IUOE set up training programs to educate its members and others to obtain higher licenses and better employment in the power industry. Apprenticeship programs were set up to bring new members into the Power Engineering craft. The IUOE locals listed run training programs to help individuals get into the Power industry.

IUOE Local 2
2929 S. Jefferson Ave.
St. Louis, MO 63118
314/865-1300
FAX: 314/865-1423

IUOE Local 6
2637 East 9th St.
Kansas City, MO 64124
816/483-9444

International Union of Operating Engineers   Chapter 5

IUOE Local 30
115-06 Myrtle Ave.
Richmond Hill, NY 11418
718/847-8484 x205
FAX: 718/805-2172

IUOE Local 39
2280 Palou Street
San Francisco, CA 94124
415/285-3939 TR site
415/861-1135
FAX: 415/861-5264

IUOE Local 68
14 Fairfield Place
PO Box 534
W. Caldwell, NJ 07006
201/227-6426 x210
FAX: 201/227-8373

IUOE Local 70
970 Raymond Ave., Ste 103
St. Paul, MN 55114
612/646-4566
FAX:612/646-2858

IUOE Local 94
331-337 West 44th St.
New York City, NY 10036
212/245-7040
FAX: 212/245-7886

IUOE Local 99
2461 Wisconsin Ave., NW
Washington, DC 20007
202/337-0099
FAX: 202/625-7982

IUOE Local 280
P.O. Box 807
Richland, WA 99352
509/946-5101

IUOE Local 399
763 W. Jackson Blvd.
Chicago, IL 60661
312/372-9870
FAX: 312/372-7055

IUOE Local 501
2501 West Third St.
Los Angeles, CA 90057
213/385-2889
FAX: 213/385-7324

IUOE Local 501
313 Deauville St.
Las Vegas, NV 89106
702/385-5005

IUOE Local 547
Stationary Engineers Education Center
24270 W. Seven Mile Rd.
Detroit, MI 48219
313/532-5345
FAX: 313/532-7306

IUOE Local 882
4333 Ledger Ave., Rm. 304
Burnaby, BC CANADA V5G3T3
604/294-5266
FAX: 604/294-0694

IUOE Local 900
P.O. Box 4548
Oak Ridge, TN 37831
615/523-5147

Chapter 6

License Requirements for States and Cities

Alabama

No State or City Licenses

Alaska

State of Alaska

SUMMARY OF QUALIFICATIONS:

1. Exam Location, Address and Phone Number

Alaska Department of Labor
Mechanical Inspection
P.O. Box 107020
3301 Eagle Street, Suite 203
Anchorage, Alaska 99510-7020
907/269-4925
Fairbanks 907/451-2887
Juneau 907/465-4842
Kenai 907/283-3778
Sitka 907/747-6380

Exams can be taken in any Alaskan community provided there is a Mechanical Inspection office, an employment center, or a magistrate present.

2. Exam Days and Times

Any Magistrates office in the State of Alaska can administer the examination. The State will send the test to the Magistrate to administer and then it will be sent to the Department of Labor for grading. No set dates or times. The exam can also be

Chapter 6 — License Requirements for States and Cities

taken at a Mechanical Inspection office, or an employment center.

3. Examination Cost

Free

4. License Cost and Renewal

Free. The license is valid for three years and is renewable upon request by contacting the Anchorage Office of the Department of Labor.

5. Type of Test

Fireman— True and False

Third Class Stationary Engineer— multiple choice and short answer

Second Class Stationary Engineer—multiple choice and short answer

First Class Stationary Engineer— multiple choice and short answer

6. Grades of Licenses

Fireman—Apprentice

Third Class Stationary Engineer

Second Class Stationary Engineer

First Class Stationary Engineer

7. Experience and Education Requirements

Fireman—Apprentice-No experience

Third Class Stationary Engineer—6 months experience or training. Boiler capacity not to exceed 3,500 pounds of steam an hour or 3,500,000 british thermal units per hour for high temperature or High Pressure water boilers.

Second Class Stationary Engineer—1 year experience or held 3rd Class Stationary Engineer's license for one year. Boiler capacity not to exceed 100,000 pounds of steam an hour or 100,000,000 british thermal units per hour for high temperature or high pressure water boilers.

First Class Stationary Engineer—2 years experience or held 2nd Class Stationary Engineer's license for one year Unlimited boiler capacity

8. U.S. Citizenship Required?

No

9. Local Residency Required?

No

10. Recognition of Licenses from other locales

Yes

CODES for State of Alaska

Alaska Statutes

Title 18. Chapter 60. Article 3.

Boilers.

Alaska Statute 18.60.395 states:

LICENSING OF BOILER OPERATORS.

(a) The Department of Labor shall promulgate regulations for the licensing of boiler operators. The regulations shall conform to the generally accepted nationwide standards and practices established for boiler operators.

Chapter 6 License Requirements for States and Cities

(b) Operators' licenses shall be provided in the following categories:

- (1) Fireman - Apprentice,
- (2) Third Class - Boiler capacity not to exceed 3,500 pounds of steam an hour or 3,500,000 British thermal units per hour for high temperature or high pressure water boilers,
- (3) Second Class - Boiler capacity not to exceed 100,000 pounds of steam an hour or 100,000,000 British thermal units per hour for high temperature or high pressure water boilers,
- (4) First Class - Unlimited.

(c) This section does not require a person to be licensed in order to be a boiler operator.

Title 8 of the Alaska Administrative Code, Chapter 80, Section 130 states:

REQUIREMENTS FOR BOILER OPERATOR LICENSE.

(a) An applicant for a boiler operator license must show documentation of qualifications for the respective license category as follows:

- (1) Fireman - No experience required;
- (2) Third Class - Six months of experience in the trade or six months of boiler training;
- (3) Second Class - Third class license for one year or experience in the trade for at least one year;

- (4) First Class - Second class license for one year or experience in the trade for at least two years.

(b) An examination will be given upon approval of an application. If the applicant fails the examination, the applicant must wait 30 days before reexamination.

(c) A license is valid for three years and is renewable upon request by contacting the Anchorage office of the department.

(Eff. 6/21/84, Register Authority: AS 18.60.180; AS 18.60.395

REQUIREMENTS FOR DOCUMENTATION:

Type of License:

- First Class: 2 years experience in the trade or held a 2nd class license for one year.
- Second Class: 1 year in the trade or held a 3rd class license for 6 months.
- Third Class: 6 months experience in the trade or 6 months of boiler training - no experience necessary.
- Fireman: No experience required.

Possession of the Alaska license will be verified by Mechanical Inspection. Experience is documented by submitting letters of verification on letterhead stationary from prior employers stating actual duties performed and dates and hours employed. Documentation of boiler training shall show dates and hours of training.

This form, when properly completed and submitted with the necessary supporting documents will be reviewed by an inspector or designee. Applicants whose documentation is found to

| Chapter 6 | License Requirements for States and Cities |

be in accordance with the requirements of the law will be admitted for examination.

Exams can be taken in any Alaskan community provided there is a Mechanical Inspection office, an employment center, or a magistrate present. City to Test: There is a four-hour time limit on all exams. Exams are graded in a Mechanical Inspection Office.

Upon successful completion of the exam, a license will be issued from the Anchorage office. A license is valid for 3 years and is renewable upon request by contacting the Anchorage office.

Arizona

No State or City License

Arkansas

State of Arkansas

SUMMARY OF QUALIFICATIONS:

1. Exam Location, Address and Phone Number

Arkansas Department of Labor
Boiler Division
10421 West Markham
Little Rock, Arkansas 72205
501/682-4513

2. Exam Days and Times

1st Monday of every month

3. Examination Cost

$16.00

4. License Cost and Renewal

Initial license included in cost of exam.

$12.00 for renewal

5. Type of Test

All Written - fifty multiple choice questions

Up to the inspectors discretion - if the applicant has difficulty reading or cannot read he may give an oral examination.

6. Grades of Licenses

High Pressure Boiler Operator

Low Pressure Boiler Operator

7. Experience and Education Requirements

High Pressure Boiler Operator - Six months practical experience

Chapter 6 License Requirements for States and Cities

Low Pressure Boiler Operator - Six months practical experience

8. U.S. Citizenship Required?

Yes

9. Local Residency Required?

No

10. Recognition of Licenses from other locales

Yes - They count military and previous work experience

CODES for State of Arkansas

Rules & Regulations

Governing the Inspection, Installation and Operation of Steam Boilers, Hot Water Heating and Supply Boilers and Unfired Pressure Vessels Including Anhydrous Ammonia Containers and Equipment

81-507

Section V

BOILERS TO BE ATTENDED AT ALL TIMES

(a) All power boilers subject to the provisions of this act shall be under regular attendance by some responsible person whenever they are in use for any purpose. Boilers that are fired up are considered in use whether steam is being withdrawn or not.

(b) Boilers 50 horsepower and over, as rated by the manufacturer, and boilers used in hospitals, hotels, schools, theaters and office buildings, but not limited to, must be under regular attendance by a licensed Operator who holds a certificate of competency issued by the Boiler Inspection Division.

BOILER OPERATORS' FEES

(1) The Boiler Inspection Division shall conduct examinations for each applicant seeking a Boiler Operator's license. This examination may be either written or oral.

(2) Each applicant shall pay a fee of sixteen ($16.00) dollars for the examination and first license. Each license must be renewed annually. The annual fee shall be twelve ($12.00) dollars.

(3) If licenses are not renewed within two (2) to five (5) years after the expiration date, the renewal fee shall be sixteen ($16.00) dollars. Any Operator who shall allow his license to lapse for a period of over five years will be required to participate in a written examination before license may be reissued. An average of 70 percent shall be required for a passing grade.

(4) Any Operator found operating a boiler without a certificate issued by the Boiler Inspection Division, or operating a boiler knowing it to be defective, shall have his license revoked at once. Any person found operating a boiler without an Operator's license shall be guilty of a misdemeanor and upon conviction thereof shall be punished by a fine of not less than Twenty-Five ($25.00) Dollars and not more than One Hundred ($100.00) Dollars, and in addition may be imprisoned for not more than two (2) years or both.

Chapter 6   License Requirements for States and Cities

California

Los Angeles, California

SUMMARY OF QUALIFICATIONS:

1. Exam Location, Address and Phone Number

City of Los Angeles
Department of Building and Safety
Licensing Section - Room 495, City Hall
200 N. Spring Street
Room 460-P
Los Angeles, California 90012
213/485-3787

2. Exam Days and Times

Wednesday at 5 p.m.

3. Examination Cost

$43.40

4. License Cost and Renewal

Initial license included in cost of exam.

$22.20 for Renewal

5. Type of Test

Written and Oral

6. Grades of Licenses

Boiler Operator-35 h.p.

Steam Engineer-500 h.p.

Steam Engineer-Unlimited

Steam Engineer-Unlimited-with Turbine Endorsement

7. Experience and Education Requirements

License Requirements for States and Cities   Chapter 6

BOILER OPERATOR-35 h.p. - No previous experience required

STEAM ENGINEER-500 h.p. - One year full time experience as a Steam Engineer in charge of the operation of one or more steam boilers of over 35 h.p. rating or three years experience as an assistant Engineer to the person in charge of the operation of one or more steam boilers of over 35 h.p.

STEAM ENGINEER-UNLIMITED - One year full time experience as a Steam Engineer in charge of one or more steam boilers of over 500 h.p. each or over 500 h.p. total if connected for joint operation or three years experience as an assistant Engineer to the person in charge of the operation of one or more steam boilers of over 500 h.p.

STEAM ENGINEER UNLIMITED and TURBINE ENDORSEMENT - One year of experience in charge of a steam turbine and possession of current unlimited Steam Engineer license or three years of experience as assistant to the person in charge of the operation of a steam turbine and possession of a current unlimited Steam Engineer license.

8. U.S. Citizenship Required?

No

9. Local Residency Required?

No

10. Recognition of Licenses from other locales

Yes

CODES for Los Angeles, California

Examination for Boiler Operator - 35 H.P.

| Chapter 6 | License Requirements for States and Cities |

1. DUTIES: Operation of Steam Boilers of not to exceed 35 horsepower capacity, at a specified address only, as shown on the license.

2. REQUIREMENTS: No previous experience required.

3. FILING FEE: $43.40 Annual Renewal: $22.20

4. EXAMINATION:

- WHEN HELD: Applicants are scheduled for examination at the time application is filed.
- WHERE HELD: Room 495, City Hall, Use Main Street entrance.
- TIME: Examination begins at 5:30 p.m.
- SCOPE: Examination, in general, consists or questions on the following subjects, together with an oral interview:

    1. Precautions to be observed and valves to be opened when starting a cold boiler.

    2. Precautions to be observed when shutting down a boiler for the night.

    3. Function of various valves on the boiler.

    4. Importance of maintaining proper water level in the boiler.

    5. Blowing down a boiler, frequency and amount of blowdown.

    6. Foaming in a boiler and why this condition is hazardous.

7. Boiler priming and why this condition may prove hazardous.

8. Causes and correction of foaming and priming in a boiler.

9. Importance of maintaining a clean boiler and treatment of the water.

10. Cause of excess smoke from a boiler.

11. Other subjects pertinent to operating boilers.

- MINIMUM SCORE: Seventy percent is the minimum passing total score. Applicants who fail to attain at least one-half of the required passing score in the written examination shall not be given an oral examination.

- TIME LIMIT: Failure to appear for examination within six months after filing application shall void the application, and the applicant shall file a new application and pay filing fee before another examination will be given.

5. RE-EXAMINATION: Applicants who fail to pass the first examination are entitled to a second examination not less than two weeks thereafter. No additional fee is required for such second examination.

6. Applicants who fail to pass the second examination are required by ordinance to wait at least six months thereafter before they may file another application for examination, at which time a new application and filing fee is required.

7. Boiler Operator - 35 h.p. license issued as a result of passing the examination is valid for a period of one year from date

Chapter 6 License Requirements for States and Cities

of issuance but may be renewed from year to year by the payment of the annual renewal fee or late fee, if it is within a year from the expiration date.

Examination Information for Steam Engineers

1. FILING FEE: $43.40 Annual Renewal Fee: $22.20

   EXCEPTION: No charge for Turbine Endorsement when taken in conjunction with Unlimited Steam Engineers Examination. If taken at a later date, a separate application and filing fee will be required.

2. EXAMINATION:
   - WHEN HELD: Applicants are scheduled for examination at the time application is filed.
   - WHERE HELD: Room 495, City Hall - 200 N. Spring St. Los Angeles, CA. 90012 (Use City Hall Main Street Entrance at 201 N. Main St.)
   - NOTE: Please wait in the hallway outside of Room 485, City Hall. The Examiner will call you into Room 495 at 5:30 p.m.
   - TIME: Examination begins at 5:30 p.m.
   - SCOPE: Steam Engineer Examinations consist of fifty multiple choice questions pertaining to the safe operation of steam boilers and steam-driven equipment together with an oral interview to determine the applicants practical knowledge of such equipment. In general, the Steam Engineer's exam may contain questions in the following categories:

License Requirements for States and Cities   **Chapter 6**

1. Pumps and compressors
2. Boiler operation
3. Boiler maintenance
4. Controls
5. Boiler safety
6. Physics and chemistry
7. Water treatment
8. Steam traps
9. Safety valves

- The Turbine Endorsement examination consists of 25 multiple choice questions.
- MINIMUM SCORE: 70% is the minimum passing total score. Applicants who fail to attain at least 35 points in the written examination shall not be eligible for an oral examination.

3. TIME LIMIT: Failure to appear for examination within six months after filing an application shall void the application, and the applicant shall file a new application and pay the required filing fee before another examination will be given to him/her.

4. REEXAMINATION: Applicants who fail to pass the first examination are entitled to a second examination not less than two weeks thereafter. No additional fee is required for the second examination.

5. Applicants who fail to pass the second examination are required by ordinance to wait at least six months before they

may file another application for examination, at which time a new application and filing is required.

Steam Engineer - 500 H.P.

1. DUTIES: Operation of one or more steam boilers rated to 500 h.p. If multiple boilers are connected for joint operation, the total horsepower rating cannot exceed 500 h.p.

2. REQUIREMENTS: (a) One year full time experience as a Steam Engineer in charge of the operation of one or more steam boilers of over 35 horsepower rating; or (b) three years experience as an assistant Engineer to the person in charge of the operation of one or more steam boilers of over 35 horsepower rating. Such positions generally are identified as Fireman, water tender, boiler tender or assistant Engineer.

Steam Engineer Unlimited

1. DUTIES: Operation of steam boilers of any size.

2. REQUIREMENTS: (a) One year full time experience as a Steam Engineer in charge of one or more steam boilers of over 500 horsepower each or over 500 horsepower total if connected for joint operation; or (b) three years experience as an assistant Engineer to the person in charge of the operation of one or more steam boilers of over 500 horsepower.

Steam Engineer - Turbine Endorsement (Optional at time of filing)

1. APPLICABILITY: The turbine endorsement is only applicable to the unlimited license.

2. REQUIREMENTS: (a) One year of experience in charge of a steam turbine and possession of current unlimited Steam

License Requirements for States and Cities — Chapter 6

Engineer license or (b) three years of experience as assistant to the person in charge of the operation of a steam turbine and possession of a current unlimited Steam Engineers license.

3. RETROACTIVITY: Unlimited licenses issued prior to Dec. 22, 1980 (when unlimited tests included turbine questions) may be returned for endorsement.

Colorado

Denver, Colorado

SUMMARY OF QUALIFICATIONS:

1. Exam Location, Address and Phone Number

Contracting Licenses & Certificates
200 W. 14th Ave.
Denver, Colorado 80204
303/640-5903

2. Exam Days and Times

Twice a month - Usually 2nd & 4th week

Date varies - Always on a Wednesday - Twice a day at 9 a.m. and 1:30 p.m.

3. Examination Cost

$15.00 filing fee

4. Certificate of Qualification and Renewal

Stationary Engineer and Boilermaker Supervisor........$25.00

Boilermaker Journeyman $10.00

Boiler Operator A & B..$10.00

5. Type of Test

All Written - multiple choice

6. Grades of Licenses

Stationary Engineer

Boiler Operator Class A

Boiler Operator Class B

Boilermaker - Journeyman

License Requirements for States and Cities   Chapter 6

Boilermaker - Supervisor

7. Experience and Education Requirements

STATIONARY ENGINEER:

Four years in Stationary Engineering work consisting of a minimum of three (3) years operating High Pressure steam boilers producing over 100 hp and steam pressure over 100 p.s.i. or water heating boilers when the water temperature exceeds 250 degrees F in the system, and a minimum of one (1) year operating refrigeration equipment requiring an Operator per chapter 2, Denver Building Code.

BOILER OPERATOR-CLASS A:

Three (3) years in boiler operating work as a trainee or apprentice on equipment producing over 100 hp and steam pressure over 100 p.s.i. and water heating systems when the water temperature exceeds 250 degrees F. in the system.

BOILER OPERATOR-CLASS B:

Three (3) years in boiler operating work as a trainee or apprentice on equipment producing a total of between 10 and 100 h.p. and steam pressure between 15 and 100 p.s.i. and water heating systems when the water temperature exceeds 250 degrees F in the system.

BOILERMAKER-JOURNEYMAN:

Four (4) years boilermaker field installation experience in an apprenticeship or on-the-job training program completing a minimum of 7,000 hours 2. Experience as a journeyman, supervisor or contractor under a recognized licensing authority may be accepted on a year for year basis.

BOILERMAKER-SUPERVISOR:

Chapter 6 License Requirements for States and Cities

Eight (8) years in boilermaker work consisting of the following: 1. Four (4) years field installation experience in an apprenticeship or on-the-job training program consisting of a minimum of 7,000 hours and 2. Four (4) years experience as a Designer or Supervisor.

8. U.S. Citizenship Required?

Not specified

9. Local Residency Required?

No

10. Recognition of Licenses from other locales

Yes, counts for military experience

Codes for Denver, Colorado

Building Code for the City and County of Denver

Based Upon the Uniform Codes of the International Conference of Building Officials (ICBO)

Division 1 - Amendments to the 1988 Edition of the Uniform Building Code

Chapter 2

LICENSING, CERTIFICATION, REGISTRATION, BOARDS OF STANDARDS

SECTION 210. CERTIFICATES.

(a) Definition.

A Certificate of Qualification is authority to perform certain skills and is issued by the Department on the successful completion of an examination. This certificate is not transferable. The term "Certificate" means Certificate of Qualification.

(b) Temporary Certificate.

The Department may issue a temporary Certificate when the applicant has previously exhibited his skills to the satisfaction of the Department and the applicant's qualifications are acceptable. The Department shall determine the period of validity of the temporary Certificate.

(c) Certificate Application.

Every applicant for a Certificate shall be required to complete a form provided by the Department and to pay an application fee of $15.00 at the time of the filing. The fee shall not be refundable and shall not apply to the Certificate fee. The payment of the fee shall entitle the applicant to one examination only. If the applicant is reexamined for any reason, a new application and fee shall be required.

(d) Successful Applicants.

If an applicant who has successfully passed the examination given by the Department fails to procure this Certificate within 90 days after notification, the Certificate shall be declared to be null and void and a new application and fee shall be filed.

(e) Failure to Pass Examination.

When an applicant has failed to pass the examination, he shall be notified in writing by the Department.

(f) Certified Supervisors.

(1) Every supervisor required for a particular license shall be examined by the Department, and if qualified, shall be issued a Supervisor Certificate of Qualification. The Certificate holder shall be entitled to perform and supervise the work in the particular skill for which he is qualified and certified. The Certificate is personal to that holder and shall not be construed to be a license.

Chapter 6　　　License Requirements for States and Cities

(2) The Certificate holder shall actively supervise the workmen of the licensee by whom he is employed in accordance with Section 202(d).

SECTION 211. CLASSIFICATION OF SUPERVISOR CERTIFICATE OF QUALIFICATION.

A Supervisor Certificate for the particular work to be performed shall permit the holder to be a Supervisor under the licenses listed in Table 2-B.

SECTION 212. CLASSIFICATION OF JOURNEYMAN AND OPERATOR CERTIFICATE OF QUALIFICATION.

(a) General.

Unless otherwise provided for in this Section or in this Building Code, all journeymen and operators required to be certified shall perform that work permitted under the provisions of licenses for a particular type of work. The work permitted by the certification shall be performed in the employ of the licensee as hereinafter set forth.

(c) Stationary Engineer and Operator Certificates.

It shall be unlawful to operate any of the following equipment without the personal attendance of a properly Certified Stationary Engineer or a properly Certified Operator.

(1) Any steam boiler and appurtenances thereto, steam pumps, steam turbines, and steam engines where the steam pressure is in excess of 15 psi working pressure and where the equipment produces a total of 10 boiler horsepower or more at Denver altitude.

(2) Water heating systems when the water temperature exceed 250 degrees F. in the system.

(3) Composite grouping of refrigeration machines where machines are 25 tons in capacity and parallel to a common refrigerant piping system. The total charge in the entire system shall determine the capacity of the system.

(4) Refrigeration systems utilizing Group 2 or 3 refrigerants as defined in the Mechanical Code and which contains a charge of 200 lbs. or more.

(5) Refrigeration systems having manual or semi-automatic control with charges of 1500 lbs. or more of Group 1 refrigerants as outlined in the Mechanical Code.

(6) Refrigeration systems with fully automatic controls with charges of 1500 lbs or more of Group 1 refrigerants.

NOTE: As used in this Section, semi-automatic shall mean plants or systems which are provided with automatic safety controls by manual load proportioning controls requiring other than seasonal adjustments.

(d) Stationary Engineer Certificate.

Permits the holder to take charge of and operate all steam boilers and appurtenances hereto, steam pumps, steam turbines, steam engines and mechanical refrigeration systems.

(e) Boiler Operator Certificate, Class A.

Permits the holder to take charge of and operate all steam boilers and appurtenances, steam pumps, steam turbines and steam engines.

(f) Boiler Operator Certificate, Class B.

Permits the holder to take charge of and operate all steam boilers and appurtenances, steam pumps, steam turbines and steam engines containing a steam pressure between 15 and 100 psi and where the equipment produces a total of between 10 and 100 horsepower at Denver altitude.

SECTION 213. APPRENTICES AND TRAINEES.

(a) General.

This Section shall govern the requirements for apprentices and trainees and shall be limited to the crafts listed in this Chapter where a Journeyman Certificate holder is required.

(b) Requirements.

Apprentices and trainees shall not be required to possess a Certificate, but shall be permitted to work as prescribed in other Sections of this Chapter.

(c) Definition.

(1) An apprentice shall mean any person who has entered into an apprentice agreement which provides for participation in a program of training through employment and education in related and supplementary subjects.

(2) A trainee shall mean any person working at the trade under the direct supervision of a certified Journeyman or Supervisor.

(d) Work.

An apprentice or trainee may perform any work which is distinctive to a specific craft, but only under the direction and supervision of a Certified Supervisor or Journeyman of the craft, during working hours. Persons working on tasks not distinctive to any specific craft shall not be classed as an apprentice.

(e) Employment of Apprentices.

Contractors may employ apprentices or trainees for the licensed crafts or trades. The ratio of apprentices and trainees to Journeyman employed shall not exceed one apprentice or trainee to one Journeyman.

(f) Employer.

All apprentices or trainees shall be in the employ of the licensed crafts where Journeymen Certificate holders are required.

SECTION 214. CERTIFICATE FEES.

(a) Annual Fees.

Annual Certificates of Qualification fees shall be paid the Department as follows:

Supervisor Certificate............$25.00

Journeyman Certificate.........$10.00

Engineer Certificate...............$10.00

Operator Certificate...............$10.00

EXCEPTION: The certificate fee for employees of the City shall be waived when performing work for the City or when employed by the Department.

(b) Certificate Fee Refund.

Certificate fees are not refundable.

SECTION 215. CERTIFICATE RENEWAL.

Certificates shall be renewed annually and expire on the date specified on the Certificate. No work shall commence or continue after the date of expiration.

SECTION 216. REISSUANCE.

Chapter 6   **License Requirements for States and Cities**

(a) General.

The Department shall have the authority to renew a Certificate, provided that the renewal is accomplished within the limits set forth herein:

(1) The Certificate may be reissued without a new application, provided that such reissuance is accomplished within one year after the Certificate has expired.

(2) If the Certificate holder reapplies within 3 years of the date of expiration, re-examination shall not be required.

(3) If the Certificate holder applies more than 3 years after expiration, re-examination shall be required.

SECTION 217. CERTIFICATE HOLDER RESPONSIBILITY.

(a) General.

All Certificate holders shall be responsible to insure that the work performed by the licensee is in accordance with the requirements of this Building Code, without limitation, and to:

(1) Have in possession at all times a Certificate.

(2) Present a Certificate when requested by the Department.

(3) Faithfully construct without departure from or disregard of approved drawings and specifications.

(4) Obey any order issued under authority of this Building Code.

(5) Pay any fee assessed under the authority of this Building Code.

(6) Observe the safety requirements of this Building Code.

License Requirements for States and Cities — Chapter 6

(7) Actively supervise and oversee all work performed by or for the licensee by whom he is employed.

(8) Be responsible for all permits being issued prior to the beginning of work.

(9) Maintain a current local mailing address and accept all mail so addressed.

(10) Notify the Department within 3 days whenever he leaves the employ of licensee.

(11) Provide minimum safety measures and equipment to protect workmen and the public.

(12) Faithfully construct, without departure from drawings and specifications filed and approved by the Department and permit issued for same, unless changes are approved by the Department.

(13) Complete all work authorized by the permit issued under the authority of the Department, unless the cause of incomplete work is determined by the Department to be not the fault of the Certificate holder.

(14) Obtain inspection services where required by the Department.

SECTION 218. SUSPENSION OR REVOCATION OF CERTIFICATE.

(a) Authority.

The Director may suspend or revoke a Certificate for any one or more of the following acts or omissions.

(1) Incompetence.

Chapter 6 — License Requirements for States and Cities

(2) Misuse of the Certificate.

(3) Violation of any of the provisions of this Building Code.

(4) Failure to comply with any of the Certificate holder responsibilities outlined in Section 217.

(5) Knowingly conspire with a person to permit a license to be used by another person.

(6) Act as agent, partner, associate or in any capacity with persons to evade the provisions of this Building Code.

(7) Willfully violate or disregard any of the provisions of this Building Code.

(8) Intentionally fail to perform in accordance with any written contract to conduct work which is regulated by this Building Code.

(9) Create, as a result of work performed, an unsafe condition as defined in Chapter 1 of this Building Code.

(10) Intentionally or fraudulently misrepresent the condition of any structure or utility or the requirements of this Building Code.

(11) Repeatedly violate the provisions of this Code, or repeatedly fail to obey orders in a timely fashion.

Procedures.

When a Certificate holder commits any acts or omissions enumerated above and the Director deems that the Certificate shall be suspended or revoked, the action shall be as follows:

(1) Notification. The Department shall send written notice to the Certificate holder, by certified mail or by personal service, at least 7 days prior to suspension or revocation.

(2) **Request Hearing.** Upon receipt of the notice, the Certificate holder may request a hearing to show cause why a certificate should not be suspended or revoked. This request shall be in writing to the Department within 7 days after receipt of the notice.

(3) **Time of Hearing.** If a hearing is requested by the Certificate holder, the Director shall notify the Certificate holder of the time, date and place of the hearing. Suspension or revocation of the certificate shall be stayed until after the hearing. In the event the Certificate holder fails to appear, the Certificate may be suspended by the Director.

(4) **Attendance.** The Certificate holder, the Department and other interested parties may be in attendance at the hearing. Upon completion of the hearing, the Director shall take all evidence available as a result of the Department's investigation and all evidence presented at the hearing under advisement, and shall give written notice of the findings and ruling to the Certificate holder by certified mail or personal service.

(c) **Emergency Suspension or Revocation.**

If the Director finds that cause exists for emergency suspension or revocation of a Certificate, and that continued work under the Certificate could be hazardous to life or property, an order may be entered for immediate suspension or revocation of the Certificate, pending further investigation. The Certificate holder may, upon notice of the emergency suspension or revocation, request an immediate hearing before the Department. The hearing shall be conducted in the manner prescribed herein.

(d) **Delegation of Authority.**

The Director may appoint a qualified member of the Department to sit in his stead as the Hearing Officer to conduct the hearing. The final decision shall be rendered by the Director.

(e) Right to Appeal.

The Board of Appeals may review the suspension or revocation under Section 121.

STEPS IN APPLYING FOR A CERTIFICATE

(1) Read the Standards for the Certificate.

(2) If you feel you meet the minimum requirements, fill out the application. Be sure to cover in detail the duties performed in any job that you are listing to meet the experience requirement. It is not sufficient to state just "apprentice" or "journeyman" or "see resume."

(3) Experience required for the Certificate must be verified by letters (from persons other than the applicant). The letters should be in the following format and original letters submitted with the application:

- Be on stationery with the company letterhead.
- Be notarized.
- Have exact listing of duties or projects performed or equipment operated.
- Have exact dates of employment (month & year started: month & year ended).

(4) Please make copies of reference letters if needed. They can not be returned to you once they are in our files and copies can not be made.

(5) Applicants will be scheduled for examination in the order in which applications are received and approved. Check with this office a week after submitting application to be sure everything is in order.

(6) The Examination:

- You will be notified one week before the exam as to time and place.
- You will be tested on the Denver Building Code and the knowledge and skills necessary for your trade.
- Be on time for the examination.
- Read all instructions given with the examination.
- Read questions carefully.
- Guess if you are not sure of the answer.
- All exams are closed book. You may not take any notes or code book into the exam.
- You may use a calculator or slide rule in the exam.
- Please bring your notification slip and some picture ID to the exam.

(7) Do not telephone us for results of the exam. You will be notified by mail as soon as possible. It will generally take two weeks.

(8) A $15.00 money order or check is required with each application. (Make payable to Manager of Revenue, City and County of Denver). No cash can be accepted with an application.

(9) Notify us of any change of address or telephone number.

Chapter 6 License Requirements for States and Cities

Standards for Stationary Engineer Certificate

A. The Application

(1) The application form shall be filled out completely and shall be in the form of a sworn statement.

(2) Examinations will be scheduled only upon:

- (a) Completion and receipt of the application by the Building Inspection Division.

- (b) Receipt of all necessary employment and experience record verifications by the Building Inspection Division.

- (c) Recommendation by the Director, Building Inspection Division, or his authorized delegate(s) that applicant is qualified to take the examination.

B. Experience

Four (4) years in Stationary Engineering work consisting of a minimum of three (3) years operating High Pressure steam boilers producing over 100 hp and steam pressure over 100 p.s.i. or water heating boilers when the water temperature exceeds 250 degrees F. in the system, and a minimum of one (1) year operating refrigeration equipment requiring an Operator per Chapter 2, Denver Building Code.

C. The Examination

(1) The applicant will be required to pass the written examination with a grade of at least 70% out of a possible 100%.

(2) Areas of inquiry

License Requirements for States and Cities   Chapter 6

A. Denver Building Code

Chapter 1: Title, Scope and General

Chapter 2: Licensing, Certification, Registration Boards of Standards

Chapter 3: Permits, Plans, Inspection, Certificate of Occupancy

Chapter 4: Definitions and Abbreviations

Uniform Mechanical Code and amendments

B. Technical and Related Job Knowledge

*There may be other chapters or references that apply to this certificate.

Standards for Boiler Operator Certificate

A. The Application

(1) The application form shall be filled out completely and shall be in the form of a sworn statement.

(2) Examinations will be scheduled only upon:

- (a) Completion and receipt of the application by the Building Inspection Division
- (b) Receipt of all necessary employment and experience record verifications by the Building Inspection Division
- (c) Recommendation by the Director, Building Inspection Division, or his authorized delegate(s) that applicant is qualified to take the examination.

B. Experience

Chapter 6   **License Requirements for States and Cities**

(1) Class A - Three (3) years in boiler operating work as a trainee or apprentice on equipment producing over 100 hp and steam pressure over 100 p.s.i. and water heating systems when the water temperature exceeds 250 degrees F. in the system.

(2) Class B - Three (3) years in boiler operating work as a trainee or apprentice on equipment producing a total of between 10 and 100 hp and steam pressure between 15 and 100 p.s.i. and water heating systems when the water temperature exceeds 250 degrees F. in the system.

C. The Examination

(1) The applicant will be required to pass the written examination with a grade of at least 70% out of a possible 100%

Standards for Journeyman Certificate - Boilermaker

A. The Application

(1) The application form shall be filled out completely and shall be in the form of a sworn statement.

(2) Examinations will be scheduled only upon:

- Completion and receipt of the application by the Building Inspection Division

- Receipt of all necessary employment and experience record verifications by the Building Inspection Division

- Recommendation by the Director, Building Inspection Division, or his authorized delegate(s) that applicant is qualified to take the examination.

B. Experience

(1) Four (4) years boilermaker field installation experience in an apprenticeship or on-the-job training program completing a minimum of 7,000 hours.

(2) Experience as a journeyman, supervisor or contractor under a recognized licensing authority may be accepted on a year for year basis.

Standards for Supervisor Certificate - Boilermaker

The Application

(1) The application form shall be filled out completely and shall be in the form of a sworn statement.

(2) Examinations will be scheduled only upon:

- Completion and receipt of the application by the Building Inspection Division
- Receipt of all necessary employment and experience record verifications by the Building Inspection Division.
- Recommendation by the Director, Building Inspection Division, or his authorized delegate(s) that applicant is qualified to take the examination.

B. Experience

Eight (8) years in boiler maker work consisting of the following:

(1) Four (4) years field installation experience in an apprenticeship or on-the-job training program consisting of a minimum of 7,000 hours; and

Chapter 6   License Requirements for States and Cities

(2) Four (4) years experience as a designer or Supervisor.

C. The Examination

(1) The applicant will be required to pass the written examination with a grade of at least 70% out of a possible 100%.

Pueblo, Colorado

SUMMARY OF QUALIFICATIONS:

1. Exam Location, Address and Phone Number

Pueblo Regional Building Department
316 W. 15th Street
Pueblo, Colorado 81003
719/543-0002

2. Exam Days and Times

Monday - 1:30 p.m.

Tuesday & Thursday - 8:30 a.m. and 1:30 p.m.

Friday - 8:30 a.m.

3. Examination Cost

$10.00

4. License Cost and Renewal

C-Class - Boiler Operator $10.00

B-Class - Operating Engineer $15.00

A-Class - Chief Engineer $20.00

5. Type of Test

All Written

6. Grades of Licenses

Class C - Boiler Operator

Class B - Operating Engineer

Class A- Chief Engineer

7. Experience and Education Requirements

Class C - Boiler Operator .... 1 year practical experience

Chapter 6   License Requirements for States and Cities

Class B - Operating Engineer.. 2 or more years practical experience

Class A - Chief Engineer.......... 5 or more years practical experience

8. U.S. Citizenship Required?

No

9. Local Residency Required?

No

10. Recognition of Licenses from other locales

No - Must take the test

CODES for Pueblo, Colorado

Building Regulations

Supplement #49 - 12-31-83

4-1-34: STATIONARY ENGINEERS

(a) (1) It shall be unlawful for any person to have charge of, or operate, any steam boiler or steam engine, either stationary or portable, except an automatically fired low pressure hot water and low pressure steam boiler installed under provisions of this code, within the City without first having obtained the appropriate license therefor from the Plumbing Board of Review of the Pueblo Regional Building Department or of any owner or user to put any person in charge of a steam boiler or steam engine, either stationary or portable, unless such person put in charge is a duly licensed Engineer. Engineers and boiler tenders operating locomotives under the Interstate Commerce Commission regulations, shall be exempt from the requirements and provisions of this section.

(2) For purposes of this Section, a low pressure hot water and low pressure steam boiler is a boiler furnishing hot water at temperatures not exceeding 250 degrees F., or steam at pressures not more than 15 pounds per square inch.

(b) For the purpose of providing for the regulation and licensing of Stationary Engineers, such Engineers are divided into classes as follows:

(1) Class A - Chief Engineer's License

Applicants for Class A licenses shall be required to pass a written examination indicating ability to supervise the operation, installation and maintenance of any plant in the City and must have had not less than five years practical experience in the operation and supervision of such plants, boilers, compressors or other apparatus. In order to qualify for such licenses, the applicant shall pass such examination with a grade of not less than eighty percent.

(2) Class B - Operating Engineer's License

Applicants for Class B license shall be required to pass a written examination indicating ability to operate any plant in the City and must have had not less than two years' experience in the operation of such plants, boilers, compressors or other apparatus. In order to qualify for such license, the applicant shall pass such examination with a grade of not less than seventy-five percent.

(3) Class C License - Boiler Operator

Class C licenses shall permit the holder thereof to operate High Pressure boilers without machinery, except the pumps necessary to feed the boilers and the necessary equipment attached thereto. Applicants for a Class C license shall have at least one year of practical experience in the operation of such

Chapter 6 License Requirements for States and Cities

boilers and shall be required to pass a written examination with a grade of not less than seventy-five percent.

(c) Every license issued under this section shall be displayed in some conspicuous place near the boiler, engine, or plant where it may be readily seen at all times, and any failure or neglect to comply with this section shall be deemed a violation of this code.

(d) Any license issued hereunder may be suspended or revoked by the Plumbing Board of Review for failure of the licensee to properly operate any boiler in accordance with industry standards, or for carrying higher steam or other pressure than allowed by law, for failure to properly post the license, or other neglect or incapacity.

(e) All persons holding Class A or Class B licenses under this section shall assist the State Boiler Inspector in his inspections of any boiler under their charge and shall point out defects and imperfections known to them in the boilers and machinery. It shall also be the duty of persons holding Class A or Class B licenses before vacating any position as Engineer, to give to his employer at least one week's notice in advance of such intention and to give the same notice to the Building Official. Any failure to comply with the provisions of this section may result in suspension or revocation of the license of the person failing to comply.

License Requirements for States and Cities   Chapter 6

Connecticut

Bridgeport, Connecticut
SUMMARY OF QUALIFICATIONS:

1. Exam Location, Address and Phone Number
City of Bridgeport, City Hall, City Clerk
45 Lyon Terrace
Bridgeport, Connecticut 06604
203/576-7081
City Clerks Office for application, Council Chambers for exam

2. Exam Days and Times
Varies - they wait until they get 12 applicants

3. Examination Cost
$10.00

4. License Cost and Renewal
Power Engineer.................................$5.00

Boiler or Water Tender.....................$3.00

Low Pressure Boiler Operator............$3.00

5. Type of Test
50 questions, true-false, multiple choice, and short answer

6. Grades of Licenses
Boiler Tender

No Seal - Low Pressure Boiler Operator

Red Seal - Limited on Boilers and Engines

Blue Seal - Unlimited Boilers - Limit on Engine HP

Gold Seal - Chief Engineer - Unlimited

Chapter 6 License Requirements for States and Cities

7. Experience and Education Requirements

Under Codes for Bridgeport

8. U.S. Citizenship Required?

Yes

9. Local Residency Required?

You must be employed in the City of Bridgeport, or promised employment which would qualify them for an interim certificate until such time as a test is given.

10. Recognition of Licenses from other locales

Yes, for experience requirements only

CODES for Bridgeport, Connecticut

An Ordinance Revising The Ordinances Of The City of Bridgeport Relative To The Licensing Of Power Engineers, Boiler Tenders, Or Water Tenders.

Be it ordained by the Common Council of the City of Bridgeport:

Section 41.1. Board of Power Engineers.

There shall continue to be a board to be known as the Board of Examiners of Power Engineers which shall consist of three members, each of whom shall be a resident of the city and two of whom shall be practical power Engineers having at least ten years' experience in the operation of steam boilers, steam turbines and steam engines. Annually during the month of December the Mayor shall appoint, as a member of said board for a term of three years, a person who shall come within the classification of the member whose term shall have expired. Said board shall meet at such times as it may determine or upon the

License Requirements for States and Cities Chapter 6

call of its chairman for the purpose of holding examinations for the licenses provided in this ordinance.

Section 41-2. Licenses Required.

No person shall be the Engineer of, or shall have charge of, or operate, or perform the duties of boiler tender or water tender for any steam boiler, steam turbine or steam engine operating with more than fifteen pounds gauge pressure or more than twenty five horsepower without having procured a license therefor from said board, except as provided in this ordinance.

Section 41-3. Application for a License.

Application for such license shall be made to the board upon such forms as shall be prescribed by it. The applicant shall clearly indicate thereon the character of the license, for which application is made and the maximum gauge pressure and horsepower of the apparatus which the applicant intends to operate under his license. In applying for said license, the applicant may apply for an interim certificate authorizing him to act as a power engineer, boiler tender or water tender in a specified power plant under supervision as hereinafter provided in the interim between such application and the date upon which the board shall pass upon his application. No such interim certificate shall be issued unless the licensed Power Engineer, boiler tender or water tender under whose supervision the applicant is to act under said certificate joins in requesting its issuance. Such certificate may be issued by the board or by its chairman when the board is not in session. Under such certificate, the applicant shall be privileged to act as Power Engineer, boiler tender or water tender in a specified power plant under the supervision of a licensed power Engineer if the applicant seeks to be licensed as a Power Engineer or under the supervision of a licensed Power Engineer, boiler tender or water tender if the applicant seeks to be licensed as a boiler tender or

Chapter 6   **License Requirements for States and Cities**

water tender. Such certificate shall expire and be of no effect sixty days after the date of its issuance or at such earlier date as the board shall have passed upon the applicant's license application, and shall be surrendered to the board upon its expiration.

Section 41-4. Licenses, Issuance and Renewal.

No such license shall be issued to any person until he shall have been examined by said board and demonstrated that he is qualified by training and experience to be licensed by it. The examination may include a written test, oral test, practical test or any combination of such tests as the board may determine to be necessary to test the qualification of applicants.

Although licenses shall not be necessary for the operation of low pressure boilers, the board shall examine applicants for low pressure boiler operator's license and shall issue licenses to such applicants as it shall determine to be qualified to operate steam boilers operating with less than fifteen pounds gauge pressure.

No person shall be eligible for examination by said board, and no license shall be issued to any person unless he shall have attained the age of twenty-one years, shall be of temperate habits, shall be able to read and write the English language and unless he shall possess the following educational and experience qualifications:

(a) To be licensed as a Power Engineer:

the applicant (1) shall have served as a journeyman boiler maker or machinist engaged in the construction or repair of steam boilers or steam engines for a period of not less than four years and shall have had one year's experience in the operation and maintenance of stationary steam power plants; or (2) shall have had not less than two years of study at an Engineering school and one year's experience in the operation and maintenance of stationary steam power plants; or (3) shall possess a Power or Stationary Engineer's or marine engineer's certificate issued by the United States or a Power or Stationary Engineer's certificate issued by any state or government subdivision thereof, or (4) shall have been employed as a boiler tender or water tender for not less than three years and shall have had experience in the operation and maintenance of boilers and engines.

(b) To be licensed as a boiler tender or water tender:

the applicant (1) shall have the foregoing qualifications provided for the licensing of Power Engineers; or (2) shall have been employed as a boiler tender, water tender, oiler, or assistant to a Power Engineer, boiler tender or water tender operating stationary steam plants having in excess of fifteen pounds gauge pressure or twenty five horsepower for a period of not less than one year; or (3) shall have had theoretical and practical training in a technical school for boiler or water tenders for not less than one year together with not less than six months practical experience in the operation of steam boilers.

(c) To be licensed as a Low Pressure Boiler Operator:

the applicant (1) shall have the foregoing qualifications provided for the licensing of Power Engineers or boiler or water tenders; or (2) shall have had at least six months' experience as a boiler tender or water tender or as an assistant to a qualified Power Engineer, boiler tender or water tender; or (3) shall have had not less than six months theoretical and practical training in a technical school for boiler or water tenders; or (4) shall have had not less than one year's experience in the operation of low pressure boilers.

Section 41-5. Licenses Fees.

Chapter 6   License Requirements for States and Cities

No application for any of the foregoing licenses shall be received until the applicant shall have paid to said board for the use of the City an examination fee of ten dollars for a Power Engineer's examination, five dollars for a boiler tender's or water tender's examination and five dollars for a low pressure boiler operator's examination. The examination fee shall not be refunded to any applicant who fails to pass said examination or who fails to present himself for examination at such time within sixty days of the application as said board shall assign for the examination to the applicant. Said fee, however, shall be refunded to the applicant who is not examined by the board because of his failure to meet the qualifications requisite for examination. A like examination fee shall be paid to the board at the time of filing any subsequent application for a license which requires an examination by the board. No license shall be issued by the board until the following fees have been paid to it: a fee of five dollars for a Power Engineer's license and a fee of three dollars for a boiler tender's, water tender's or low pressure boiler operator's license. Any unrevoked license shall be renewed without examination for a further period of one year on payment of an annual renewal fee of five dollars for Power Engineer and three dollars for boiler tender, water tenders and low pressure boiler operators within thirty days after its expiration. The board may, in its discretion, renew any license within sixty days after its expiration provided the applicant justifies his failure to renew the same within said thirty day period. Any application for a license which is made after a period of sixty days after the expiration of any former license possessed by the applicant shall be treated as a new application. If any applicant shall fail to pass an examination, he may re-apply for a license and to be examined therefor after a lapse of thirty days from the date of his original examination. If an applicant shall fail to pass an examination a second or more times, he may not reapply for a license and for re-examination until a lapse of six months from the date of the last examination.

Section 41-6. License, Posting of.

Each license shall designate the duties for which it is issued and the maximum gauge pressure and horsepower which the license may operate thereunder. Said license shall be framed and hung in a conspicuous place in the plant, or upon or near the equipment being operated under said license.

Section 41-7. Revocation of License; Appeal.

Said board shall have the power to revoke any license, upon hearing held after not less than five days notice to the licensee, if any license shall have been obtained from the board through fraud or misrepresentation or if the holder of the license shall have been found guilty by the board or by a court of competent jurisdiction of any fraud, deceit, gross negligence, incompetency or misconduct in his duties. Any person aggrieved by the action of the board in refusing to grant a license or in revoking a license issued under the provisions of this ordinance may appeal to the Mayor by filing such appeal with the Mayor within ten days after receiving notice of such action from said board and the Mayor shall thereupon appoint three disinterested persons who shall be residents of the city, two of whom shall be practical Power Engineers having at least ten years' experience in the operation of steam boilers, steam turbines and engines, who shall meet within one week following said appointment to examine the person aggrieved and to confirm or reverse the decision of said board. If they shall find the applicant qualified and entitled to the license in question, then such license shall be issued forthwith. A new license, to replace any license which has been lost, destroyed, or mutilated, may be issued subject to the rules of the board and upon the payment of two dollars for the same.

Section 41-8. Prohibition Against Unlicensed Operation.

Except as provided in this ordinance, no owner, lessee, agent or any other person having control of any premises shall permit the control, management or operation of any boiler, engine or turbine of more than fifteen pounds gauge pressure or more than twenty five horsepower to be entrusted to any person other than a Power Engineer, boiler tender or water tender licensed hereunder. No engine, boiler or turbine over one hundred horsepower shall be operated except by, or under the supervision of, a Power Engineer.

Section 41-9. Exceptions.

Nothing in this ordinance shall apply to the operation of the locomotives of any railroad, nor to the operation of any steamboat by persons duly licensed by authority of the Federal Government.

Section 41-10 Penalty.

Any person who shall violate any provision of Section 2 or Section 11 of this ordinance shall be fined not more than one hundred dollars or imprisoned not more than thirty days or both.

Section 41-11 Enforcement.

This ordinance shall be enforced by the Inspector of Combustibles of the Fire Department.

Section 41-12. Payment of Fees for Renewals of License of Power Engineers, Boiler Tenders, and Water Tenders serving in the Military or Naval Forces of the United States.

During the period of time that any Power Engineer, boiler tender or water tender shall be a member of the Military or Na-

License Requirements for States and Cities   **Chapter 6**

val forces of the United States no license fee shall be required to be paid to the city as provided by ordinance for the renewal of Power Engineer, boiler tender or water tender licenses of such persons. If, upon the expiration of any existing Power Engineer, boiler tender or water tender license, it shall appear to the satisfaction of the Board of Examiners of Power Engineers that any Power Engineer, boiler tender or water tender is a member of the Military or Naval Forces of the United States, such licenses shall be renewed by said Board even though no application for such renewal shall have been made to it.

Adopted December 1, 1952.

Attest: John M. Brannelly, City Clerk

Chapter 6 — License Requirements for States and Cities

Connecticut

New Haven, Connecticut

SUMMARY OF QUALIFICATIONS:

1. Exam Location, Address and Phone Number

City Hall
200 Orange Street
New Haven, Connecticut 06517
203/787-8346

2. Exam Days and Times

1st and 3rd Tuesday of the Month, (July and August - 1st Tuesday)

3. Examination Cost

Free

4. License Cost and Renewal

Initial license.............$25.00

Renewal of license.....$10.00

5. Type of Test

All Oral

6. Grades of Licenses

Fireman & Engineer

Refrigeration

Fireman Engineer

7. Experience and Education Requirements

None

8. U.S. Citizenship Required?

No

9. Local Residency Required?

No

10. Recognition of Licenses from other locales

Yes, reciprocal and for experience

CODES for New Haven, Connecticut

Chapter 26

Steam, Electrical and Refrigeration Equipment

Article 1. Operators

Sec. 26-1. License required.

No person shall be the Engineer of, or shall have charge of, or shall operate any steam boiler, steam engine, electric generator, or refrigerating machinery who shall not have a license certificate authorizing him to have charge of, or operate such boiler, engine, generator, or refrigerating machinery, from the board of examiners of Engineers hereinafter constituted, except as provided in section 26.3.

Sec. 26-2. Board of examiners established.

The Mayor shall appoint a board, to be known as the board of examiners of Engineers, which board shall consist of three (3) members, one of whom shall be a practical Engineer who has had at least ten (10) years' experience in the operation of steam boilers or steam engines and refrigerating machinery, one of whom shall be a practical Engineer who has had at least ten (10) years' experience in the operation of steam boilers or engines and who has a practical working knowledge of electricity and electric generators, and one of whom shall be a manufacturer doing business in said city. All members shall hold office for a period of two (2) years or until their successors are

Chapter 6 License Requirements for States and Cities

appointed and qualified. The board shall grant licenses as hereinafter provided. Said board shall meet on the first and third Tuesdays of each month except in July and August, when it shall meet on the first Tuesdays of said months. Each member of said board shall be entitled to receive as compensation, twenty-five dollars ($25.00) for each meeting of said board which he shall attend and as shall appear by its records. The member of the board designated by the mayor as its secretary shall, in addition, receive five hundred dollars ($500.00) per year.

Sec. 26-3. Applications for licenses.

Before any person shall be employed as an Engineer of, or have charge of, or shall operate any such boiler, engine, electric generator, or refrigerating machinery, he shall make a written application to such board of examiners of Engineers for the license hereinbefore mentioned, and shall specify in such application the particular boiler, engine, generator, refrigerating machinery, or plant consisting of some of all of the above equipment, which he desires to operate or take charge of, which application shall be accompanied by references as to his character and ability, and the filing of such application with said board shall be considered a compliance of this chapter for fifteen (15) days thereafter, or until his application shall have been acted upon by said board, and said applicant, after filing of such application, shall have the right to operate and have charge of any such steam boiler, steam engine, electric generator, or refrigerating machinery, or plant, until his application shall have been passed by the board of examiners of engineers.

Sec. 26-4. Scope of license examinations.

Each person who shall satisfy the board of examiners by a practical and thorough examination that he is a safe and competent person to operate or have charge of the equipment or plant specified in his application shall, upon payment of the fee specified in section 17-20(25), receive a license permitting him to operate same for a period of one year. Said licensee shall thereafter pay yearly a further sum as required by section 17-20(25) for the renewal of said license. Said license shall apply only to one location at which such equipment or plant is installed. Before taking charge of plant or equipment at any other location, the licensee shall apply for another license for the plant or equipment at such other location. For any such additional license there shall be an additional fee charged in the amount shown in section 17-20(25). For the renewals above mentioned there shall be no additional examination required unless, in the judgment of the board of examiners, it shall be necessary. Said license must be displayed in a conspicuous place in the plant, or upon or near the boiler, engine, electric generator or refrigerating machinery operated or attended by the person to whom such license has been issued.

Sec. 26-5. Effect of refusal to grant license.

If said board shall refuse to grant an applicant a license, no license shall be issued him for the next six (6) months following, but after said time, said applicant may make another application, and if found qualified shall be granted a license.

Sec. 26-6. Appeals from refusals to grant licenses.

Whenever the board of examiners of Engineers shall refuse to grant a license, it shall give immediate notice of such action to the applicant or licensee. If the applicant or licensee wishes to contest such action on the part of said board, he or she shall within ten (10) days after receiving such notice file with said board a request for a hearing. Such hearing, if so requested, shall be held by said board within fifteen (15) days after receipt of such request and shall be conducted in accordance with the procedures set forth in Sections 4-177 to 4-180 inclusive of the general statutes.

Sec. 26-7. Duty of owner to assure compliance with licensing requirements.

The owner of a steam boiler, steam engine, electric generator, or refrigerating machinery, other than such of the above mentioned equipment as hereinafter excepted, shall not operate, or cause to be operated any such equipment unless the person in charge of or operating it has been duly licensed by said board to operate such equipment, provided, however, that said owners or user may employ any person without a license for a period of fifteen (15) days upon notification by said owner or user to the clerk of said board of examiners of such employment, in the event that said person so employed has made application for a license, and that such application, at the time of such employment, has not been acted upon by said board.

Sec. 26-8. Revocation of licenses.

The board of examiners of Engineers may at any time revoke a license issued on account of inebriety, incompetency, or negligence of the licensee, or for any other good cause, and no license shall be issued to such person whose license has been revoked, during the next six (6) months following, after which time the license revoked may again be issued if, in the judgment of the board, the cause of the revocation no longer exists.

Sec. 26-9. Abandoning running machinery.

Any licensed person in charge of or operating any steam boiler, steam engine, electric generator or refrigerating machinery, who shall abandon it while in operation, without leaving in charge of the same another licensed person, or an applicant for such license within the fifteen-day period specified in section 26-7, and any other person who shall procure, counsel, instruct or require such person to abandon such machinery and equipment, shall be guilty of a misdemeanor, and upon conviction shall be punished by a fine of not less than one hundred dollars ($100.00) or more than three hundred dollars ($300.00) for each offense. Each day during which such abandonment continues may be treated as a separate offense under this section.

Section 26-10. Machinery excluded from regulations.

This article shall not apply:

- (a) To railway locomotives nor to Engineers employed thereon, nor to vessels subject to inspection and licensing by any agency or instrumentality of the United States government, and the Engineers employed thereon; nor

- (b) To boilers used for heating or steam generating purposes in which the maximum pressure does not exceed fifteen (15) pounds per square inch gauge pressure; nor

- (c) To refrigerating machinery in which the amount of refrigerant used in such system does not exceed three hundred (300) pounds of such refrigerant.

Sec. 26-12. Penalty.

Except as otherwise herein provided, every person who shall violate any of the provisions of this article, or who shall aid, counsel, abet, require or procure the violation thereof, shall be guilty of a misdemeanor, and upon conviction shall be punished by a fine of not less than fifty dollars ($50.00) not more than one hundred dollars ($100.00) for each violation.

Chapter 6   License Requirements for States and Cities

Delaware

Wilmington, Delaware

SUMMARY OF QUALIFICATIONS:

1. Exam Location, Address and Phone Number

City County Building
800 French Street
6th Floor Conference Room
Wilmington, Delaware 19801-3537
302/571-4220

2. Exam Days and Times

3rd Monday evening of each month (excluding July) at 7:00 p.m. Excluding holidays.

3. Examination Cost

$30.00

4. License Cost and Renewal

Initial license..........................$30.00

Renewal (every two years)......$50.00

5. Type of Test

All Written

6. Grades of Licenses

Fireman

Third Class Engineer

Second Class Engineer

First Class Engineer

7. Experience and Education Requirements

Fireman.........................Six months experience

License Requirements for States and Cities   Chapter 6

Third Class Engineer......Two years experience

Second Class Engineer...Two years experience as Third Class Engineer

First Class Engineer ....Two years experience as Second Class Engineer

8. U.S. Citizenship Required?

No

9. Local Residency Required?

No

10. Recognition of Licenses from other locales

U.S. Merchant License - Equivalent license will be given with an oral exam. Other Licenses - must take written exam for equivalent.

CODES for Wilmington, Delaware

General Requirements

Fireman

Minimum age 18

Must have knowledge and experience pertaining to powerhouse boiler equipment and operation, with 6 months experience under direction of an Engineer. Requires knowledge of combustion and heat transfer, safety controls and fuels. Must pass a written examination for black seal Fireman's license.

Third Class Engineer

Must have two years' experience or equivalent beyond Fireman. Requires knowledge and experience in powerhouse equipment and facilities including boilers, pump, turbines, die-

Chapter 6 License Requirements for States and Cities

sel engines, air conditioning, fuels, condensers, lubrication, pressures and water treatment. Must pass a written examination for red seal Third Class Engineers license.

Second Class Engineer

Must have two years' experience or equivalent beyond Third Class Engineer. Requires knowledge and experience pertaining to codes, tests, turbine operating and maintenance, refrigeration, air conditioning, electrical principles, generators, motors and controls. May act as operation Engineer under supervision of a properly licensed Chief Engineer. Must pass written examination for blue seal Second Class Engineers license.

First Class Engineer

Must have two years' experience or equivalent beyond Second Class Engineer. Requires extensive knowledge and experience in all phases and equipment in powerhouses. Must possess ability in direction of powerhouse personnel. May act as Chief Engineer in any plant. Must pass written examination for gold seal First Class Engineers license.

Presenting a license from elsewhere:

- (1) U.S. Merchant Marine License - Equivalent license will be given with an oral exam
- (2) Other Licenses - Must take written exam for equivalent.
- (3) U.S. Navy or Coast Guard - Must take written exam - Grade of license will be determined by the Board from the application.

Examinations

- (1) Original exam cost $30.00 - if you fail.

A. One retake in 30 days for an additional $30.00 - if you fail.

B. Exam can be retaken again after 6 months with $30.00 fee.

- (2) License will be issued for 1 year for an additional fee of $30.00 when exam is passed.

License Renewal:

- (1) All licenses are renewable on July 1
- (2) Licenses may be renewed by mail

License Fees:

- (1) All fees to be paid by check or money order payable to "Board of Examining Engineers, City of Wilmington".
- (2) $30.00 fee for all 1 year renewals
- (3) $50.00 fee for a 2 year renewal
- (4) Those with expired licenses, prior to April 1, 1990, must pay $8.00 per year for the number of years the license was invalid.
- (5) Those with licenses that expired after April 1, 1990, will have to pay $30.00 for each year the license was invalid.
- (6) After 3 years of having a non-valid license, the person must pay the $30.00 fee for the back years. In addition, the person will have to retake the exam and pay the examination fee of $30.00. The back fee may be waived if the person can show proof of living outside of the general area.

Chapter 6   License Requirements for States and Cities

Please address all mail and draw checks in favor of:
Board of Examining Engineers
Louis J. Redding City/County Building
800 French Street
Sixth Floor
Wilmington, Delaware 19801

Florida

Hillsboro County-Tampa, Florida
SUMMARY OF QUALIFICATIONS:

1. Exam Location, Address and Phone Number
Hillsboro County Building Dept.
800 E. Twigs
Room 102
Tampa, Florida 33601
813/272-5652

2. Exam Days and Times

There is no set time every week or month

3. Examination Cost

$145.00

4. License Cost and Renewal

Initial license (2 years).......$10.00

Renewal (2 years)...............$10.00

5. Type of Test

All Written - Have to apply to take an oral test.

6. Grades of Licenses

First Class Unlimited Engineer

First Class Steam Engineer

Second Class Steam Engineer

High Pressure Fireman

Low Pressure Fireman

7. Experience and Education Requirements

Chapter 6 — License Requirements for States and Cities

Under Codes for Hillsboro County

8. U.S. Citizenship Required?

Not Specified

9. Local Residency Required?

No

10. Recognition of Licenses from other locales

Not Specified

CODES for Hillsboro County, Florida

Tampa Code

Sec. 44-51. Licenses to be signed by board chairman, attested by secretary.

All licenses granted by the board of examiners of stationary engineers shall be signed by the chairman and attested by the secretary.

Sec. 44-52. Classification of licenses for engineers, operators and firemen.

Licenses for engineers, operators and firemen shall be classified as follows:

(1) First Class Unlimited Engineers' licenses:

Shall qualify the holder to take charge of and operate any steam plant without limitation as to character or size, and all refrigerating or generating machinery driven by steam, diesel engines, internal combustion or electricity, and shall be issued to such persons as shall qualify therefor by passing the necessary examinations and who have had the following experience:

- (a) Any person who has been employed as an oiler or as a general assistant under the instructions of a licensed First Class Engineer for a period of not less than three (3) years.

- (b) Any person who has served as fireman or general assistant to the engineer on a steamboat or railway locomotive for a period of not less than three (3) years and shall have been employed for not less than one (1) year as an assistant under a licensed First Class Engineer.

- (c) Any person who has learned the trade of machinist or boilermaker and has worked at such trade two (2) years exclusive of the time served as apprentice or while learning such trade, and has not had less than two (2) years' experience as an assistant under a First Class Engineer.

- (d) Any person who is a graduate mechanical engineer of a recognized school of technology, and has had not less than one year's experience under a licensed first class engineer.

- (e) Any holder of a second class engineer's license as issued by the board of examiners of the city and who has actively performed the duties as such for a period of not less than two (2) years.

- (f) Any person holding a first grade or first class engineer's license issued by any city in the United States having a population of one hundred thousand (100,000) or over, or holding Steam Engineer's license issued by the United States Government or marine service.

(2) First Class Steam Engineers' licenses:

Chapter 6 — License Requirements for States and Cities

Shall qualify the holder to take charge and operate any steam plant, without limitation as to character or size.

(3) First Class Refrigeration Engineers' licenses:

Shall qualify the holder of such license to operate all refrigeration machinery, driven by diesel, internal combustion or electricity.

(4) Second Class Steam and Refrigeration Engineers' licenses:

Shall qualify the holder to take charge of and operate steam plants up to one hundred and seventy-five (175) horsepower, and refrigeration and internal combustion engines and shall be issued to such persons as shall qualify therefor, pass the necessary examination and have had at least two (2) years' experience as oiler, Fireman or assistant under a licensed First Class Engineer; or, any person who is a graduate of a recognized school of technology and has had at least one year's experience as an Oiler, Fireman, or assistant under a licensed First Class Engineer.

(5) Second Class Steam Engineers' licenses:

Shall qualify the holder to take charge of and operate steam plants, up to one hundred and seventy-five (175) horsepower.

(6) Second Class Refrigeration Engineers' licenses:

Shall qualify the holder of such license to operate all refrigeration machinery, driven by diesel, internal combustion or electricity up to one hundred and seventy-five (175) horsepower.

(7) Third Class Hoisting and Portable Engineers' licenses:

Shall qualify the holder to take charge of and operate portable boilers with engines and machinery other than a boiler

License Requirements for States and Cities Chapter 6

feed-pump in operation and internal combustion engines and shall be issued to any person who has had at least six (6) months' experience in the operation of such machinery and equipment and shall pass the necessary examination therefor.

The application and operator's license shall describe generally the type of machinery and equipment for which applicant desires to and is licensed to operate.

(8) Fireman's High Pressure license:

Shall qualify the holder to take charge of and operate boilers of from ten (10) to fifty (50) horsepower, and shall be issued to such persons as may qualify therefor, pass the necessary examination and have had at least six (6) months' experience in the care and operation of boilers as Fireman or assistant under a licensed First or Second Class Engineer.

(9) Fireman's low pressure license shall qualify the holder to take charge of and operate boilers carrying the pressure of less than fifteen (15) pounds per square inch, and shall be issued to such persons as may qualify therefor, and pass an examination consisting of safety questions pertaining to the actual physical operation of Low Pressure heating boilers, safety valves and oil burning systems. No definite period of experience shall be required but the applicant shall satisfy the examiners that he is capable of operating safely the boiler for which he is applying for a license.

Georgia
No State or City License

Hawaii
No State or City License

Idaho
No State or City License

Chapter 6   License Requirements for States and Cities

Illinois

Chicago, Illinois

SUMMARY OF QUALIFICATIONS:

1. Exam Location, Address and Phone Number

Department of Buildings
Bureau of Licensing and Registration
Room 405, 4th Floor
320 North Clark Street
Chicago, Illinois 60610
312/744-3895

2. Exam Days and Times

The test is given twice a year; March and August

3. Examination Cost

$70.00

4. License Cost and Renewal

Initial license (for four years).....$60.00

Renewal (for four years).............$60.00

5. Type of Test

All Written - multiple choice

6. Grades of Licenses

Stationary Engineer

7. Experience and Education Requirements

Stationary Engineer - 2 years experience on boilers over 15 psi

8. U.S. Citizenship Required?

Yes

9. Local Residency Required?

Chicago, Illinois   137

No

10. Recognition of Licenses from other locales

No

IN ORDER TO PROCESS THE APPLICATION IT IS NECESSARY THAT YOU PROVIDE THE FOLLOWING:

- An application form completely filled out with notarized signature;
- Examination fee of $70.00; make check payable to Chicago Department of Revenue
- An original letter on company stationery, signed by a Chief Engineer and including the Chief Engineer's license number; letter must identify the nature of your experience, the pounds of pressure of steam of the boilers and the dates of your experience. Have this letter notarized.

Chapter 6   License Requirements for States and Cities

Elgin, Illinois
SUMMARY OF QUALIFICATIONS:
1. Exam Location, Address and Phone Number
City Hall
150 Dexter Ct.
Elgin, Illinois 60120-5555
708/931-5650
2. Exam Days and Times
1st Tuesday of every month at 7 p.m. (3 hr test)
3. Examination Cost
Engineers..................$10.00

Boiler Fireman..........$ 6.00
4. License Cost and Renewal
Initial license included in cost of exam.

Stationary Engineer-Renewal of License.... $8.00

Boiler Fireman-Renewal of License..........$4.00
5. Type of Test
Written and Oral (50 questions)
6. Grades of Licenses
Boiler Tender and Fireman

Stationary Engineer
7. Experience and Education Requirements
Boiler Fireman............None

Stationary Engineer ...1 year practical experience
8. U.S. Citizenship Required?

Yes

9. Local Residency Required?

No

10. Recognition of Licenses from other locales

No

CODES for Elgin, Illinois

BOARD OF EXAMINERS OF STATIONARY ENGINEERS -STATIONARY ENGINEER LICENSING

3.56.010. Definitions

(A) "Boiler pressure" means the pressure in a boiler as determined by a gauge and shall not be interpreted as absolute pressure.

(B) "Fireman" means any person whose duty it is to keep water in a steam boiler and fire or regulate the firing of such boiler whether under the supervision of a licensed Stationary Engineer or not, and who is qualified to operate boilers, boiler feed pumps, injectors, and turbine-driven pumps.

(C) "High Pressure Boiler" means any steam boiler carrying more than fifteen pounds per square inch of pressure.

(D) "Low Pressure Boiler" means any steam boiler carrying fifteen pounds or less per square inch of pressure.

(E) "Stationary Engineer" means a person who has had not less than one year of practical experience in the operation of steam engines, boilers, pumps, and turbines. The word "Engineer" shall not include persons operating steam boilers carrying less than fifteen pounds of pressure per square inch.

Chapter 6   License Requirements for States and Cities

(F) "Steam boiler" means a pressure vessel under steam pressure, carrying a steam pressure of more than fifteen pounds per square inch gauge regardless of intended use.

(G) "Steam engine" means any engine or turbine operated by steam pressure, not including however any steam locomotive or steam-propelled vehicle operated upon the highway.

3.56.020. Stationary Engineer - License required.

It is unlawful for any person to operate a steam engine, boiler or turbine in the city unless such person has first obtained a license as a Stationary Engineer as provided in this chapter.

3.56.030. Fireman - License required.

It is unlawful for any person to tend a steam boiler unless such person has first obtained a license as a Fireman as provided in this chapter, provided that no license shall be required for any person tending Low Pressure boilers, and provided further that a licensed Stationary Engineer shall be authorized to perform services as a Fireman.

3.56.040. Duty of owners - Licensed operators required.

It is unlawful for any person owning a boiler or stationary engine or turbine to cause the same to be operated unless such boiler or stationary engine or turbine is in the charge of a person having a license to operate the same and such person is in actual attendance in his duties during the operation of such stationary engine or turbine or boiler.

3.56.050. Board of Examiners of Stationary Engineers - Created.

There is created a board of examiners of Stationary Engineers (hereinafter called the board). The board shall consist of

Elgin, Illinois

three members who shall be licensed Stationary Engineers, belong to the N.A.P.E., and be well informed as to the construction, operation and theory of steam boilers and steam engines. Members of the board shall be appointed by the city council. They shall take the oath required of city officers. The board shall elect a chairman. The city clerk or a designated representative of that office shall serve as secretary of the board.

3.56.060. Board of examiners of Stationary Engineers - Terms and meetings.

A. The terms of the members of the board of examiners of Stationary Engineers shall expire on the first Monday in May succeeding the regular biannual municipal election, provided however, all appointments shall remain in effect until successors have been appointed.

B. Regular meetings of the board shall be held in the city hall on the first Tuesday of each month between the hours of seven p.m. and ten p.m. Special meetings of the board may be called by the chairman or any two members, either by actual notice or by written notice mailed not later than the day preceding the day upon which the special meeting is to be held.

3.56.070. Board of examiners of stationary engineers - Rules and regulations.

The board shall adopt rules and regulations governing meetings of the board and for the purpose of carrying out the requirements contained in this chapter. Such regulations shall be reduced to writing and filed with the secretary of the board. Such regulations may be amended. The board may adopt and use an official seal.

3.56.080. Board of Examiners of Stationary Engineers - Duties:

A. Applications. Applicants for examination shall pay the required original fee to the city clerk who shall furnish the applicant with proper forms for application for examination.

B. Examinations. The board shall examine the qualifications of all applicants for licenses as Stationary Engineers or Firemen. Such examinations may be either written or oral or both. The board shall require that such applicant be qualified to perform the duties required for the safe operation of steam boilers and stationary engines and shall inquire into the habits of the applicant as to sobriety and faithfulness to duty. After examination of the applicant, the board may grant a license to the applicant.

C. Renewals. Renewals of licenses for Stationary Engineers and Firemen shall be made without examination, provided that the board may require applicants for renewals to submit themselves for examination. No license shall be renewed or shall be renewable after March 31st succeeding the calendar year for which the license was last in effect.

D. Temporary Permit. The board may issue temporary permits for Engineers and Firemen upon affidavit by the applicant showing that he is operating as an apprentice under a licensed Fireman or Engineer. Such temporary permits shall be for not more than one year and are not to be renewed.

E. Revocation. The board may revoke any license granted by it for violation of this chapter. Hearings on such violations shall be after not less than three days' written notice to the license holder.

3.56.090. License - Issuance.

The licenses herein required shall be issued by the City Clerk upon recommendation of the board and upon the payment of the fee set forth in Section 3.56.100. All licenses shall expire on the thirty-first day of December of each year and shall be prorated as to the first or the second half of the calendar year.

3.56.100. License - Fees.

All fees shall be payable to the city upon filing application whether for original appointment or for renewal. No license or permit shall be issued until such fee has been deposited with the city clerk.

License Fees

Original license as a stationary engineer.........$10.00

Annual license fee for renewal........................$ 8.00

Original license as Boiler Fireman..................$ 6.00

Annual license fee for renewal........................$ 4.00

Not more than two re-examinations shall be permitted on any application.

3.56.110. Penalty for violation.

Any person violating any of the provisions of this chapter shall be fined not more than two hundred dollars. Each day any offense continues shall be deemed to be a separate offense.

Chapter 6 — License Requirements for States and Cities

Evanston, Illinois

In order to be licensed as a Stationary Engineer in the City of Evanston, you must first have an existing license from N.I.U.L.P.E. or another locale such as the City of Chicago. You must also have a letter from an employer stating that you work at his location in Evanston. You may also apply for a license in Evanston if you have a letter from an employer stating you have three years experience operating boilers at the employer's business. If you qualify under one of these two categories, then you go to the Water and Sewer Department at 555 Lincoln Street and pick up an application. Call the Director of the Water and Sewer Department at 1-708-866-2942 to set up an appointment to take a safety examination before the license is issued. There is no cost to take the exam. The exam consists of 8 true and false and 33 multiple choice questions about safety around boilers. You must have at least a 70% score to pass. If you pass, you go to the Building and Zoning Department where the license will be issued. The first issue costs $44.00 and the renewal is $36.00.

License Requirements for States and Cities   Chapter 6

Peoria, Illinois
SUMMARY OF QUALIFICATIONS:
1. Exam Location, Address and Phone Number
Peoria City Hall
419 Fulton
Peoria, Illinois 61602
309/672-8588
Exam Location is in Room 400

2. Exam Days and Times
Second and fourth Wednesday of each month at 7:00 p.m.

3. Examination Cost
Boiler Tender..............$25.00

Stationary Engineer....$50.00

4. License Cost and Renewal
Initial license included in cost of exam.

Boiler Tender...$15.00

Stationary Engineer..$20.00

5. Type of Test
All Oral

6. Grades of Licenses
Boiler Tender

Second Class Engineer

First Class Engineer

7. Experience and Education Requirements
Boiler Tender....................................1 year experience

Second Class Stationary Engineer.....2 years experience

Chapter 6   License Requirements for States and Cities

First Class Stationary Engineer..........3 years experience

8. U.S. Citizenship Required?

No

9. Local Residency Required?

No

10. Recognition of Licenses from other locales

Yes - for experience

CODES for Peoria, Illinois

Chapter 35

Stationary and Hoisting Engineers

Article I. In General

Sec. 35-2 - Same - Applications.

Every application for a certificate or license under this article must be made on the printed blanks furnished by the board of examiners, and, for an Engineer, either stationary, hoisting or portable, must be accompanied by a fee of fifty dollars ($50.00), and, for a boiler or water tender, must be accompanied by a fee of twenty-five dollars ($25.00). (Code 1940, 103-2, Ord. No. 6411, 2; Ord. No. 10770, 1, 1-6-81)

Sec. 35-3 - Same - Grades.

The certificates or licenses for Stationary Engineers shall consist of two (2) Grades: First and Second Grades, but only one Grade of licenses for hoisting or portable Engineers shall be granted by the board of examiners of stationary and hoisting Engineers. (Code 1940, 103-2, Ord. No. 6411, 2; Ord. No. 10770, 1, 1-6-81)

Sec. 35-4 - Same - Qualifications of applicant.

An applicant for a boiler tender's license must be a person who has a thorough knowledge of the construction and management of steam boilers. Each applicant must state upon the blank the extent of his experience. (Code 1940, 103-3; Ord. No. 10770, 1, 1-6-81)

Sec. 35-5 - Same - Duration; fees.

If an applicant under this article is found qualified by a majority of the members, the members of the examining board of stationary and hoisting Engineers shall sign a certificate of qualification, which shall be delivered to such applicant. The certificate shall expire one year from the date of issue. Reissues shall be made upon the payment of a fee of twenty dollars ($20.00) for Engineers, either stationary, hoisting or portable, and fifteen dollars ($15.00) for boiler tenders' licenses. (Code 1940, 103-1, ord. No. 6411, 1; Ord. No. 10770, 1, 1-6-81)

Sec. 35-6 - Same - Attestations.

Every certificate or license issued by the board of examiners shall be signed by a majority of the board and shall bear the imprint of the seal of the board and be attested to by the city clerk. (Code 1940, 103-5; Ord. No. 10770, 1, 1-6-81)

Sec. 35-7 - Same - Display.

All Engineers regulated by this article shall display their certificate of qualification at all times in a conspicuous place in the boiler or engine room (Code 1940, 103-7; Ord. No. 10770, 1, 1-6-81)

Sec. 35-8 - Same - Suspension or revocation.

A majority of the board of examiners shall have power to suspend or revoke the license of any person in charge of steam boilers or engines for an absence from his post of duty; or, for any violation of the provisions of this article, or other neglect

Chapter 6 — License Requirements for States and Cities

or incapacity; provided, however, that no license shall be suspended or revoked without first giving the accused party an opportunity to be heard in his own defense.

When a license shall be suspended or revoked, no license shall be reissued to the same person for the first offense for thirty (30) days thereafter; for the second offense for ninety (90) days thereafter; and for any offense thereafter, his license shall be permanently revoked, and then only reissued upon a full compliance with the conditions and provisions prescribed for an original license. In case of a suspension or revocation of license, the fee therefore shall be forfeited to the city. (Code 1940, 103-4; Ord. No. 10770, 1, 1-6-81)

License Requirements for States and Cities   Chapter 6

Indiana

Terre Haute, Indiana

SUMMARY OF QUALIFICATIONS:

1. Exam Location, Address and Phone Number

City of Terre Haute
City Hall, 3rd Floor, Conference Room
17 Harding Ave.
Terre Haute, Indiana 47802
812/232-3375

2. Exam Days and Times

1st and 3rd Monday of every month from 6 p.m. to 7:30 p.m.

3. Examination Cost

$10.00

4. License Cost and Renewal

Initial license................$10.00

Renewal of license.......$10.00

5. Type of Test

All Written - 40 to 75 questions

6. Grades of Licenses

3rd Class Engineer - up to 50 h.p.

2nd Class Engineer - up to 200 h.p.

1st Class Engineer - unlimited

7. Experience and Education Requirements

YOU MUST BE RECOMMENDED BY 2 LICENSED ENGINEERS.

Chapter 6   License Requirements for States and Cities

All of the grades of licenses require 1 year practical experience under a licensed engineer. Their grade of license depends on their test score.

60% correct - 3rd Class Engineer

75% correct - 2nd Class Engineer

90% correct - 1st Class Engineer

8. U.S. Citizenship Required?

Yes, and 21 years of age.

9. Local Residency Required?

No

10. Recognition of Licenses from other locales

Yes - they have people with licenses from Ohio, Illinois & Indiana.

Iowa

Des Moines, Iowa

SUMMARY OF QUALIFICATIONS:

1. Exam Location, Address and Phone Number

City of Des Moines
Building Department
Armory Building
602 E. 1st Street
Des Moines, Iowa 50309
515/283-4946

2. Exam Days and Times

1st and 3rd Tuesday of every month at 5 p.m.

Except for June, July and August.

3. Examination Cost

First Class Engineer.................$20.00

Second Class Engineer.............$15.00

Third Class Engineer................$15.00

First Class Fireman..................$10.00

Second Class Fireman.............$10.00

4. License Cost and Renewal

First Class Engineer.................$20.00 - Initial license & renewal

Second Class Engineer............$15.00 - Initial license & renewal

Third Class Engineer................$15.00 - Initial license & renewal

Chapter 6   License Requirements for States and Cities

First Class Fireman..................$10.00 - Initial license & renewal

Second Class Fireman..............$10.00 - Initial license & renewal

5. Type of Test

All Written - multiple choice - based on practical experience

6. Grades of Licenses

First Class Engineer

Second Class Engineer

Third Class Engineer

First Class Fireman

Second Class Fireman

7. Experience and Education Requirements

First Class Engineer.................Five years experience

Second Class Engineer............Three years experience

Third Class Engineer...............Two years experience

First Class Fireman..................Two years experience

Second Class Fireman..............One year experience

8. U.S. Citizenship Required?

No

9. Local Residency Required?

No

10. Recognition of Licenses from other locales

Yes - for experience

Des Moines, Iowa

CODES for Des Moines, Iowa

Subchapter 5, Des Moines City Code

8-365. POWERS AND DUTIES.

(a) The board of examiners shall hold regular stated meetings at least twice monthly for the purpose of examining into and determining the qualifications of applicants for license as Engineers or Firemen and shall make and post such rules and regulations not inconsistent with the provisions of this subchapter, other provisions of this code or the statutes of the state as shall be necessary and proper for carrying into effect the provisions relative to examinations and hearings.

(b) The board of examiners shall have the power to issue licenses, as provided for in this subchapter, to those found to be eligible after due examination.

(c) After giving the accused licensee an opportunity to be heard, the board of examiners shall have the power to suspend or revoke the licenses provided in this subchapter for the following enumerated reasons:

- (1) For carrying a higher steam pressure than authorized by the senior mechanical inspector.
- (2) For intoxication or the drinking of intoxicating liquors while on duty.
- (3) For any unauthorized absence from post of duty.
- (4) For defacing or obstructing a license certificate.
- (5) For any negligence, incompetence or incapacity that may endanger life or property.

- (6) For any violation of the provisions of this subchapter.

8-366. APPEAL FROM ACTION OF BOARD.

(a) Any person questioning the action of the board of examiners in refusing to grant a license because of failure to pass the prescribed examination, or for other cause, or who feels himself aggrieved by an order of revocation or suspension by the board of examiners may, within ten days from the board action complained of, appeal his or her case to the city council. The council shall then appoint a special board of examiners consisting of three persons holding First class licenses as Engineers in the city, which shall review the action complained of together with other evidence or facts pertaining to the action in question, after which they shall submit a finding of facts to the city council together with a recommendation in the premises.

(b) After the report and recommendation has been filed, the city council shall either, affirm or reverse the action of the board of examiners.

(c) Compensation for the special board of examiners shall be at the same rate as for the regular board of examiners.

8-367. SENIOR MECHANICAL INSPECTOR APPOINTED.

The mechanical division of the department of building shall be under the supervision of a senior mechanical inspector appointed by the city manager.

8-368. INSPECTOR'S POWERS AND DUTIES.

(a) The senior mechanical inspector shall be responsible for the enforcement of the provisions of this subchapter.

(b) He or she shall have authority to inspect all steam or power generating equipment for conformance with the provisions of the city mechanical code.

(c) He or she shall supervise the licensing of Boiler Firemen and Boiler Engineers in accordance with the provisions of this subchapter and in accordance with the rules and regulations adopted by the board of examiners.

8-369. OPERATOR'S LICENSE REQUIRED.

No person shall operate, control, or assume responsible supervision of any stationary or portable steam engine, any stationary or portable steam boiler or other steam generating apparatus, or any appurtenance thereto, unless and until he or she is properly licensed as provided in this subchapter. No owner, user, or agent of any owner or user of any such engine, boiler or other steam generating apparatus or appurtenance thereto shall cause, permit, or allow the same to operate or be operated without first having determined that the person operating and the person in responsible charge thereof are in possession of a proper and valid license for that purpose.

8-370. APPLICATION.

(a) Any person desiring to act as a stationary or portable Engineer or Fireman shall apply to the board of examiners for license to do so.

(b) The application referred to in subsection (a) of this section shall be in writing, on forms furnished for that purpose, shall be accompanied by the required fee and shall set out the applicant's name, age, place of residence, nationality and present status of citizenship, present place and position of employment and a complete record of his or her experience as an Engineer or Fireman, all of which information shall be vouched for

Chapter 6 License Requirements for States and Cities

by two citizens of the city or may be verified under oath by the applicant.

8-371. FALSE APPLICATION STATEMENTS.

A willfully false statement in the application referred to in section 8-370 of this code shall be sufficient cause for revocation of any license issued by reason thereof.

8-372. QUALIFICATIONS OF APPLICANT.

To be eligible for examination in any classification, an applicant for a license, under the provisions of this subchapter, shall be a citizen of the United States, or have declared intention to become such, not less than 18 years of age, of temperate habits and not addicted to the use of drugs or the excessive use of intoxicating liquor, and be able to meet the requirements of the particular class of license applied for.

8-373. PERSONS EMPOWERED TO ADMINISTER AFFIRMATION.

Any member of the board of examiners or the Inspectors shall be empowered and qualified to administer any affirmation required by this subchapter.

8-374. CLASSES OF LICENSES.

Licenses required under this subchapter shall be of the following classifications: First Class Engineer; Second Class Engineer; Third Class Engineer; Portable Engineer; First Class Fireman; and Second Class Fireman.

8-375. WORK WHICH MAY BE DONE.

By holder of First Class License:

A license as First Class Engineer, unless restricted in any manner by the board of examiners, shall entitle the holder to take charge of any plant referred to in this subchapter.

By holder of Second Class License:

A license as Second Class Engineer, unless restricted in any manner by the board of examiners, shall entitle the holder to take charge of any plant not exceeding 200 plant horsepower, or to act as a shift Engineer in a First class plant under the supervision of a First Class Engineer in charge.

By holder of license as Portable Engineer.

A license as Portable Engineer shall entitle the holder to operate any hoisting or portable engine, boiler or other portable equipment.

By holder of Third Class License:

A license as Third Class Engineer, unless restricted in any manner by the board of examiners, shall entitle the holder to take charge of and operate any plant not exceeding 125 plant horsepower, except plants in which steam engines are operated, or to act as a shift Fireman under the immediate supervision of as shift Engineer holding First or Second Class license.

By holder of First Class Fireman's License:

A license as a First Class Fireman, unless restricted in any manner by the board of examiners, shall entitle the holder to take charge of and operate any Low Pressure heating plant or not more than 75 plant horsepower; unless a qualified Engineer is in charge and on duty.

By holder of Second Class Fireman's License:

A license as a Second Class Fireman, unless restricted in any manner by the board of examiners, shall be limited to Low Pressure heating plants of not more than 50 plant horsepower.

8-376. LIMITED OR RESTRICTED LICENSES.

If, after examination, the board of examiners finds that an applicant is qualified to operate a type of plant or a specific plant within the general classifications, set out in section 8-374 but is not qualified to hold an unrestricted license in any of the foregoing classifications, the board may issue a limited license within any of these classifications, upon which the restriction shall be noted.

8-377. EXPERIENCE REQUIREMENTS OF APPLICANT.

No person shall be granted a license as a First Class Engineer until he or she furnishes the board of examiners with the following satisfactory proof:

(1) For a First Class Engineer, that he or she has had five years' experience in Steam Engineering or refrigeration plants and has had experience in the operation of heating ventilation and electric apparatus.

(2) For a Second Class Engineer, that he or she has had three years' experience in Steam Engineering and knowledge of refrigeration, heating, ventilation and electric apparatus.

(3) For Third Class Engineer, that he or she has had two years' experience in Steam Engineering as Fireman or helper around a boiler plant.

(4) For a First Class Fireman's license that he or she has had two years' experience as Fireman or helper around a boiler plant.

(5) For a Second Class Fireman's license that he or she has had one year's experience as Fireman's helper around a boiler plant.

8-378. EXAMINATIONS.

(a) Notice shall be given to applicants of the time and place of examination at least three days prior thereto.

(b) Examinations shall be designed to test fairly the applicant's knowledge of Engineering matters and to determine his or her competence and fitness to hold the grade of license applied for.

(c) All examinations shall be in writing or oral, by the question and answer method, and shall be graded on a percentage basis.

(d) If, after examination, it shall appear to the board of examiners that the applicant is not qualified to serve in the classification for which he or she has taken examination, the board shall refuse to issue the applicant a license in that classification.

8-379. FEE.

An examination fee in the sum of $20.00 for First Class Engineer, $15.00 each for Second and Third Class Engineer, and $10.00 each for First and Second Class Fireman, shall accompany each application for a license and shall be a condition precedent to taking an examination.

8-380. LICENSE FEES.

An annual license fee for the license required by this subchapter shall be paid to the city as follows:

First Class Engineer. $20.00

Second Class Engineer....................$15.00

Chapter 6 License Requirements for States and Cities

Third Class Engineer......................$15.00
First Class Firemen.........................$10.00
Second Class Firemen....................$10.00
Duplicate License...........................$ 5.00

8-381. DISPLAY OF LICENSE.

(a) Each Engineer and Fireman licensed under the provisions of this subchapter shall at all times keep his or her license posted under glass in a conspicuous place at the plant in which he or she is employed. He or she shall report at once to the mechanical division any change in his or her employment or in his or her place of residence.

(b) The owner, agent or lessee shall post the current certificate of inspection of the boiler in a conspicuous place near the boiler.

8-382. EXPIRATION AND PRORATION.

All licenses required by this subchapter shall expire on January 1 of each year. Original license fees shall be prorated quarterly to the following January 1. Licenses expiring other than on January 1 shall be prorated on the quarterly basis to the following January 1, at which time they shall be renewed on the annual basis. Any license not renewed within 30 days from the date of expiration shall be considered void and its holder shall be required to pass another examination to operate within the city.

Sioux City, Iowa

SUMMARY OF QUALIFICATIONS:

1. Exam Location, Address and Phone Number

Application Site

Community Development Department
Inspection Services Department
City Hall, Room 110
P.O. Box 447
Sioux City, Iowa 51102-0447
712/279-6120

Examination Site

Briarcliff College
3303 Rebecca St.
Sioux City, Iowa 51102-0447
Exam Time: 8:00 a.m.
712/279-6310

2. Exam Days and Times

Quarterly- on the second Saturday of the month (January, April, July, October) Unless it falls on a holiday, then it will be moved up one week.

3. Examination Cost

$45.00

4. License Cost and Renewal

1st Class Hydronic Engineer....$35.00 - Initial license and renewal

2nd Class Hydronic Engineer...$30.00 - Initial license and renewal

3rd Class Hydronic Engineer....$25.00 - Initial license and renewal

| Chapter 6 | License Requirements for States and Cities |

4th Class Hydronic Engineer....$20.00 - Initial license and renewal

If license has not been renewed by January 31st the license fee renewal will be double the cost.

5. Type of Test

All Written - multiple choice

6. Grades of Licenses

4th Class Hydronic Engineer Low Pressure 51-100 HP High Pressure 0-25 HP

3rd Class Hydronic Engineer - Low Pressure 101-200 HP High Pressure 26-100 HP

2nd Class Hydronic Engineer - Low Pressure 201-400 HP High Pressure 101-200 HP

1st Class Hydronic Engineer - Low Pressure 401 HP and over High Pressure 201 HP and over

7. Experience and Education Requirements

4th Class Hydronic Engineer - one years experience - education in engineering or technical college may count toward part of this experience.

3rd Class Hydronic Engineer - two years experience - education in engineering or technical college may count toward part of this experience

2nd Class Hydronic Engineer - three years experience - education in engineering or technical college may count toward part of this experience

1st Class Hydronic Engineer - four years experience - education in engineering or technical college may count toward part of this experience

8. U.S. Citizenship Required?

Not Specified

9. Local Residency Required?

Not Specified

10. Recognition of Licenses from other locales

Application considered on individual basis - Up to Board

CODES for Sioux City, Iowa

20.28.040. LICENSE REQUIRED.

(a) No person shall operate any gas, oil or electric hydronic heating or Group II refrigerants equipment in a system nor shall any owner or user of any such heating/refrigeration equipment permit the operation of same without having determined that the person or persons operating, controlling or assuming supervision thereof are in possession of a proper and valid license for that purpose, which license or copy thereof shall at all times be displayed in a conspicuous place adjacent to the equipment supervised, protected from damage. The type of license required shall be in accordance with the following schedule:

Chapter 6 License Requirements for States and Cities

(1) Hydronic Heating Engineer

Hydronic Heating Engineer		
LICENSE	LOW PRESSURE	HIGH PRESSURE
None	0-50 HP	0
4th Class	51-100 HP	0-25 HP
3rd Class	101-200 HP	26-100 HP
2nd Class	201-400 HP	101-200 HP
1st Class	401 HP and over	201 HP and over

20.28.050. GAS, OIL, OR ELECTRIC HYDRONIC HEATING EQUIPMENT.

All gas, oil, or electric hydronic heating equipment shall be operated in accordance with and subject to the requirements of this chapter; except when an inspection and remote monitoring system (plans) have been submitted to and approved by the manager of inspection services division and after the approved monitoring system has been installed, then said hydronic heating equipment shall be exempt from requiring an operator present at all times to operate and control such equipment; but such equipment shall be operated under assumed responsible supervision as set forth in Section 20.28.030(2), Definitions.

Owners or operators of gas, oil, or electric hydronic heating systems shall submit to the manager of inspection services division a detailed plan of all inspection and monitoring procedures for review and approved by the manager of inspection services division. The owner or operator of such equipment shall, upon demand, produce all records required to be kept by this section indicating the dates, times, and findings of such inspections made pursuant to such approved inspection plans.

All High Pressure fossil fuel boilers are not included in this Section.

20.28.060. CLASS OF LICENSES.

Licenses shall be issued by the city as follows:

First-Class Hydronic Engineer.

Every applicant for a license certifying competency and proficiency as a First-Class Hydronic Engineer shall have had at least four year's practice in the management, operation, or construction of steam engines, boilers, turbines, absorption units, pumping equipment and appurtenances to the foregoing or equal as determined by the board. Education in Engineering or technical college may be considered by the board as part of the four year's required practice. An examination must be passed for this license.

A First-Class Hydronic Engineer license shall entitle the holder to operate any hydronic heating equipment and appurtenances used in conjunction therewith except as said license may be restricted by the board.

Second-Class Hydronic Engineer.

Every applicant for a license certifying competency and proficiency as a Second-Class hydronic Engineer shall have had at least three years' practice in the management, operation or construction of steam engines, boilers, turbines, absorption units and/or pumping equipment and appurtenances to the foregoing or equal as determined by the board. Education in Engineering or technical college may be considered by the board as part of the three years' required practice. An examination must be passed for this license.

Chapter 6   License Requirements for States and Cities

A Second-Class Hydronic Engineer license shall entitle the holder to operate any hydronic heating equipment and appurtenances used in conjunction therewith of not more than four hundred horsepower of a Low Pressure system and two hundred horsepower of a High Pressure system except as said license may be restricted by the board.

Third-Class Hydronic Engineer.

Every applicant for a license certifying competency and proficiency as a Third-Class Hydronic Engineer shall have had at least two years' practice in the management, operation, or construction of steam engines, boilers, turbines, absorption units, and/or pumping equipment and appurtenances to the foregoing, or the equivalent as determined by the board. Education in Engineering or technical college may be considered by the board as part of the two years' required practice. An examination must be passed for this license.

A Third Class Hydronic Engineer license shall entitle the holder to operate any hydronic heating equipment and appurtenances used in conjunction therewith of not more than two hundred horsepower of a Low Pressure system and one hundred horsepower of a High Pressure system except as said license may be restricted by the board.

Fourth-Class Hydronic Engineer.

Every applicant for a license certifying competency and proficiency as a Fourth-Class Hydronic Engineer shall have had at least one year's practice in the management, operation or construction of steam engines, boilers, turbines, absorption units and/or pumping equipment and appurtenances to the foregoing or equal as determined by the board. Education in Engineering or technical college may be considered by the board as part of the one year's required practice. An examination must be passed for this license.

A Fourth-Class Hydronic Engineer license shall entitle the holder to operate any hydronic heating equipment and appurtenances used in conjunction therewith of not more than one hundred horsepower of a Low Pressure system and twenty-five horsepower of a High Pressure system except as said license may be restricted by the board.

20.28.090. APPLICATION FOR EXAMINATION.

Any person who desires to be licensed as a Hydronic Engineer and/or Refrigeration Engineer shall make application to the board for an examination by the second Monday of each month. The inspector shall provide application forms for this purpose. The completed forms shall include the name of the applicant, his home address, business address, a brief resume of his training and experience, the date of the application, the classification of license for which he is applying, and any other information as required by the board.

20.28.100. EXAMINATION FEES.

Every person who applies for examination or reexamination for a hydronic Engineer's or refrigeration Engineer's license shall first pay the following nonrefundable examination fee to the City Treasurer and a receipt therefor shall be attached to the application:

First-Class Hydronic Engineer..................$35.00

Second-Class Hydronic Engineer..............$30.00

Third-Class Hydronic Engineer.................$25.00

Fourth-Class Hydronic Engineer...............$20.00

First-Class Refrigeration Engineer.............$35.00

Second-Class Refrigeration Engineer........$30.00

Third-Class Refrigeration Engineer...........$25.00

20.28.110. WHEN EXAMINATIONS ARE CONDUCTED.

The board shall normally conduct an examination on the second Tuesday of each calendar month, and at such other times as the board or manager may deem necessary; if additional meetings are necessary to administer examinations, the board shall schedule them as needed. The time and place for each examination shall be designated by the board. The examination shall be practical, written, and shall be of such a nature as to test the capabilities of all applicants uniformly within the scope of the desired license. The applicant shall clearly demonstrate to the board his qualifications for being licensed.

20.28.120. REEXAMINATION.

If an applicant fails to pass an examination, he may reapply for reexamination for the same position at the next regularly scheduled test upon payment of another examination fee.

20.28.130. ISSUANCE OF LICENSE.

The board shall certify and record the names of all applicants who successfully complete the examination and are qualified to be licensed. The secretary shall issue a license (and subsequent annual renewal licenses) to each applicant who has been certified by the board; provided, that each applicant shall have paid a license fee in accordance with Section 20.28.140 of this chapter.

20.28.140. LICENSE FEES AND EXPIRATION DATE.

License Fee. A license fee shall be paid for each original license and annual license renewal as follows:

First-Class Hydronic Engineer................$35.00

Second-Class Hydronic Engineer............$30.00

Third-Class Hydronic Engineer...............$25.00

Fourth-Class Hydronic Engineer.............$20.00

First-Class Refrigeration Engineer..........$35.00

Second-Class Refrigeration Engineer......$30.00

Third-Class Refrigeration Engineer........$25.00

Apprentice Engineer..............................$10.00

2. Expiration Date. Each license shall expire on the next December 31st after the date of issue and may be renewed annually upon request for such renewal by the licensee. Any license renewal which has not been filed with the inspector prior to December 31st, shall expire on that date and may not be renewed without examination except on recommendation of the board. If the license is renewed without an examination, the fee shall be doubled.

Exception: Only the highest fee shall be required of a licensee who applies on the same date for renewal of licenses he holds in more than one classification.

20.28.150. LICENSE NOT TRANSFERABLE.

It is unlawful for any license holder to transfer a license to another or to allow it to be used directly or indirectly by another person.

20.28.160. SUSPENSION OR REVOCATION OF LICENSE.

Chapter 6 License Requirements for States and Cities

(a) The board shall have the authority to suspend or revoke any license issued under this chapter or renewal thereof for the following reasons:

(1) Misrepresentation in obtaining the license;

(2) Failure to comply with any of the ordinances of the city of Sioux City;

(3) Failure to obtain the necessary permits and/or pay for the same;

(4) Carelessness or lack of skill, or operation of any equipment in an unsafe manner;

(5) Upon a showing that the licensee is not physically or mentally competent to perform the functions for which he is licensed.

(b) Suspension or revocation proceedings shall be commenced by the filing of a written complaint with the board, signed by the complainant.

(c) The board may conduct a preliminary investigation prior to a public hearing. If the board finds it advisable after a preliminary investigation or after the filing of the written complaint, a hearing shall be set and the holder of the license shall be given a ten day notice sent by certified mail of the time and place of the hearing. The notice shall also advise the holder of his right to appear at the hearing in person or by counsel for the purpose of defending the allegations made in the complaint. The hearing shall be open to the public and all interested persons shall be given an opportunity to be heard.

(d) The board, may, by majority vote, revoke or suspend all licenses, place the licensee on probation, or dismiss the charges.

(e) A person whose license has been revoked shall not apply for a new license within one year after the revocation aforesaid. Said application shall be accompanied by an examination, application fee, and shall be processed in the same manner as for an original application. The suspension or revocation of a license shall not entitle the holder to a refund or any part of fee which he may have paid.

20.28.170. APPEAL.

In the event any person feels aggrieved by any action of the board, he may appeal from such action to the city council by filing written notice of his appeal within ten days from the date of such action. The council shall give the appealing party and the examining board five days' written notice by certified mail of the date, time and place of hearing. All interested persons shall be given an opportunity to be heard at such hearing and the city council may affirm, modify or overrule the action of the board.

Kansas

No State or City License

Kentucky

No State or City License

Chapter 6　　　　　License Requirements for States and Cities

Louisiana

New Orleans, Louisiana

SUMMARY OF QUALIFICATIONS:

1. Exam Location, Address and Phone Number

New Orleans Municipal Training Academy
401 City Park Avenue
New Orleans, Louisiana 70112
504/565-6107

2. Exam Days and Times

Twice a month on Tuesday at 7 p.m.

3. Examination Cost

Third Class and Second Class Operating Engineer..$13.00

First Class Operating Engineer................................$23.00

4. License Cost and Renewal

Third Class Operating Engineer's Certificate-Initial license and renewal..$28.00

Second Class Operating Engineer's Certificate-Initial license and renewal..$28.00

First Class Operating Engineer's Certificate-Initial license and renewal..$53.00

Special Operator's Certificate-Initial license and renewal..$28.00

Handling fee of $3.00 has been added to license cost.

5. Type of Test

All Written

6. Grades of Licenses

License Requirements for States and Cities Chapter 6

Third Class Operating Engineer

Second Class Operating Engineer

First Class Operating Engineer

Special Operator's Certificate

7. Experience and Education Requirements

Third Class Operating Engineer - 2 years experience; with degree; no experience

Second Class Operating Engineer - 3 years experience; with degree; 6 months experience

First Class Operating Engineer - 4 years experience; with degree; 1 year experience

Special Operator's Certificate - 2 years experience as an apprentice

8. U.S. Citizenship Required?

Yes - except if the board grants a special certificate

9. Local Residency Required?

Yes - 6 months

10. Recognition of Licenses from other locales

Yes for experience but they still need to take exam

CODES for New Orleans, Louisiana

The New Orleans Mechanical Code

110.9 - EQUIPMENT AND SYSTEMS REQUIRING OPERATORS AND CLASSIFICATION OF CERTIFICATES

Chapter 6   **License Requirements for States and Cities**

110.9.1 - A First Class Operating Engineer's Certificate is required:

For an operator to operate the following:

- Boilers - Unlimited.
- Steam engines - Unlimited.
- Internal combustion engines - Unlimited.
- Refrigeration and/or air conditioning systems B 1,2,3,4,5, using Group 1,2,3 Refrigerants - Unlimited
- Equipment involving pressure or speed - Unlimited

110.9.2 - A Second Class Operating Engineer's Certificate:

Is required for an operator to act as an assistant to and under a First Class Operating Engineer, or to operate the following without the supervision of a First Class Operating Engineer:

- Boilers up to 150 horsepower
- Steam engines up to 150 horsepower
- Internal combustion engines up to 150 horsepower
- Refrigeration systems and air conditioning system B2,3,4,5, using Group 1,2,3 Refrigerants; a total of 150 horsepower, but not more than 15 systems
- Air Conditioning systems B1 using Group 1 Refrigerant; a total of 200 horsepower, not over 15 systems, maximum size of any individual unit 60 horsepower.

110.9.3 - A Third Class Operating Engineer's Certificate:

Is required for an operator to act as an assistant to and under a Second Class Operating Engineer, or to operate the following without the supervision of a Second Class Operating Engineer:

- Boilers up to 75 horsepower
- Steam Engines up to 75 horsepower
- Internal combustion engines up to 75 horsepower
- Refrigeration systems and air conditioning systems B 1,2,3,4,5, using Group 1,2,3 Refrigerants, a total of 75 horsepower.

110.9.4 - A Special Operator's Certificate is required:

For an operator to act as an assistant to and under a Third Class Operating Engineer, or to operate the following without the supervision of a Third Class Operating Engineer:

- Boilers containing more than 30 gallons of water and up to 120 square feet of boiler heating surface any pressure above atmospheric pressure.
- Refrigeration or air conditioning systems using Group 2 or 3 Refrigerants from 5 to 20 horsepower.

110.10 - ANNUAL COST OF CERTIFICATE AND EXAMINATION FEES

110.10.1

In order to defray the expenses of the Board in conducting the necessary examinations, every applicant for a Certificate of Competency and Proficiency, either upon his original application or for a renewal thereof, shall be required to pay the following schedule of cost:

First Class Operating Engineer's Certificate..........$50.00

Second Class Operating Engineer's Certificate......$25.00

Third Class Operating Engineer's Certificate.........$25.00

Chapter 6 — License Requirements for States and Cities

Every applicant applying for a Special Certificate of Competency and Proficiency as provided in 110.13 relative to portable equipment and hoisting machinery and being duly certified.

Hoisting and Portable Engineer's Certificate.........$25.00

Every applicant applying for a Special Certificate of Competency and Proficiency for those classes of operators provided for in 110.9 relative to boilers under 120 square feet of boiler heating surface and refrigeration systems up to 20 tons, and being duly certified.

Special Operators Certificate................................$25.00

Special Engineer's Certificate

Boilers (not to exceed 19 H.P.)............................$25.00

Refrigeration (not to exceed 20 tons)...................$25.00

110.10.2 -

An examination fee will be charged on all written examinations. Fee must be paid in advance. In the event the applicant fails to pass the examination, he shall be permitted to take a second examination after the disqualification period without paying the additional fee as provided in the examination fee schedule which follows:

Examination Fee Schedule:

Stationary Engineer's Test

Written or Oral will read as follows:

Per Test

1st Class: Stationary...$20.00

2nd Class: Stationary................................$10.00

3rd Class: Stationary................................$10.00

Hoisting & Portable Engineer's Test.......$10.00

Special Engineer's Test

Boilers (not to exceed 19 H.P.)................$10.00

Refrigeration (not to exceed 20 tons).......$10.00

110.11 - MULTIPLE ISSUANCE OF CERTIFICATES

The Board shall issue an additional certificate to a successful applicant upon request and payment of an amount equal to the cost of the first issuance of the certificate. Such additional certificates shall expire on the same day the first issuance of the certificate expires.

110.12 - RENEWAL OF CERTIFICATE

110.12.1

Each certificate evidencing the necessary satisfactory qualifications of the applicant, according to the classifications stated therein shall expire one year from the date of its issuance, and must be annually renewed in order to permit the holder thereof to engage in the profession, trade or calling of Operating Engineer in the City of New Orleans, according to the classification stated in said certificate. The applicant shall submit a passport photograph, at the time of applying for certificate renewal. The photograph shall have been taken within three months of the renewal date.

110.12.2

Every applicant under this article who fails to make timely application for renewal of his certificate shall pay an additional fee of $1.00 for each month or part thereof that the certificate has been expired.

110.13 - HOISTING AND PORTABLE CERTIFICATES

A special form of Certificate of Competency and Proficiency shall be issued by the Board to those Engineers who have passed a proper examination before the Board, and who operate, or who are in charge of all hoisting or portable equipment such as boilers, steam or internal combustion engines, cranes, derricks, hoists, air compressors, or other power driven equipment.

110.14 - QUALIFICATIONS OF APPLICANTS

Any citizen in the United States, over the age of 18 years, and a resident of the City of New Orleans for at least 6 months prior thereto, may apply for an examination as heretofore set forth. The applicant, however, must have served the necessary apprenticeship, and must comply with the qualifications hereinafter set forth. The Board in its discretion may grant a special certificate to one not a citizen of the United States.

110.15 APPRENTICESHIP REQUIREMENTS

110.15.1

Every applicant for a Certificate of Competency and Proficiency as a First-Class Engineer shall be an Operating Engineer, with 4 years practice in the management, operation or construction of steam engines, boilers, internal combustion engines, refrigeration, or air conditioning equipment/except that those who are graduates of a recognized technical college or trade school are required to have 1 year's practical experience. Every applicant for a Certificate of Competency and Proficiency as a Second-Class Engineer shall be an Operating Engineer, having at least 3 years practice in the management, operation or construction of steam engines, boilers, internal combustion engines, refrigeration or air conditioning equipment; except that those applicants who are graduates from a recognized technical college or trade school shall be required to have 6 months practical experience.

110.15.2

Every applicant for a Certificate of Competency and Proficiency as a Third-Class Engineer shall have at least 2 years experience as an apprentice or junior Engineer or Fireman, etc., except that graduates of a recognized technical college or trade school are not required to have any practical experience.

110.15.3

Every applicant for a Special Certificate of Competency and Proficiency to operate boilers, steam or internal combustion engines specified in 110.9 and 110.13 shall be an Operating Engineer, having at least 2 years practical experience as an apprentice in the operation of steam engines, boilers, or internal combustion engines. Applicants for special certificates coming under the classifications of 110.9 and 110.13 must, to the satisfaction of the Board, establish their familiarity with the operation of the equipment in their care; these applicants must furnish three letters of reference, at least one of which should be from the applicant's present employer or former employer, attesting to the applicant's experience and character.

110.16 - VALIDITY OF CERTIFICATES

Special Certificates of Competency and Proficiency as provided for in 110.13 to be valid must bear at least two signatures, namely, the Chairman or Acting Chairman of the Board, and one other member of the Board. Certificates must bear at least five signatures, namely, the Chairman or Acting Chair-

Chapter 6 — License Requirements for States and Cities

man and four other members of the Board and countersigned by the Secretary. The official seal of the Board shall be placed on all certificates issued.

110.17 - REVOCATION OF CERTIFICATE

A certificate may be suspended indefinitely or be revoked, in addition to other lawful reasons, for any of the following reasons: intemperance, incompetency, neglect of duty, failure to keep the apparatus in his charge in safe working order, or leaving a refrigeration, air conditioning, steam system, etc., in operation without providing a substitute holding the proper Certificate of Competency and Proficiency from this Board. No Certificate shall be suspended or revoked without first giving the accused person an opportunity to be heard by the Board in his own defense.

Appeal from the decision of the Board of Examiners of Operating Engineers may be taken within 10 days to the Board of Building Standards and Appeals through the Director of Safety and Permits.

110.18 - RIGHT TO ENTER PREMISES

Any member, or the Secretary, of the Board is empowered, upon showing his proper credentials, to visit any plant or place of business in charge of an Engineer or operator holding a certificate of competency and proficiency, or any other plant where the Engineer or operator thereof does not hold such certificate during the operating hours of such plant or place.

110.19 - CERTIFICATES DISPLAYED

Certificates of competency and proficiency of all successful applicants before the Board must be displayed in a conspicuous place in the engine or boiler room or mechanical equipment room, with an up to date passport type photograph, attached to the license, for identification purposes. Engineers who have qualified for a special class certificate for the operation of hoisting or portable equipment must carry the certificate on their person while in charge of such hoisting or portable equipment.

110.20 - EXCEPTIONS

110.20.1

This code does not apply in any way to Engineers in charge of locomotives of franchised railroads, nor to Engineers employed by a person, firm or corporation operating under Certificates of Convenience and Necessity, franchise or indeterminate permits, nor to Engineers employed by the State of Louisiana or any of its Boards, Commissions or agencies, nor to Engineers in charge of steamboats or steamships.

110.22 - PENALTY TO ENGINEER

110.22.1

Any person engaged in the practice, calling, or profession as an Engineer, engaged in the operation of any boiler, steam engine, internal combustion engine, refrigeration system, air conditioning system, or equipment involving pressure or speed, under the terms of this Code, without being at the time the holder of a proper and valid Certificate of Competency and Proficiency issued by the Board provided herein, shall, upon conviction thereof before any court of competent jurisdiction, be penalized as heretofore set forth in this Code.

110.22.2

It shall be a misdemeanor for any person to alter, reproduce, transfer, lend or rent his certificate or to use a license not his own.

110.22.3

| Chapter 6 | License Requirements for States and Cities |

It shall be unlawful for an Engineer holding a certificate from this Board to be in charge of more than one plant, unless there are Engineers on duty with the proper certificates.

110.22.4

An engineer holding a Certificate of Competency and Proficiency from this Board who knowingly gives false statements on Voucher One and Two of the application form as the applicants' experience, qualifications, residence requirements, etc. is subject to having his certificate revoked and also prosecuted.

Maine

State of Maine

SUMMARY OF QUALIFICATIONS:

1. Exam Location, Address and Phone Number

Exam Location:
Augusta Armory
Western Avenue
Augusta, Maine

Office Location:
Maine Boiler Division
Station 45
Augusta, Maine 04337
207/624-6420

2. Exam Days and Times

Examinations are held in Augusta at the State Armory at 9:00 a.m. and scheduled after 60 to 120 people have applied. They are held four times a year on the 2nd Wednesday of the month in March, June, September and December.

3. Examination Cost

$10.00

4. License Cost and Renewal

Low Pressure Boiler Operator.........$20.00 for three years Initial license and renewal

Stationary Engineers and High Pressure Boiler Operator... $30.00 for three years-Initial license and renewal

5. Type of Test

All Written -multiple choice

6. Grades of Licenses

Chapter 6   License Requirements for States and Cities

   Heating Boiler Operator-Low Pressure

   High Pressure Boiler Operator

   Fourth Class Stationary Engineer-50,000 pounds of steam an hour

   Third Class Stationary Engineer-100,000 pounds of steam an hour

   Second Class Stationary Engineer-200,000 pounds of steam an hour

   First Class Stationary Engineer-Unlimited

7. Experience and Education Requirements

   Heating Boiler Operator-Low Pressure-6 months experience

   High Pressure Boiler Operator-6 months experience

   Fourth Class Stationary Engineer-1 year experience

   Third Class Stationary Engineer-1 year with 4th Class license

   Second Class Stationary Engineer-2 years with 3rd Class license

   First Class Stationary Engineer-2 years with 2nd Class license

8. U.S. Citizenship Required?

   Not Specified

9. Local Residency Required?

   No

10. Recognition of Licenses from other locales

Those holding a U.S. Coast Guard Marine license will be issued a State of Maine stationary license equivalent to the stated steam rating on the Coast Guard license.

CODES for State of Maine

Department of Labor

Bureau of Labor Standards

Boiler Division

ENGINEERS AND BOILER OPERATORS EXAM

The Maine Boiler Operator and Engineers licensing law became effective October 1973. The purpose of this law was to upgrade the quality of boiler room Operators and Engineers and to encourage training and provide a standardized statewide license program.

Examinations are given to applicants based on practical operation, the Maine Boiler Code and the American Society of Mechanical Engineers, Boiler and Pressure Vessel Code and Engineering Practices.

The examination fee for Engineers and Operators is $10.00.

Examinations are given quarterly in March, June, September and December. Examinations are given in the Augusta Armory on Western Avenue in Augusta, Maine at 9:00 a.m. The exam application and $10 examination fee must be submitted to this office thirty (30) days prior to the examination date. Those holding a U.S. Coast Guard Marine license will be issued a State of Maine stationary license equivalent to the stated steam rating on the Coast Guard license. General boiler experience and navy experience will be evaluated for experience

Chapter 6 License Requirements for States and Cities

qualification when an application has been submitted to the Chief Boiler Inspector.

Licenses are issued for a 3 year period. Maine has no reciprocal agreement with any other state.

LICENSE GRADING

Boiler Operator Permits are issued to persons in training for a twelve month period, this allows the applicant to operate under the direct supervision of the licensed engineer in charge while learning boiler operations.

There are two grades of Boiler Operator's licenses and four classes of Engineer's licenses. The license grading is as follows:

(A) The holder of a Low Pressure Boiler Operator's license:

May operate a heating plant covered by these statutes with steam boilers not exceeding 15 psi or hot water and hot water supply boilers not exceeding 160 psi or 250 degrees Fahrenheit.

(B) The holder of a Boiler Operator's license:

May operate, supervise or have charge of a heating plant having a capacity of not more than 20,000 #/hr or operate, or supervise a plant up to the capacity of the license of the engineer in charge of the plant in which he is employed. The applicant for a boiler operator's license shall have 6 months operating experience prior to examination under a permit. The board shall issue a permit for the purpose of gaining such experience. Such permit shall be limited to a specified plant and shall be limited to one year.

(C) The holder of a 4th Class Engineer's license:

License Requirements for States and Cities   Chapter 6

May have charge of a plant of not more than 50,000 #/hr or operate or supervise a plant up to the capacity of the license of the engineer in charge of the plant in which he is employed. Applicant for a 4th-class engineer's license shall be a high school graduate or have had equivalent education and shall have at least one year of operating or supervising experience under a duly licensed engineer having charge of a plant.

(D) The holder of a 3rd Class Engineer's license:

May have charge of a plant having a capacity of not more than 100,000 #/hr or operate, or supervise a plant up to the capacity of the license of the Engineer in charge of the plant in which he is employed. Applicants for a 3rd Class Engineer's license shall have had at least 1 year operating or supervising experience as a 4th Class Engineer.

(E) The holder of a 2nd Class Engineer's license:

May have charge of a plant having a capacity of not more than 200,000 #/hr or operate, or supervise a plant up to the capacity of the license of the Engineer in charge of the plant in which he is employed. Applicants for a 2nd Class Engineer's license shall have had at least 2 years operating or supervising experience as a 3rd Class Engineer.

(F) The holder of a 1st Class Engineer's license:

May operate, supervise or have charge of a plant of unlimited steam capacity. Applicants for a 1st Class Engineer's license shall have had at least 2 years operating or supervising experience as a 2nd Class Engineer.

(G) One year of schooling in the field of boiler operation in a school approved by the board shall be equivalent to 6 months of operating experience.

Chapter 6   **License Requirements for States and Cities**

(H) In the event of a lack of qualified personnel in the plant in which the applicant is employed, the committee may waive the operating experience requirements of the applicant for examination for the next higher grade of license and any such license issued shall be limited to that plant.

(I) Notwithstanding the provisions of this subsection, the examining committee may permit an applicant to take the examination for a license if, in the committee's opinion, the experience or education qualifications, or both, of the applicant are equivalent to the operating experience required by this subsection.

FEES.

The fees charged for examination and for licenses issued pursuant to this section shall be as follows:

(A) License and license renewal fee for Stationary Steam Engineers....$30.00[6]

(B) License and license renewal fee for High Pressure Boiler Operators...$30.00

License and license renewal fee for Low Pressure Boiler Operators......$20.00

(C) The board may charge a late fee of up to $30 on all renewals for which it receives a renewal application up to 2 years after the expirations of the license under such rules as the board may adopt.

---

[6]   First Class through Fourth Class Licenses

Maryland

State of Maryland
SUMMARY OF QUALIFICATIONS:
1. Exam Location, Address and Phone Number
Towson American Legion Hall
125 York Road
Towson, Maryland 21204
410/333-6322

2. Exam Days and Times
1st Tuesday every other month

3. Examination Cost
First Grade Stationary Engineer...........$20.00

Second Grade Stationary Engineer.......$15.00

Third Grade Stationary Engineer..........$10.00

4. License Cost and Renewal
Initial license...$15.00

Renewal of license...............................$30.00

5. Type of Test
All Written - 50 multiple choice questions

6. Grades of Licenses
First Grade Stationary Engineer

Second Grade Stationary Engineer

Third Grade Stationary Engineer

7. Experience and Education Requirements
See Codes for State of Maryland below

Chapter 6　　　　　License Requirements for States and Cities

8. U.S. Citizenship Required?

Yes

9. Local Residency Required?

No

10. Recognition of Licenses from other locales

Yes - Count military and previous work experience

CODES for State of Maryland

Department of Licensing and Regulation

Board of Examining Engineers

Public Local Laws of Maryland, Article 4

Code of Maryland Regulations

4-2. Certificates of proficiency; grades.

The Board has general supervision of all Stationary Engineers within the State, except as hereinafter provided. The Board shall examine all Engineers who are 21 years old or older and apply for examination and shall give all parties so examined a certificate of proficiency if found proficient, and refuse to give a certificate if not found proficient. Each person applying for the examination shall pay the Board the following fee: $20.00 for the First Grade, $15.00 for the Second Grade, $10.00 for the third Grade and $10.00 for the Fourth Grade. Each person receiving the certificate shall pay to the Board $15.00 for the certificate issued and $30.00 for all renewals of all Grades. The certificate shall be of four Grades; a certificate of the First Grade will permit the holder to take charge of any plant machinery; the Second Grade to take charge of any plant machinery from one to five hundred horsepower; the Third Grade to take charge of any plant machinery from one to thirty

License Requirements for States and Cities   **Chapter 6**

horsepower; and the Fourth Grade to take charge of any hoisting or portable plant of machinery. The certificate shall be renewed biennially on or before the first day of February of each odd numbered year. An Engineer holding a certificate may not have charge of more than one plant of machinery at the same time unless the plant is of the same company and at one and the same place. A substitute who has not been examined and received a certificate may not be placed in charge of machinery by any Engineer who has.

4-3. Requirements for Examination.

(a) Applicants for the First Grade Stationary Engineer's examination shall possess at least 1 of the following qualifications:

(1) A Second Grade Stationary Engineer license for 1 year or more and employment for at least 12 months (1,750 working hours) in the operation of high pressure boilers;

(2) Five years or more operating experience in a plant that generates a minimum of 17,250 pounds of steam per hour;

(3) A degree in mechanical engineering from an accredited university or college, and 6 months or more of practical experience; or

(4) A valid Marine Engineer's certificate or a chief petty officer's certificate from the United States Navy, with steam boiler Engineering training.

(b) Applicants for the Second Grade Stationary Engineer's examination shall possess at least 1 of the following qualifications:

(1) Employment for at least 24 months (3,500 working hours) in a power plant under the direct supervision of a licensed First Grade Stationary Engineer;

Chapter 6   License Requirements for States and Cities

(2) Employment as a Third Grade Stationary Engineer for 1 year or more; or

(3) A valid Marine Engineer's certificate or Chief petty officer's certificate from the United States Navy.

(c) Applicants for a Third or Fourth Grade Stationary Engineer's examination are required to show experience in the operation of or experience around a plant or machinery.

(d) An applicant for any examination may substitute 2 years of education in a trade or vocational school for 1 year practical experience.

4-4. Same; applications, exceptions.

Any person 21 years old or older who desires to fill a position as a Stationary Engineer shall apply to the Board of Examining Engineers for examination and certificate of proficiency, before he can pursue his avocation as Stationary Engineer. Any person 18 years old or older who desires to fill a position as an apprentice Stationary Engineer shall apply to the Board of Examining Engineers for examination and certificate of proficiency before he can pursue his avocation as apprentice Stationary Engineer. This section does not apply to persons running engines and boilers in sparsely settled country places, where not more than twenty persons are engaged in work about such engines and boilers, to Engineers running country saw and grist mills, threshing machines and other machinery of a similar character, to Marine Engineers engaged in steamboats, ships and other vessels run by steam, or to those engaged as Locomotive Engineers of any steam railway company. If any charge is made to the Board, of any Engineer who holds certificate from them, of being intoxicated while in charge of an engine or boiler, or of the neglect of duty on the part of such Engineer or Engineers, the Board immediately shall hear the charge and, if sustained, annul the certificate. Each Engineer against whom the charge is made shall be given due notice thereof and an opportunity of being heard in person or by counsel. The certificate granted to the respective applicants, including a photograph of each respectively, must be framed and kept in a conspicuous place at such place as such person may be respectively at work. Any person violating the provisions of this subtitle of this Article is guilty of a misdemeanor, and upon conviction shall be fined not less than $25.00 or more than $50.00.

Chapter 6 License Requirements for States and Cities

Massachusetts

State of Massachusetts

SUMMARY OF QUALIFICATIONS:

1. Exam Location, Address and Phone Number

Various testing sites all over the state. Applicants go to the town they are employed in.

Office Location:
Department of Public Safety
1 Ashburton Place
Boston, Massachusetts 02108
617/727-3200

2. Exam Days and Times

Varies per testing site

3. Examination Cost

First Class Engineer.........$60.00

Second Class Engineer.....$50.00

Third Class Engineer........$50.00

Fourth Class Engineer......$50.00

First Class Fireman...........$50.00

Second Class Fireman......$50.00

4. License Cost and Renewal

Every class of license is $35.00 to renew.

Renew on birthday every two years.

5. Type of Test

License Requirements for States and Cities — Chapter 6

Written or Oral. This is up to the examiner. If they did poorly on the written test the examiner has the option of giving oral test.

6. Grades of Licenses

First Class Engineer

Second Class Engineer

Third Class Engineer

Fourth Class Engineer (portable)

First Class Fireman

Second Class Fireman

7. Experience and Education Requirements

See Codes for State of Massachusetts below

8. U.S. Citizenship Required?

Yes

9. Local Residency Required?

No

10. Recognition of Licenses from other locales

They recognize experience but you still have to take the test.

CODES for State of Massachusetts

Massachusetts General Law

Chapter 146

Section 49. Classes of Licenses.

Licenses shall be granted according to the competence of the applicant and shall be classified as follows:

Engineers' licenses: First Class:

Chapter 6 License Requirements for States and Cities

to have charge of and operate any steam plant.

Second Class:

to have charge of and operate a boiler or boilers, and to have charge of and operate engines, no one of which shall exceed one hundred and fifty horsepower, or to operate a first class plant under the Engineer in direct charge thereof.

Third Class:

to have charge of and operate a boiler or boilers not exceeding, in the aggregate, one hundred and fifty horse power when solid fuel is burned or not exceeding, in the aggregate, five hundred horse power based upon the relieving capacity of the safety valve or valves when steam is generated by the use of liquid or gaseous fuel, electric or atomic energy, or any other source of heat, and an engine or engines not exceeding fifty horse power each, or to operate a Second class plant under the Engineer in direct charge thereof.

Fourth Class:

to have charge of and operate hoisting and portable steam engines and boilers. Portable Class: to have charge of or to operate portable boilers and portable engines, except hoisting engines or steam fire engines.

Steam Fire Engineer's Class:

to have charge of or to operate steam fire engines and boilers.

Firemen's licenses: First Class:

to have charge of and operate any boiler or boilers where the safety valve or valves are set to blow at a pressure not exceeding twenty-five pounds to the square inch, or to operate high pressure boilers under the Engineer or Fireman in direct charge thereof. No Engineer or Fireman in charge of a steam plant

License Requirements for States and Cities   Chapter 6

shall be permitted to be in charge of any other steam plant, unless it is within one mile from the specific plant of which the Engineer or Fireman is designated to be in charge or unless a licensed Second or Third Class Engineer or Fireman shall be in attendance at such plant and shall perform his duties under the supervision of such Engineer or Fireman.

Second Class:

To operate any boiler or boilers under the Engineer or Fireman in direct charge thereof. A person holding a First Class Fireman's License may operate a Third Class plant under the Engineer in direct charge thereof.

Special Licenses:

A person who desires to have charge of or to operate a particular steam plant may, if he files with his application for such examination a written request signed by the owner or user of the plant, be examined as to his competence for such service and no other, and, if found competent and trustworthy, he shall be granted a license for such service, and no other; provided, that no special license shall be granted to give any person charge of or permission to operate an engine of over fifty horse power, or a boiler or boilers exceeding, in the aggregate, two hundred and fifty horse power, except that where the main power plant is run exclusively by water power, developed on the premises of such plant a major part of the year, and has auxiliary steam power for use during periods of low water; a special license may be granted to an applicant holding an Engineer's license.

Section 50. Qualifications for Licenses.

To be eligible for examination for a Second Class Fireman's license:

Chapter 6 License Requirements for States and Cities

A person must furnish evidence as to his previous training, and except students attending all day state aided vocational high schools in the Steam Engineering course or students performing the duties of a First or Second Class Fireman at said school, be at least eighteen years of age; provided, however, that no such license issued shall be used in employment by the holder thereof unless he is at least eighteen years of age.

To be eligible for examination for a First Class Fireman's license:

A person must furnish evidence as to his previous training and experience and must have been employed in a boiler or steam power plant as a Steam Engineer, Fireman, control room operator, water tender, auxiliary operator or Engineer's assistant for not less than one year, or he must have held and used a Second Class Fireman's license for not less than six months.

To be eligible for examination for a Third Class Engineer's license:

A person must be a citizen or furnish proof of having filed a declaration of his intention to become a citizen of the United States, must furnish evidence as to his previous training and experience and must have been employed in a boiler or steam power plant as a Steam Engineer Fireman, control room operator, water tender, auxiliary operator or Engineer's assistant for not less than one and one half years, or held and used an equivalent license in the United States Merchant Marine for one year or used a current Third Class steam license issued by another state for one year or must have held and used a First Class Fireman's license for not less than one year.

To be eligible for examination for a Second Class Engineer's license:

State of Massachusetts

License Requirements for States and Cities Chapter 6

A person must be a citizen or furnish proof of having filed a declaration of his intention to become a citizen of the United States, must furnish evidence as to his previous training and experience and must have been employed as an Engineer in charge of or operating a steam plant or plants having at least one engine or turbine of not less than fifty horse power for not less than two years or held and used an equivalent license in the United States Merchant Marine for two years, or have held and used an equivalent license from another state for two years or must have held and used a Third Class Engineer's license either as an Engineer, assistant Engineer, control room operator or a Fireman for not less than one year, or must be a person who has held and used a special license to operate a first class plant for not less than two years, except any person who is a United States citizen and served three years as an apprentice to the machinist or boiler making trade in Stationary, Marine or Locomotive engine or boiler works and who has been employed one year in connection with the operation of a steam plant, or any person who has a bachelor or science degree in Engineering from any duly recognized school of Engineering, who has been employed for one year in connection with the operation of a steam plant, shall be eligible for examination for a Second Class Engineer's license.

To be eligible for examination for a First Class Engineer's license:

A person must be a citizen or furnish proof of having filed a declaration of his intention to become a citizen of the United States, must furnish evidence as to his previous training and experience and must have been employed for not less than three years as an Engineer in charge of a steam plant or plants having at least one engine or turbine of over one hundred and fifty horse power, or must have held and used a Second Class Engi-

Chapter 6 License Requirements for States and Cities

neer's license in a Second class plant for not less than one and one half years, or in a First class plant as assistant Engineer for one and one half years or held and used an equivalent license in the United State Merchant Marine for three years or have held and used an equivalent license from another state for three years.

Every application for any grade of license, except a Second Class Fireman or special application granted under the provisions of chapter one hundred and forty-six, shall be endorsed by an Engineer or Fireman holding the same or higher grade of license. The endorsee shall state his name, address, grade of license currently in force and that he has personal knowledge of the applicant's experience and trustworthiness. The license of an endorsee making a wilful falsification shall be suspended or revoked. District Engineering inspectors of the department of public safety shall not endorse applications, except when applying the oath when the applicant is being examined.

Section 51. Posting of License; Daily Record.

An Engineer's or Fireman's license shall be so placed in the engine or boiler room of the plant operated by the licensee as to be easily read. The person in charge of a stationary steam boiler upon which the safety valve is set to blow off at more than twenty-five pounds pressure to the square inch, except boilers in private residences, boilers in apartment houses of less than five apartments, boilers under the jurisdiction of the United States, boilers used for agricultural purposes exclusively, and boilers of less than nine horse power, shall keep a daily record of the boiler, its condition when under steam, and of all repairs made and work done on it, upon forms to be obtained upon application to the department. These records shall be kept on file and shall be always accessible to the Chief and inspectors of the division.

License Requirements for States and Cities   Chapter 6

Section 54. License to be Carried on the Person.

A license to operate such hoisting machinery shall be carried on the person of the holder thereof when operating the same.

Section 55. Penalty.

Whoever violates any provision of section forty-two to fifty-four, inclusive, or any rule made thereunder, or prevents or attempts to prevent an inspector from entering on any premises in the discharge of his duty shall be punished by a fine of not less than ten nor more than three hundred dollars or by imprisonment for not more than three months.

Section 57. Applications for Licenses.

Each application for a license as an Engineer or Fireman of a class specified herein or as an operator of hoisting machinery not run by steam shall be made upon a blank furnished by the department, signed and sworn to by the applicant, and shall show the total experience of the applicant. Each such application for a First Class Engineer's license, Second Class Engineer's license or for a special license; for a Third Class, Fourth Class or Portable Class Engineer's license or a Steam Fire Engineer's license; for a First Class or Second Class Fireman's license; and for a license for operating hoisting machinery not run by steam shall be accompanied by an examination fee to be determined annually by the commissioner of administration under the provision of section three B of chapter seven. Each such application shall entitle the applicant to one examination only, except in case of an appeal under section sixty-six; provided, however, that no person shall make application hereunder for a license of any particular class oftener than once in ninety days. The fee for an examination on appeal shall be determined annually under the aforementioned chapter seven provision.

Chapter 6   License Requirements for States and Cities

Section 58. Applicants May Have a Person Present.

In all examinations or appeals, the applicant may have one person present who may take notes if he so desires. In case of applicants for certificates of competency to inspect boilers such person shall be a representative of an insurance company employing the applicant or wishing to do so.

Section 59. Revocation of Licenses (Expires Dec. 31, 1982)

A certificate of competency to inspect boilers shall be revoked and a license as Engineer or Fireman or operator of hoisting machinery shall be suspended or revoked for incompetence or untrustworthiness of the holder thereof. A wilfully false statement in the application shall be sufficient cause for revocation at any time. If a certificate or license is lost or destroyed a new certificate shall be issued without examination upon satisfactory proof thereof. The fee for such new certificate shall be determined annually by the commissioner of administration under the provision of section three B of chapter seven.

Section 64. Application for License as Engineer or Fireman; Examination.

Whoever desires to act as an Engineer or Fireman shall apply for a license to the department. The applicant shall be examined by the inspector assigned to the district in which the applicant resides or is employed. The examinations shall be uniform throughout the commonwealth in a form approved by the Chief, and shall be given in two parts, the first part of which shall be written and the second of which shall be oral; provided, however, that any applicant may be given only an oral examination if he so requests in writing, stating the reasons for such request and if such reasons are deemed valid by the Chief or his designee. A mark of seventy percent shall be considered passing on any such written examination and such passing mark shall remain valid for a period of one year. The failure of any applicant taking a written examination to achieve such a passing mark shall result in the removal of his application from any further consideration. The time allotted for the written examination shall be four hours.

Upon the successful completion of the examination process, the applicant if found competent and trustworthy, shall receive a license graded in accordance with the merits of his examination. An applicant for the First or Second Class Engineer's license shall be examined by a board consisting of three district Engineering inspectors of the department, or two such inspectors and the Chief, if such Chief is a qualified examiner, and, if the applicant is employed, one member of said board shall be the department inspector of boilers assigned to the area where the applicant is so employed, and the decision of said board shall be final. An applicant for a license as an Engineer of any other class or as a Fireman, or for a special license, shall be examined by one inspector of the division, from whose decision there shall be an appeal as provided in section sixty-six.

Section 65. Application For License to Operate Hoisting Machinery.

Whoever desires to act as an operator of hoisting machinery shall apply to the inspector of the division for the town where he resides or is employed. He shall be given a practical examination by a department inspector, and if found competent and trustworthy shall receive a license to operate such machinery. He shall be examined only as to his ability to use the particular machinery, whether internal combustion, compressed air, electric engine or otherwise, which he desires to operate, and the license shall be limited to that particular kind of machinery; but if he so requests in his application the applicant may be examined as to his proficiency in the various kinds of machinery

Chapter 6 License Requirements for States and Cities

used for hoisting, and the license shall include those kinds of machinery in respect to which he is found competent.

Section 66. Appeal.

A person aggrieved by the action of a single examiner in refusing, suspending or revoking a license to act as Engineer, Fireman or operator of hoisting machinery may, within one week, appeal therefrom to the Chief, who shall appoint three inspectors of the division, or himself and two inspectors, to act together as a board of appeal. The decision of a majority of the members of the board of appeal shall be final.

Section 67. Time of Expiration of Licenses and Renewals; Exceptions; Fees.

A license shall continue in force until the date of birth of the licensee occurring more than twelve months but not more than twenty-four months after the effective date of such license unless suspended or revoked for incompetence or untrustworthiness of the licensee, except that a special license shall not continue in force after the holder thereof ceases to be employed in the plant specified in the license. If any such license or the renewal thereof expires in an even year, any subsequent renewal shall expire on the next anniversary of the licensee's date of birth occurring in an even year. If any such license or renewal thereof expires in an odd year, any subsequent renewal shall expire on the next anniversary of the licensee's date of birth occurring in an odd year. A license issued to a person born on February twenty-ninth shall, for the purposes of the section, expire on March first. The fee for the renewal of a license shall be determined annually by the commissioner of administration under the provision of section three B of chapter seven for the filing thereof. Licenses not renewed at expiration date shall become void, and shall after one year be reinstated only by re-examination of the licensee. A notice of the date of expiration of a license shall, at least thirty days prior to such date, be sent to the licensee. The inspector of the division for the town where a licensee resides may issue a renewal license. A person whose license is suspended or revoked shall surrender his license to the Chief or an inspector of the division. If new license of a different grade is issued, the old license shall be destroyed by the examiner.

Michigan

Dearborn, Michigan

SUMMARY OF QUALIFICATIONS:

1. Exam Location, Address and Phone Number
City Hall
Personnel Department
4500 Maple
Dearborn, Michigan 48126
313/943-2151

2. Exam Days and Times

3rd Monday of every month

3. Examination Cost

Filing Fee..$ 5.00

Examination Fee..$20.00

4. License Cost and Renewal

1st Class Engineer..$15.00

2nd Class Engineer.. $13.00

3rd Class Engineer...$11.00

4th Class Engineer...$ 9.00

High Pressure Boiler Operator............................$ 8.00

Low Pressure Boiler Operator............................$ 8.00

Refrigeration Engineer..$13.00

Turbine and Reciprocal Engine Operator..............$ 8.00

Renewal of any of the licenses listed above.....Same as issuance of initial license

License Requirements for States and Cities   Chapter 6

5. Type of Test

All Written - multiple choice

6. Grades of Licenses

First Class Engineer

Second Class Engineer

Third Class Engineer

High Pressure Boiler Operator

Low Pressure Boiler Operator

Portable Steam Equipment Operator

7. Experience and Education Requirements

See Codes for Dearborn below

8. U.S. Citizenship Required?

Yes

9. Local Residency Required?

No

10. Recognition of Licenses from other locales

Yes, count military experience

CODES for Dearborn, Michigan

Ordinance No. 44-266

High Pressure Boiler Code

Chapter 4. License

Section 4.1. LICENSE REQUIRED, WHEN, AGE, LIMITATION.

Chapter 6   **License Requirements for States and Cities**

It shall be unlawful for any person to be in charge of or to operate, to supervise or to be in charge of persons operating, any apparatus or equipment governed by this Code in the city without a license therefor as hereinafter set forth; provided, however, that no license shall be issued to any person who is not a citizen of the United States, and unless he shall have attained the age of 18 years. "Person to be in charge of" shall mean the immediate supervisor of an employee at the time of the operation of the apparatus or equipment governed by this Code when relating to said person being in charge of or having supervision over persons in this section.

Section 4.2. CLASSIFICATION OF LICENSES.

There shall be issued in the City, the following classifications of licenses involving apparatus and equipment covered by the within Code:

(a) Low Pressure Boiler Operator.

There shall be a classification of license known as Low Pressure Boiler Operator which shall be issued to persons properly qualified for the operation of boilers where steam pressure of 15 pounds per square inch or less is used or hot water boilers carrying 30 pounds per square inch or less is used. Such license shall be issued to any person who has had one year's experience in the boiler room as fireman, coal passer, or general helper and passes a satisfactory examination showing that he is familiar with and can operate such boilers with safety to life and property.

(b) High Pressure Boiler Operator.

There shall be a classification of license known as High Pressure Boiler Operator which shall be issued to persons properly qualified for the operation of boilers of which steam pressure in excess of 15 pounds per square inch or hot water boilers carrying 30 pounds per square inch is in use. Such license shall be issued to any person who has had two years experience in a high or low pressure steam or hot water plant as fireman, oilers, coal passer or general helper and passes a satisfactory examination showing that he is familiar with and can operate such boilers with safety to life and property.

(c) Fourth Class Engineer.

There shall be a classification of license known as Fourth Class Engineer which shall be issued only to persons who operate hoisting portable boilers and track locomotive equipment excluding locomotive boilers used on railroad tracks. Each person to whom this classification of license is issued must have had at least two years experience in the operation of such mechanical equipment working under the direction of a licensed Engineer, and must pass a satisfactory examination that he has reasonable skill and familiarity in the operation of such apparatus. The license issued for this classification shall be in card form and shall be retained at all times in the possession of the operator when engaged in the operating of such hoisting and portable apparatus. It shall be unlawful for any person to operate hoisting, portable apparatus without such license, except a person who holds a First Class Engineer's license may operate this equipment.

(d) Third Class Engineer.

There shall be a classification of license known as Third Class Engineer which shall be issued to persons who have had three years experience in the operation of power boilers or prime movers and shall be limited to an aggregate of 100 Horsepower. The applicant shall possess a High Pressure Boiler Operator's license and shall have had twelve months actual experience as such an operator. Such persons shall pass a

Chapter 6 — License Requirements for States and Cities

satisfactory examination showing he is competent and understands the operation of high pressure boilers and prime movers.

The Chief Safety Engineer, with the approval of the Board, may waive the length of experience for this classification, if the applicant submits documentary evidence of equivalent education.

(e) Second Class Engineer.

There shall be a classification of license known as Second Class Engineer which shall be issued to persons who have had four year's experience operating high pressure boilers or prime movers. The applicant shall possess a Third Class Stationary Engineer's license and shall have had twelve months actual experience as such an Engineer.

Anyone who holds a Mechanical Engineer's degree and has had one year's experience operating high pressure boilers or prime movers. Such a license shall be limited to an aggregate of 250 horsepower. Any applicant shall pass a satisfactory examination showing they are competent and understand the operation of high pressure boilers and prime movers. The Chief Safety Engineer, with the approval of the Board, may waive the length of experience for this classification if the applicant submits documentary evidence of equivalent education.

(f) First Class Engineer.

There shall be a classification of license known as First Class Engineer which shall be unlimited as to horsepower of boilers, engines and pressures and shall include all types of refrigeration, such license shall be granted to any person who passes a satisfactory examination showing he is competent and understands the operation of any and all equipment covered by the

License Requirements for States and Cities　　　Chapter 6

within Code if such person has had experience in any one of the following fields of work:

1. Any person who has been employed as an oiler, general assistant, foreman or general Engineer in any steam plant for a period of not less than five years providing such person has in force a Second Class Engineer's license for at least one year.

2. Any person who has served as fireman or general Engineer in any steam plant in any other city or state having comparable requirements and qualifications if such person holds a First Class Engineer's license from said city or state.

3. Any person who is a graduate Mechanical Engineer from a recognized college or university , after such person has had two years experience in any steam plant providing he has had a Second Class Engineer's license one year.

Provided however, the experience requirements may be waived if the applicant holds a Mechanical Engineering degree or is a registered Mechanical Engineer and submits to the Board evidence of such knowledge and experience.

(g) Refrigeration Engineer.

There shall be a class of license known as Refrigeration Engineer which shall be limited to the operation of refrigeration plants of any kind and using any refrigerant. This license shall be granted to any person who has had two years experience as an oiler or general assistant under the instruction and direction of a licensed Refrigeration Engineer and who has passed a satisfactory examination. Provided, however, the experience requirement shall be waived if the applicant holds a Mechanical Engineering degree or is a registered Mechanical Engineer and submits to the Board evidence of such knowledge and experience.

(h) Turbine and Reciprocal Engine Operator.

There shall be a classification of license known as turbine and reciprocating engine operator. Each person to whom this classification of license is issued, shall have at least four years of experience in the operation of such equipment and must pass a satisfactory written and oral examination showing that he understands the operation of such apparatus.

(i) Post of Duty-Low Pressure Boiler Operators and Engineers.

In Low Pressure boiler plants, the post of duty of the licensed operator in responsible charge shall be in the boiler rooms which contains the auxiliaries pertinent to the operation of said boiler or boilers providing such rooms are accessible for passage, visibility or audibility. Post of Duty - Need not apply when "banked operations" are met.

(j) Post of Duty - High Pressure Boiler Operators and Engineers.

In high pressure boiler plants or stationary prime mover plants where licensed personnel are required by the Code to be in attendance, the post duty of such operator or Engineer shall be in the room or rooms where such boilers, engines, turbines or auxiliaries are located pertinent to the plant operation if such rooms are readily accessible for passage, visibility or audibility.

(k) Post of Duty - Refrigeration Operators.

In all plants where systems of one hundred twenty-five (125) tons or less are used in other than institutional occupancies the post of duty of the licensed refrigeration operator at all times when the refrigerating systems are in operation shall be on the premises in which the refrigerating systems are located. In all other plants the post of duty of the licensed refrigeration operator at all times when the systems are in operation shall be on the premises within easy reach and hearing distance of where the equipment which he is licensed to operate is located, provided that the operator may be absent for a period of not more than fifteen (15) consecutive minutes during any one hour for the purpose of inspecting or adjusting parts, auxiliaries, or appurtenances of the system which he is licensed to operate and provided further that the Department shall be permitted to modify, by written notice to the owner, any of the above limitations where special conditions warrant such action, and on systems where in the opinion of the Chief Safety Engineer the refrigeration plant design is such that the operation does not endanger the life or health of individuals. The post of duty of the operator when the systems are in operation shall be on the premises in which the refrigeration systems are located.

Section 4.3. LICENSE AND APPLICATION FEES.

The following fees shall be paid directly to the City Treasurer before the issuance of any classification of license hereinbefore set forth:

1st Class Engineer..$15.00

2nd Class Engineer...$13.00

3rd Class Engineer..$11.00

4th Class Engineer..$ 9.00

High Pressure Boiler Operator............................$ 8.00

Low Pressure Boiler Operator.............................$ 8.00

Refrigeration Engineer...$13.00

Turbine and Reciprocal Engine Operator............$ 8.00

Renewal of any of the licenses Above Listed.....Same as issuance

Chapter 6 License Requirements for States and Cities

Examination fees and license fees for conditional licenses will be the same as those charged for comparable classifications.

At the time of filing the application for examination for any license, a fee of $2.00 shall be paid to the Treasurer.

Section 4.4. APPLICATION AND EXAMINATION.

Application forms approved by the Board shall be supplied to any applicant by the Department for all classifications of licenses herein provided. All information and data pertaining to the applicant set forth on said application form must be answered in full, sworn to before a Notary Public and upon payment of the filing fee, filed with the Secretary of the Board. Falsification of any material fact upon the application shall be grounds for revocation of any license granted. Regular examinations as prescribed by the Board shall be held at least once each month at the time and place specified by the Board in its rules and regulations. Applicants will be notified of the time and place where such examinations will be held. Special examinations may be held in the discretion of the Director and at such time and place as designated by the Board. When any applicant for a given classification has passed the required examination prescribed by the Board, the Safety Engineer shall certify such fact to the City Clerk and upon the payment to the City Treasurer of the respective fee, as above set forth, the City Clerk shall issue the particular license specified by the Safety Engineer to which such applicant is entitled.

Section 4.5. POSTING LICENSE.

Every license granted or issued to any person shall be hung in a frame under glass in the boiler or engine room where such person operates the mechanical apparatus, except that of Fourth Class Engineer.

Section 4.6. LOWER CLASSIFICATION, ISSUANCE OF.

Any person applying for a particular classification of a license and failing to pass the examination required for that classification may in the discretion of the Board be issued a license of lower classification than that applied for to which his qualification and his examination may entitle him, but such person shall not make a re-application for the license, the examination for which he failed to pass, or for a higher classification, until after a period of at least three months has expired.

Section 4.7. EXPIRATION AND RENEWAL.

Engineers and operators licenses issued under this Code shall expire at midnight following the first Tuesday of May after issuance, and shall be renewable annually at the Office of the City Clerk upon payment of the renewal fee above set forth. If any licensee fails to apply for a renewal of his license within one month after the expiration thereof, such license will be deemed to have lapsed. Any person allowing his license to lapse will be subject to re-examination before obtaining a renewal thereof. Before any license shall be issued after an examination or re-examination of the licensee, as provided in this section, such licensee shall pay the fee as prescribed above for the original issuance of the license.

Section 4.8. MINIATURE BOILERS AND UNFIRED PRESSURE VESSELS EXCEPTED.

It shall not be necessary for the operator of any miniature boiler, as defined by the within Code and covered thereby, to obtain a license to operate same, but all other provisions of the Code shall apply to such boilers; likewise it shall not be necessary for the operator of any unfired pressure vessel, except refrigeration plants covered by the Code, to obtain a license to

Chapter 6 License Requirements for States and Cities

operate same, but all other provisions of the Code shall apply to such pressure vessels.

Detroit, Michigan

SUMMARY OF QUALIFICATIONS:

1. Exam Location, Address and Phone Number
City of Detroit
Buildings and Safety Engineering Department
Fourth Floor, City-County Building
Detroit, Michigan 48226
313/224-3184

2. Examination Days and Times

Applicant is interviewed between 8 a.m. and 9 a.m. The examination begins at 9:30 a.m. If the written examination is passed, then you will be scheduled for an oral interview.

3. Examination Cost

$30.00 for written examination (increasing to $35.00 as of July 31, 1993)

$30.00 for oral examination (increasing to $35.00 as of July 31, 1993)

4. License Cost and Renewal

License cost...............$20.00

5. Type of Test

Oral with drawings for Portable, Minature, Low Pressure and High Pressure Boiler Operators examinations

Written - multiple choice with an oral examination after sucessful completion of the written for 3rd Class to 1st Class Stationary Engineers Licenses

6. Grades of Licenses

License Requirements for States and Cities   Chapter 6

First Class Stationary Engineer

Second Class Stationary Engineer

Third Class Stationary Engineer

High Pressure Boiler Operator

Low Pressure Boiler Operator

Miniature Boiler Operator

Portable Steam Equipment Operator

7. Experience and Education Requirements

See Codes for Detroit below

8. U.S. Citizenship Required?

Yes

9. Local Residency Required?

No

10. Recognition of Licenses from other locales

Not Specified

CODES for Detroit, Michigan

City of Detroit

Stationary Engineer, Boiler Operator and

Refrigeration Operator Licensing Code

Ordinance No. 706-G

Article VI

Qualifications of Applicants for License

Sec. 6.0. Age.

Chapter 6 — License Requirements for States and Cities

Applicants for all licenses regulated by this Ordinance shall have attained a minimum of nineteen (19) years of age.

Sec. 6.1. Health Requirements.

The applicant for a license or its renewal shall possess such physical and mental capacities as to perform the duties covered by his license without endangering life and property. The Department shall have the authority to require medical certification of the health of the applicant.

Sec. 6.2. Substantiation of Experience.

The Department shall have the authority to require substantiation of professed experience by means of affidavits from present or previous employers or by other acceptable documentary evidence.

Sec. 6.3. Applicants Holding Non-Detroit Licenses.

Applicants for Detroit licenses holding active recognized licenses as Boiler Operator, Portable Steam Equipment Operator, Stationary Engineer or Refrigeration Operator, issued by any other governmental subdivision or agency, may qualify for examination for a class or grade of license commensurate with the experience requirements for, and limitations of, the non-Detroit license which they hold, provided all other requirements are complied with.

Sec. 6.4. Experience Requirements for Miniature Boiler Operator License.

An applicant for Miniature Boiler Operator license shall have had at least one (1) month experience in the operation of any type of high-pressure boiler.

Sec. 6.5. Experience Requirements for Low Pressure Boiler Operator License.

An applicant for Low Pressure Boiler Operator license shall have been actively employed in connection with the operation or maintenance of low or High Pressure boilers, steam prime movers or their auxiliaries for a minimum of one (1) year.

Sec. 6.6. Experience Requirements for High Pressure Boiler Operator License.

An applicant for High Pressure Boiler Operator license shall have been actively employed in connection with the operation or maintenance of low or High Pressure boilers, steam prime movers or their auxiliaries for a minimum of two (2) years, or shall possess a Low Pressure Boiler Operator license and have been actively employed as such an operator for a minimum of one (1) year

Sec. 6.7. Experience Requirements for Portable Steam Equipment Operator license.

An applicant for Portable Steam Equipment Operator License shall have had at least one (1) year of experience in the operation of high pressure boilers or steam prime movers.

Sec. 6.8. Experience Requirements for Third Class Stationary Engineer License.

An applicant for Third Class Stationary Engineer License shall meet one the following experience requirements:

(a) He shall possess a High Pressure Boiler Operator License and shall have had at least one (1) year of experience as such an operator, or

(b) He shall possess a Low Pressure Boiler Operator License with at least one (1) year of experience as such an operator, together with at least one (1) year of maintenance experience on high-pressure boilers and boiler auxiliary apparatus.

Chapter 6   **License Requirements for States and Cities**

(c) He shall possess a High Pressure Boiler Operator License, and his experience shall include at least one (1) year in the operation of a high pressure boiler plant having an aggregate heating surface of more than 4,000 square feet, as a boiler maintenance man or as an apprentice in an approved training program, or

(d) He shall have at least three (3) years of experience in the actual operation of boilers in a high pressure boiler plant having an aggregate heating surface of more than 4,000 square feet, the foregoing experience to be established by documented evidence from present and previous employers, or

(e) He shall have at least one (1) year of experience in the actual operation of boilers in a high pressure boiler plant having an aggregate heating surface of more than 4,000 square feet, plus sufficient experience in the actual operation of steam prime movers in excess of 10 horsepower to make a total of at least three (3) years, the foregoing experience to be established by documented evidence from present and previous employers, or

(f) He shall have at least one (1) year of experience in the actual operation of boilers in a high pressure boiler plant having an aggregate heating surface of more than 4,000 square feet, plus sufficient experience in the actual operation of boilers in a High Pressure boiler plant having an aggregate heating surface of 4,000 square feet or less to make a total of at least three (3) years, the foregoing experience to be established by documented evidence from present and previous employers.

Sec. 6.9. Experience Requirements for Second Class Stationary Engineer License.

An applicant for Second Class Stationary Engineer License shall meet one of the following experience requirements:

License Requirements for States and Cities Chapter 6

(a) He shall possess a Third Class Stationary Engineer license and shall have had at least one (1) year of experience as such a Stationary Engineer, or

(b) He shall be an engineering graduate of a college of recognized standing and have been actually employed in the engineering or research division of a steam electric power generating plant for at least one (1) year, or

(c) He shall have at least four (4) years of experience in the actual operation of boilers in a high pressure boiler plant having an aggregate heating surface of more than 7,500 square feet, the foregoing experience to be established by documented evidence from present and previous employers, or

(d) He shall have at least one (1) year of experience in the actual operation of boilers in a high pressure boiler plant having an aggregate heating surface of more than 7,500 square feet, plus sufficient experience in the actual operation of steam prime movers in excess of 100 horsepower to make a total of at least four (4) years, the foregoing experience to be established by documented evidence from present and previous employers, or

(e) He shall have at least one (1) year of experience in the actual operation of boilers in a high pressure boiler plant having an aggregate heating surface of more than 7,500 square feet, plus sufficient experience in the actual operation of boilers in a high pressure boiler plant having an aggregate heating surface of more than 4,000 square feet to make a total of at least four years, the foregoing experience to be established by documented evidence from present and previous employers.

Sec. 6.10. Experience Requirements for First Class Stationary Engineer License.

Chapter 6 License Requirements for States and Cities

An applicant for First Class Stationary Engineer license shall meet one of the following experience requirements:

(a) He shall possess a Second Class Stationary Engineer license and shall have had at least two (2) years of experience as such a Stationary Engineer, or

(b) He shall have at least six (6) years of experience in the actual operation of boilers in a high pressure boiler plant having an aggregate heating surface of more than 20,000 square feet, the foregoing experience to be established by documented evidence from present and previous employers, or

(c) He shall have at lest two (2) years of experience in the actual operation of boilers in a high pressure boiler plant having an aggregate heating surface of more than 20,000 square feet, plus sufficient experience in the actual operation of steam prime movers in excess of 200 horsepower to make a total of at least six (6) years, the foregoing experience to be established by documented evidence from present and previous employers, or

(d) He shall have at least two (2) years of experience in the actual operation of boilers in a high pressure boiler plant having an aggregate heating surface of more than 20,000 square feet, plus sufficient experience in the actual operation of boilers in a high pressure boiler plant having an aggregate heating surface of more than 7,500 square feet of heating surface to make a total of at least six (6) years, the foregoing experience to be established by documented evidence from present and previous employers.

City of Detroit
Building and Safety Engineering Department
Fourth Floor, City-County Building
Detroit, Michigan 48226

Memorandum

Subject:

Procedure and Subject Matter that an Applicant shall be familiar with, when applying for a Boiler Operator License.

PREREQUISITES:

The City of Detroit Ordinance No. 706-G requires that an applicant for a Boiler Operator License shall have a minimum of boiler plant experience in accordance with the following:

Miniature Boiler Operator - One month experience in the operation of any type high pressure boiler.

Low Pressure Boiler Operator - One year actively employed in connection with the operation or maintenance of low or high pressure boilers; or have completed an approved low pressure boiler operator program as established by completion letters or certificates.

High Pressure Boiler Operator - Two years actively employed in connection with the operation or maintenance of low or high pressure boilers, steam prime movers or their auxiliaries; or have completed an approved high pressure boiler operator program as established by completion letters or certificates.

Application and Documentation:

Application Forms will be obtained from this Department. Completed applications and letters of documentation or other documentary evidence shall be submitted by the applicant to the examination office for evaluation, verification and scheduling of examination dates. Notarized letters from company officials on company stationary per the attached samples are acceptable. Photocopies will be accepted for scheduling purposes only. The originals must be presented for verification on the

Chapter 6   License Requirements for States and Cities

date of the appointment. Failure to do so, will void the application and cancel the examination. Documentary material will be retained with the application and will not be returned. Applications that are not accepted, will be returned.

Examination:

The oral examinations for these licenses are given by appointment only, on Tuesdays and Thursdays between 6:00 a.m. and 2:00 p.m., upon payment of the appropriate fee. Applicants who pass the examination may secure their license upon payment of the required fee. Failure to appear for an appointment will result in voiding the application. A new application and documentation will be required to make a new appointment.

The applicant is reminded that if he/she fails the examination, he/she is not eligible to reapply for the same grade of license for at least ninety (90) days.

Subject Matter:

The examination will include questions on the following subjects:

MINIATURE BOILER OPERATOR

Boiler appliances and auxiliaries; feedwater pumps; fusible plugs; low water alarms; water column; gage glass; tricocks; types of valves; pressure regulator; steam gage; safety valves; low water and over pressure procedure; post of duty; etc.

LOW PRESSURE BOILER OPERATOR

That information required for a miniature boiler operator license plus: Low water cutoff; automatic combustion controls; preparation of boilers for inspection; starting and shutting down a boiler; smoke control; flame sensing devices; fuels; fuel burning equipment; injector; venting of boiler and furnace.

HIGH PRESSURE BOILER OPERATOR

That information required for a low pressure boiler operator license plus: Feedwater regulators; steam header; feedwater header and blow-off valve arrangements and operation; soot blowers; injectors; safety valve requirements and testing.

City of Detroit
Buildings and Safety Engineering Department
Fourth Floor, City-County Building
Detroit, Michigan 48226
Mechanical/Electrical Inspection Division
Examination Section - Room 404
Telephone: 313/224-3184

Chapter 6   License Requirements for States and Cities

Memorandum

Subject:

Subject Matter that an Applicant should be familiar with when applying for a Third Class Stationary Engineer's License.

The City of Detroit Ordinance No. 706-G requires that an applicant for a Third Class Stationary Engineer's License possess a High Pressure Boiler Operator's license and have one year of actual experience as such an operator or otherwise qualify per Article VI, Section 6.8.

Bring pen and pencil together with a properly filled out application for the written examination, that is conducted on Monday of each week, applicant will be interviewed between 8 and 9 a.m., the examination beginning at 9:30 a.m. You will be notified of your score by mail. If the written examination is passed, an appointment will be made for the oral examination.

The subject matter of the examination will include:

Boiler and Accessories:

Types of boilers: Safety devices - their function and testing; Water Walls; Columns; Headers; Other boiler appurtenances; Low water and over pressure procedure.

Boiler Plant Auxiliaries:

Pumps; Injectors; Regulators; Feed water heaters; Valves; Traps; Separators.

Water and Steam:

Describe steam cycle; Blow downs; Basic water treatment principles; Scale prevention; Reactions under temperature and pressure; Boiler water circulation.

Fuels and Combustion:

License Requirements for States and Cities — Chapter 6

Forced and induced draft fans; Combustion control equipment and regulation for various fuels; Smoke control; Fundamentals of operation of Gas, Oil and Coal feeding equipment.

Steam Engines:

Turbines and Accessories: Types of engines and turbines; Discuss operating fundamentals; starting, stopping and governing of small heat engines; Valve settings for reciprocating engines; Lubrication.

Electricity:

Fundamental concepts of AC & DC - Ohms, volts, amperes, watts; Staring and stopping motors; Proper use of fuses; Safety precautions with electricity.

Plant Operation:

Foaming, priming, carry-over and remedies; Cutting in boilers; Water hammer; Reducing stations; Expansion joints.

General Knowledge: Preparing boiler for inspection; Scale removal; Blow-off tanks; Packing and Gasket uses; Lubricating oils and contaminants; oil preheaters; "etc."

City of Detroit
Buildings and Safety Engineering Department
Fourth Floor, City-County Building, Detroit, Michigan 48226
Mechanical/Electrical Inspection Division - Examination Section - Room 404
Telephone: 313/224-3184

Chapter 6　　　　License Requirements for States and Cities

Memorandum
Subject:

Information pertinent to First or Second Class Stationary Engineer License Applicants.

The City of Detroit Ordinance No. 278-E requires that an applicant for a First Class Stationary Engineer license shall have a valid Second Class Stationary Engineer license for two years, or otherwise qualify. An applicant for a Second Class Stationary Engineer license shall have a valid Third Class Stationary Engineer license for one year, or otherwise qualify.

Bring pen and pencil together with a properly filled out application for the written examination, that is conducted on Monday of each week. Applicants will be interviewed between 8:00 and 9:00 a.m., the examination beginning at 9:30 a.m. You will be notified of your score by mail. If the written examination is passed, an appointment will be made for the required oral examination.

The subject matter of the written examination will include multiple choice statements in the following categories.

Boiler and Accessories:

Boilers; foundations; and supports; safety devices-their function and testing; water walls; water columns; headers; drum internals; other boiler appurtenances; laying up boilers; heat absorption rates of various water surfaces.

Boiler Plant Auxiliaries:

Pumps; injectors; regulators; feedwater heaters; superheaters; de-superheaters; economizers; air preheaters; collectors; valves; traps; separators; draft fans; automatic control equipment and operation.

Detroit, Michigan

Water and Steam:

Steam cycle; blow-downs; water treatment principles and practices; scale causes and prevention; reactions under temperature and pressure; boiler water circulation; calorimeters; evaporators.

Fuels and Combustion:

Smoke; fuel and storage; flue gas analysis and interpretations; coal analysis; heating of coal in storage.

Steam Engines, Turbines and Accessories:

Types of engines and turbines; starting, stopping, lubrication, operating, valve arrangements, governing and fundamentals of heat engines; condensers, air ejectors; horsepower considerations; oil coolers and centrifuges; reduction gears; safety precautions with heat engines; routine tests and inspections; bearings and shafting.

Electricity:

Fundamental concepts of AC and DC, impedance, volts, amperes, watts, operating characteristics, starting and stopping of electric motors types; proper use of fuses; safety precautions with electricity; transformers; generators; exciters; phasing; power factor; synchronization; circuit breakers; panel type controllers.

Plant Operation:

Foaming, priming, carry-over, and remedies; cutting in boilers; water hammer; reducing stations; expansion joints; accumulation; hydrostatic, and boiler rating test; intercoolers; aftercoolers; bearings, low water and over pressure procedures, brick work maintenance; general safety precautions.

General Knowledge:

Chapter 6   License Requirements for States and Cities

Preparing boilers for inspection; scale removal; blow-off tanks; back-flow preventers; packing and gasket uses; lubricating oils and contaminants; oil preheaters; specific heats; measuring instruments; heat insulation; plant efficiency; piping and fittings, etc.

In addition, an applicant for a First Class Stationary Engineer's license will be required to answer any five problems pertaining to the above categories.

City of Detroit
Buildings and Safety Engineering Department
Fourth Floor, City-County Building
Detroit, Michigan 48226
Mechanical/Electrical Inspection Division
Examination Division - Room 204
Telephone - 224-3184
August 18, 1986

MEMORANDUM

Employers submitting Documented Evidence on behalf of applicants for Stationary Engineer license.

For your guidance, this sample letter has been developed to provide the minimum information required to substantiate an applicant's experience and therefore his qualifications for a stationary Engineer examination per Ord. 706-G, Art. VI.

It is important that the basic information necessary, that is company letterhead, inclusive dates of experience, positions held, resume of specific duties, equipment involved and proper notarization of the company official's signature, be included in the letter.

Very truly yours,

MECHANICAL/ELECTRICAL INSPECTION DIVISION

Thomas Riddering

Head Engineer

Chapter 6   License Requirements for States and Cities

SAMPLE LETTER:

Megawatt Industries
Foot of Power Drive
Detroit, Michigan 48226

August 18, 1986

Mr. Thomas Riddering - Head Engineer - Mechanical/Electrical Inspection Division
Buildings & Safety Engineering Department
408 City-County Building
Detroit, Michigan 48226

Dear Sir:

Mr. John Doe has been employed continuously in our power plant since Jan. 4, 1965 in the classifications indicated and in association with the plant equipment as indicated.

Our power plant consists of (8) B & W water tube boilers rated at 9571 square feet of heating surface each and operating at a pressure of 350# per square inch gage. These boilers are interconnected in battery and service (4) Westinghouse steam driven turbine generators rated 25,000 KVA each. We also have one B & W boiler rated 37,080 square feet of heating surface servicing one GE steam driven turbine generator rated 47,000 KVA at a throttle pressure of 1950 pounds per square inch. Our fuel system consists of bowl mill, ball mill and hammer mill pulverizers in conjunction with a #2 fuel oil ignition system.

Mr. Doe's power plant assignments have been as follows:

From 1-4-1965 to 1-25-1967

Assisted the licensed personnel in the general operation and maintenance of the power plant equipment.

From 1-25-1967 to 4-19-1971

Served as an attending operator in the boiler plant after qualifying for and being licensed as a high pressure boiler operator.

From 4-19-1971 to present.

Qualified and licensed as a Third Class Stationary Engineer, Detroit, and assigned as a relief attending operator on the turbine floor.

We hope this letter will provide the information necessary to supplement our John Doe's application for a First Class Stationary engineer's license.

Very truly yours,

J.M. Proficient
Vice President
Megawatt Industries

Sworn to before me on this_____day of_____. Notary Public, Wayne County, Michigan

RE: Mechanical Training Programs

The following educational institutions have programs which have been recognized by the City of Detroit as providing training in boiler and refrigeration operation. No endorsement is implied by this list.

E.T.I.T.
16801 Wyoming
Detroit, MI 48221
313/345-1871

Ferndale/Oak Park Adult & Community Education
881 Pinecrest
Ferndale, MI 48220
313/542-2535

Greater Opportunities Industrialization Center of Metropolitan Detroit
1565 Oakman Boulevard
Detroit, MI 48238-2882
313/883-4510

Henry Ford Community College
Searle Technical Building
13020 Osburn St.
Dearborn, MI 48126-3640
313/845-9609

International Union of Operating Engineers Local 547
24270 W. Seven Mile Rd.
Detroit, MI 48219
313/532-5345

Macomb County Community College, South Campus
14500 Twelve Mile Road
Warren, MI 48093-3896
313/445-7000

Chapter 6   License Requirements for States and Cities

Oakland Community College
Auburn Hills Campus
2900 Featherstone
Auburn Heights, MI 48304
313/340-6572

Ram Technical Center
8935 W. Eight Mile Road
Detroit, MI 48219
313/537-0505

Vocational Institute of Michigan
17421 Telegraph Rd. -#201 North
Detroit, MI 48219
313/537-6120

Wayne County Community College
801 W. Fort Street
Detroit, MI 48226
313/496-2758

William D. Ford Vocational/Technical Center
36455 Marquette
Westland, MI 48185
313/595-2135 [7]

---

7   Some Schools are for refrigeration only. The City of Detroit sets the number of classroom contact hours acceptable for credit toward experience requirements.

License Requirements for States and Cities   Chapter 6

Grand Rapids, Michigan
SUMMARY OF QUALIFICATIONS:

1. Exam Location, Address and Phone Number
City of Grand Rapids
Mechanical Inspection Dept.
345 State Street SE
Grand Rapids, MI 49503
616/456-3046

2. Exam Days and Times

Once every two months - time varies

February, April, June, August, October, December

3. Examination Cost

$15.00

4. Licence Cost and Renewal

Boiler Operator - Initial license and renewal... $20.00

Boiler Engineer - Initial license and renewal... $40.00

5. Type of Test

All Written - multiple choice

6. Grades of Licenses

Boiler Operator

Boiler Engineer

7. Experience and Education Requirements

Boiler Operator applicants must have had at least one year full-time work experience under the supervision of a licensed Boiler Engineer or equivalent.

Boiler Engineer applicants must have held a Boiler Operator's license a minimum of three (3) years under the supervi-

Chapter 6   License Requirements for States and Cities

sion of a licensed Boiler Engineer or present satisfactory evidence of at least four (4) years experience.

8. U.S. Citizenship Required?

Not specified

9. Local Residency Required?

No

10. Recognition of Licenses from other locales

Not Specified

CODES for Grand Rapids, Michigan

8.456 Boiler Engineers License and Boiler Operators Certificate.

It shall be unlawful for any person, company, or corporation to own and/or operate a High Pressure steam boiler within the City of Grand Rapids without having in its employ a licensed Boiler Engineer who shall be responsible for the maintenance and safe operation of the boiler. It shall further be unlawful for any person, company or corporation within the City of Grand Rapids to allow anyone who does not have a valid Boiler Operator's Certificate of Fitness to perform the duties of a Boiler Operator in conformity with Section M-500.4 of the BOCA Basic Mechanical Code and Boiler Act 290 of 1965. Licenses shall be issued by the Building Official as follows:

(1) A Boiler Engineer License shall be issued only to an applicant who has held a Boiler Operator License for at least three years while working full time under the supervision of a Licensed Boiler Engineer and has passed an examination prepared by the Department.

(2) A Boiler Operator Certificate shall be issued only to an applicant having at least one year full time work under the su-

License Requirements for States and Cities — Chapter 6

pervision of a licensed Boiler Engineer and having passed an examination prepared by the Department.

(3) Licenses shall be renewed annually. Licenses which have been allowed to lapse shall not be renewed unless the delinquent fees are paid up. However, if the license has lapsed for more than three years, the department may require a written examination.

(4) Any license issued hereunder may be revoked or suspended by the department after a hearing stating the reason for revocation or suspension. Supplying false information with the application for a license or failure to comply with the provisions of this Chapter, or the rules and regulations adopted thereunder, or failure to properly supervise work performed under a license hall be considered sufficient reason for revocation of a License.

(5) A fee of fifteen dollars ($15.00) shall be paid prior to taking a Boiler Engineer or Boiler Operator examination.

(6) A fee of forty dollars ($40.00) shall be paid for a Boiler Engineer License and a fee of ($20.00) shall be paid for a Boiler Operator Certificate.

(7) The Department shall establish rules for the examination of applicants desiring to obtain license or certificates of fitness. Such rules shall be available to any person wishing a copy thereof.

Rules and Regulations of the Boiler Engineer and Operator Licensing Ordinance

Applicants for a license must fill out and submit an application form provided by the Inspection Services Department. The Applicant for a Boiler Engineer License shall supply at least two (2) Certifications or letters completed by persons hav-

Chapter 6   **License Requirements for States and Cities**

ing direct knowledge of the applicant's boiler related experience. The applicant for a Boiler Operator license shall supply at least one (1) Certification or letter completed by a person having direct knowledge of the applicant's boiler related work experience.

Saginaw, Michigan

SUMMARY OF QUALIFICATIONS:

1. Exam Location, Address and Phone Number

Saginaw City Hall
City Clerk's Office
1315 South Washington Avenue
Room 102
Saginaw, Michigan 48601-2599
517/759-1480

2. Exam Days and Times

1st and 3rd Monday of the Month, 7:00 pm to 9:00 pm

3. Examination Cost

Free

4. License Cost and Renewal

Initial license....................$10.00

Renewal of license............$ 5.00

5. Type of Test

All Written - multiple choice

6. Grades of Licenses

By Horsepower:

0-15 HP

15-100 HP

100-250 HP

250-500 HP

700-1000 HP

7. Experience and Education Requirements

See Codes for Saginaw below

8. U.S. Citizenship Required?

No

9. Local Residency Required?

No

10. Recognition of Licenses from other locales

Yes

CODES for Saginaw, Michigan

Chapter 6, Section 219.

Section 219. Stationary Engineers and Firemen.

219.1

No person shall engage in the occupation of operating steam boilers or steam actuated machinery without first obtaining a license or permit therefor. No such license shall be granted except upon certification by the Board of Examiners of Stationary Engineers and Firemen that said person has been determined, by examination, to be qualified to operate said boilers or machinery. Permits may be granted as provided by this section.

219.2

Chapter 6   **License Requirements for States and Cities**

No such license shall be granted unless the applicant therefor shall have had at least two (2) years' experience as a Fireman, Engineer, Assistant Engineer or Oiler.

219.3

Any person who has been employed at one power plant consecutively for at least three (3) years and who, at the time of making application, is still employed at said power plant, shall, upon application to said board accompanied by a written request from his employer and upon satisfying said board of the fact of such consecutive employment, be certified for a permit authorizing him to take charge of and operate as Engineer or Fireman at such plant and not elsewhere, without submitting to any examination whatsoever.

Said certificate and permit shall plainly specify and designate the particular power plant at which such person is permitted to operate.

219.4

Any person who has been employed at one plant for at least two (2) years immediately preceding his application and who has had steam heating plant experience, may, without examination upon application accompanied by written request from said employer and upon satisfying said board of the fact of such employment and the sufficiency of his steam heating plant experience, be certified for a permit authorizing him to operate the steam heating plant of said employer, provided said plant does not contain engines, pumps, other steam actuated machinery or any equipment operating at steam pressures greater than fifteen (15) pounds per square inch or such pressure as the board may determine to be proper considering all of the facts and circumstances relating to the application.

219.5

Any original license or renewal thereof may be renewed without re-examination, provided application therefor is made within thirty (30) days after the expiration of such original or renewed license.

219.6

The provisions of this section shall not be construed to apply to steam boilers or other steam-generating apparatus installed and used in private dwellings in marine service, or by railroads in transportation.

Chapter 6　　　License Requirements for States and Cities

Minnesota

State of Minnesota

SUMMARY OF QUALIFICATIONS:

1. Exam Location, Address and Phone Number

Labor and Industry Building
443 Lafayette Road
St. Paul, Minnesota 55155-4304
612/296-1098

2. Exam Days and Times

Every day - Monday through Friday - 8 a.m. to 10 a.m.

3. Examination Cost

Chief A, B, C	$50.00
First A, B, C	$30.00
Second A, B, C	$25.00
Special	$20.00

4. License Cost and Renewal

Renewal Only:

Chief Engineers A, B, C	$25.00
First Class Engineers A, B, C	$20.00
Second Class Engineers A, B, C	$15.00
Special Engineer	$10.00

5. Type of Test

All written unless applicant can't read then they will give an oral test up to 2nd class. Up to the Chief Inspector.

6. Grades of Licenses

License Requirements for States and Cities Chapter 6

Chief A, B, C

First A, B, C

Second A, B, C

Special

7. Experience and Education Requirements

See Codes for Minnesota below

8. U.S. Citizenship Required?

No

9. Local Residency Required?

No

10. Recognition of Licenses from other locales

No

CODES for State of Minnesota

Laws Relating to Boiler Operating Engineers for the State of Minnesota

Minnesota Department of Labor and Industry

Section 13 - 183.501 - LICENSE REQUIREMENT.

(a) No person shall be entrusted with the operation of or operate any boiler, steam engine, or turbine who has not received a license of grade covering that boiler, steam engine or turbine. The license shall be renewed annually. When a violation of this section occurs, the division of boiler inspection may cause a complaint to be made for the prosecution of the offender and shall be entitled to sue for and obtain injunctive relief in the district courts for such violations.

Chapter 6   License Requirements for States and Cities

(b) For purposes of this chapter, "operation" shall not include monitoring of an automatic boiler, either through on premises inspection of the boiler or by remote electronic surveillance, provided that no operations are performed upon the boiler other than emergency shut down in alarm situations.

183.502 - SCHOOL ENGINEER OPERATIONAL REQUIREMENTS.

Any custodial Engineer employed by a school whose duties include the operation of a boiler shall be licensed pursuant to section 183.51, to operate the particular class of boiler used in the school.

183.505 - APPLICATIONS FOR LICENSES.

The Chief Boiler Inspector shall prepare blank applications on which applications for Engineers' licenses shall be made under oath of the applicant. These blanks shall be so formulated as to elicit such information as is desirable to enable the examiners to pass on the qualifications of applicants.

Section 15 - 183.51 EXAMINATIONS; CLASSIFICATIONS, QUALIFICATIONS.

Subdivision 1. ENGINEERS, CLASSES.

Engineers shall be divided into four classes:

(1) Chief Engineers: Grade A, Grade B, and Grade C

(2) First Class Engineers: Grade A, Grade B, and Grade C

(3) Second Class Engineers: Grade A, Grade B, and Grade C

(4) Special Engineers

Subd. 2. - APPLICATIONS.

License Requirements for States and Cities   Chapter 6

Any person who desires an Engineer's license shall make a written application on blanks furnished by the inspector. The person shall also successfully pass a written examination for such Grade of license applied for.

Subd. 3. HIGH AND LOW PRESSURE BOILERS.

For the purposes of this section and section 183.50, high pressure boilers shall mean boilers operating at a steam or other vapor pressure in excess of 15 p.s.i.g., or a water or other liquid boiler in which the pressure exceeds 160 p.s.i.g., or a temperature of 250 degrees Fahrenheit.

Low pressure boilers shall mean boilers operating at a steam or other vapor pressure of 15 p.s.i.g., or less, or a water or other liquid boiler in which the pressure does not exceed 160 p.s.i.g., or a temperature of 250 degrees Fahrenheit.

Subd. 4. CHIEF ENGINEER, GRADE A.

A person seeking licensure as a Chief Engineer, Grade A, shall be at least 18 years of age, and have experience which verifies that the person is competent to take charge of and be responsible for the safe operation and maintenance of all classes of boilers, steam engines, and turbines and their appurtenances; and, before receiving a license, the applicant shall take and subscribe an oath attesting to at least five years actual experience in operating such boilers, including at least two years experience in operating such engines or turbines.

Subd. 5. CHIEF ENGINEER, GRADE B.

A person seeking licensure as a Chief Engineer, Grade B, shall be at least 18 years of age, and have habits and experience which justify the belief that the person is competent to take charge of and be responsible for the safe operation and maintenance of all classes of boilers and their appurtenances;

Chapter 6 License Requirements for States and Cities

and, before receiving a license, the applicant shall take and subscribe an oath attesting to at least five years actual experience in operating those boilers.

Subd. 6. CHIEF ENGINEER, GRADE C.

A person seeking licensure as a Chief Engineer, Grade C, shall be at least 18 years of age, and have habits and experience which justify the belief that the person is competent to take charge of and be responsible for the safe operation and maintenance of all classes of low pressure boilers and their appurtenances, and before receiving a license, the applicant shall take and subscribe an oath attesting to at least five years of actual experience in operating such boilers.

Subd. 7. FIRST CLASS ENGINEER, GRADE A.

A person seeking licensure as a First Class Engineer, Grade A, shall be at least 18 years of age and have experience which verifies that the person is competent to take charge of and be responsible for the safe operation and maintenance of all classes of boilers, engines, and turbines and their appurtenances of not more than 300 horsepower or to operate as a Shift Engineer in a plant of unlimited horsepower. Before receiving a license, the applicant shall take and subscribe an oath attesting to at least three years actual experience in operating such boilers, including at least two years experience in operating such engines or turbines.

Subd. 8. FIRST CLASS ENGINEER, GRADE B.

A person seeking licensure as a First Class Engineer, Grade B, shall be at least 18 years of age, and have habits and experience which justify the belief that the person is competent to take charge of and be responsible for the safe operation and maintenance of all classes of boilers of not more than 300 horsepower or to operate as a Shift Engineer in a plant of unlimited horsepower. Before receiving a license, the applicant shall take and subscribe an oath attesting to at least three years actual experience in operating such boilers.

Subd. 9. FIRST CLASS ENGINEER, GRADE C.

A person seeking licensure as a First Class Engineer, Grade C, shall be at least 18 years of age and have habits and experience which justify the belief that the person is competent to take charge of and be responsible for the safe operation and maintenance of all classes of low pressure boilers and their appurtenances of not more than 300 horsepower or to operate as a Shift Engineer in a low pressure plant of unlimited horsepower. Before receiving a license, the applicant shall take and subscribe an oath attesting to at least three years actual experience in operating such boilers.

Subd. 10. SECOND CLASS ENGINEER, GRADE A.

A person seeking licensure as a Second Class Engineer, Grade A, shall be at least 18 years of age, and have experience which verifies that the person is competent to take charge of and be responsible for the safe operation and maintenance of all classes of boilers, engines, and turbines and their appurtenances of not more than 100 horsepower or to operate as a Shift Engineer in a plant of not more than 300 horsepower or to assist the Shift Engineer, under direct supervision, in a plant of unlimited horsepower. Before receiving a license, the applicant shall take and subscribe an oath attesting to at least one year of actual experience in operating such boilers, including at least one year of experience in operating such engines or turbines.

Subd. 11. SECOND CLASS ENGINEER, GRADE B.

A person seeking licensure as a Second Class Engineer, Grade B, shall be at least 18 years of age, and have habits and experience which justify the belief that the person is competent to take charge of and be responsible for the safe operation and maintenance of all classes of boilers of not more than 100 horsepower or to operate as a Shift Engineer in a plant of not more than 300 horsepower or to assist the Shift Engineer, under direct supervision, in a plant of unlimited horsepower. Before receiving a license, the applicant shall take and subscribe an oath attesting to at least one year of actual experience in operating such boilers.

Subd. 12. SECOND CLASS ENGINEER, GRADE C.

A person seeking licensure as a Second Class Engineer, Grade C, shall be at least 18 years of age, and have habits and experience which justify the belief that the person is competent to take charge of and be responsible for the safe operation and maintenance of all classes of low pressure boilers and their appurtenances of not more than 100 horsepower or to operate as a Shift Engineer in a Low Pressure plant of not more than 300 horsepower, or to assist the shift Engineer, under direct supervision, in a low pressure plant of unlimited horsepower. Before receiving a license, the applicant shall take and subscribe an oath attesting to at least one year of actual experience in operating such boilers.

Subd. 13. SPECIAL ENGINEER.

A person seeking licensure as a Special Engineer shall be at least 18 years of age and have habits and experience which justify the belief that the person is competent to take charge of and be responsible for the safe operation and maintenance of all classes of boilers and their appurtenances of not more than 30 horsepower or to operate as a Shift Engineer in a plant of not more than 100 horsepower, or to serve as an apprentice in any plant under the direct supervision of the properly licensed Engineer.

Subd. 14. CURRENT BOILER OPERATORS.

Any person operating a boiler other than a steam boiler on April 15, 1982, shall be qualified for application for the applicable class license upon presentation of an affidavit furnished by an inspector and sworn to be the person's employer or Chief Engineer. The applicant must have at least the number of years of actual experience specified for the class of license requested and pass the appropriate examination.

Subd. 15. RATING HORSEPOWER.

For the purpose of rating boiler horsepower for Engineer license classifications only: ten square feet of heating surface shall be considered equivalent to one boiler horsepower for conventional boilers and five square feet of heating surface equivalent to one boiler horsepower for steam coil type generators.

Section 15 - 183.52 REVOCATION OF LICENSE.

The Chief boiler inspector or representative may issue cease and desist orders to any person found to be in violation of sections 183.375 to 183.62 or the rules adopted thereunder, or for otherwise operating or allowing a boiler or pressure vessel to be operated under unsafe or dangerous conditions, and may petition for enforcement of the order in the district court. The department may also suspend or revoke the license of any Engineer for a violation.

Chapter 6 — License Requirements for States and Cities

Minnesota Department of Labor and Industry
Boiler Inspection Division

CLASS	NUMBER OF QUESTIONS*	EXPERIENCE REQUIRED**	APPLICATION FEE	RENEWAL FEE	EXPIRED FEE
SPECIAL	33	NONE	$20	$10	$20
SECOND C	50	1 YEAR LOW PRESSURE	25	15	25
SECOND B	60	1 YEAR HIGH PRESSURE	25	15	25
SECOND A	80	1 YEAR HIGH PRESSURE INCLUDING 1 YEAR TURBINE	25	15	25
FIRST C	100	3 YEARS LOW PRESSURE	30	20	30
FIRST B	100	3 YEARS HIGH PRESSURE	30	20	30
FIRST A	130	3 YEARS HIGH PRESSURE INCLUDING 2 YEARS TURBINE	30	20	30
CHIEF C	100	5 YEARS LOW PRESSURE	50	25	50
CHIEF B	115	5 YEARS HIGH PRESSURE	50	25	50
CHIEF A	140	5 YEARS HIGH PRESSURE INCLUDING 2 YEARS TURBINE	50	25	50

* Subject to Change

** Affidavit Required

License Requirements for States and Cities   Chapter 6

Mississippi

No State or City License

Missouri

Kansas City, Missouri

SUMMARY OF QUALIFICATIONS:

1. Exam Location, Address and Phone Number

City Hall
414 E. 12th Street
18th Floor
Kansas City, Missouri 64106
816/274-1678

2. Exam Days and Times

1st Saturday of every month

3. Examination Cost

Application fee...............................$36.00

Exam fee...$50.00

4. License Cost and Renewal

Certificate free for 1st year

Renewal fee after the 1st year........$36.00

5. Type of Test

All Written Test (computer graded - all multiple choice)

6. Grades of Licenses

Fireman

Plant Fireman

Operating Engineer

Chapter 6   License Requirements for States and Cities

Plant Operating Engineer

7. Experience and Education Requirements

Fireman - Letter of reference and 1 year of experience

Plant Fireman - Letter of reference and 1 year of experience as powerplant operator or letter from employer stating instructed and demonstrated ability to operate equipment safely.

Plant Operating Engineer - Letter of reference and 3 years experience in Stationary Engineering field (a Mechanical Engineering degree from an accredited school counts as 1 year of experience.)

Operating Engineer - Same as Plant Operating Engineer

8. U.S. Citizenship Required?

No

9. Local Residency Required?

No

10. Recognition of Licenses from other locales

No

CODES for Kansas City, Missouri

ORDINANCE 911125 - ADOPTING THE MODEL CODES

CODE OF GENERAL ORDINANCES - CHAPTER 9

Section 9.10.112. Certificate of Qualification Defined and Required:

(a) **Definition:**

License Requirements for States and Cities Chapter 6

A Certificate of Qualification is authority to perform certain skills and is issued by the Director of Codes Administration upon successful completion of a written examination. A certificate of Qualification is not transferable.

(b) Certificates Required:

Certificates of Qualification shall be required for all types of work hereinafter specified and classified.

(c) Supervisors:

Every supervisor required for a particular license shall be examined by the appropriate examining committee and if qualified, shall be issued a Certificate of Qualification and shall be entitled to perform and supervise the work in the particular skill for which he is qualified and certified. The certificate is an individual certificate and shall not be construed to be a license.

(d) Application and Fee:

Every applicant for a Certificate of Qualification shall fill out the form provided by the Director of Codes Administration and shall pay an application fee of $36.00 at the time of filing. Such fee shall not be refundable. The application fee shall not apply on the certificate fee.

EXCEPTION: The City and its departments shall be exempt.

(e) Temporary Certificate:

At the discretion of the Director of Codes Administration, he may issue a temporary Certificate of Qualification. Such certificate shall be in effect until the examination procedure is completed. The applicant shall be given consideration when:

(1) He has previously been certified by the City but not suspended or revoked; or

Chapter 6 License Requirements for States and Cities

(2) He has completed the written examination with a passing grade.

(f) Examination Fee:

Each applicant taking an examination shall be charged an examination fee of $50.00.

(g) Successful Applicants:

After an applicant has successfully passed the examination the Director of Codes Administration shall issue the Certificate of Qualification.

(h) Failure to Pass Examination:

When an applicant has failed to pass the examination, he shall be so notified in writing by the Director of Codes Administration. Every applicant who fails to pass the required examination shall not be eligible for another examination until the next regularly scheduled examination thereafter, and any applicant who shall fail to pass the second examination shall not be eligible for reexamination for twelve (12) months from the date of the second examination.

(i) Right of Appeal:

In every instance that the Director of Codes Administration or an examining committee disapproves the application of an applicant for a Certificate of Qualification, the applicant may appeal that decision to the Building and Fire Codes Board of Appeals in the manner provided in this Code.

Section 9.10.113. Classification of Certificates of Qualifications.

(c) Operator Certificates:

License Requirements for States and Cities Chapter 6

An Operator's Certificate of Qualification shall be required to operate and maintain the following equipment and shall entitle the holder to operate and maintain the equipment for which he is certified; except, that equipment and accessories used for operations, production or processing by public utilities, government agencies, manufacturing or processing plants or commercial enterprises may be operated and maintained by a regular operating and maintenance staff when supervised by a professional engineer registered by the State of Missouri. The work done under such supervision shall comply with all applicable provisions of this Code, including required permits and inspections.

(1) Operating Engineer:

An Operating Engineer Certificate shall entitle the holder to take charge of and to operate and maintain all steam generating boilers, steam engines, internal combustion engines, turbines, condensers, compressors, generators, motors, blowers, fuel-burning equipment, refrigeration systems, and all auxiliary apparatus, together with any necessary maintenance of piping used in connection therewith. The certificate is not required for operating the following:

(A) Steam generating boilers carrying less than 125 pounds pressure.

(B) Boilers carrying less than 100 pounds pressure when used for driving machinery.

(C) Portable boilers less than 10 horse power in size.

(D) A system containing a Group 1 refrigerant.

(E) A system with a capacity of 10 tons or less containing any Group 2 refrigerant.

(2) Plant Operating Engineer:

Chapter 6 — License Requirements for States and Cities

A Plant Operating Engineer Certificate shall entitle the holder to operate and maintain the same equipment and accessories as an Operating Engineer but is limited to only a designated plant or system of plants with similar equipment.

(4) Fireman:

A Fireman Certificate shall entitle the holder to operate and maintain boilers carrying less than 100 pounds of pressure for the purpose of driving machinery and to operate other steam tanks or steam boilers carrying less than 125 pounds pressure. The certificate is not required for the operation of steam tanks or steam boilers carrying 15 pounds or less.

(5) Plant Fireman:

A Plant Fireman Certificate shall entitle the holder to operate and maintain the same equipment and accessories as a Fireman but is limited to only a designated plant or system of plants with similar equipment.

Section 9.10.114. Certificate Fees.

(a) Annual Fees.

The annual fee for all Certificates of Qualification shall be $36.00.

EXCEPTION: The certificate fee for employees of the City shall be waived when performing work for the City as tradesmen or inspectors.

(b) Certificate Fee Refund:

Certificate Fees shall not be refundable.

Section 9.10.115. Certificate Renewal.

Certificates shall be valid for one year, and must be renewed annually. It shall be a violation of this Code to perform any regulated work after expiration of a certificate. If a certificate is renewed more than one year after its issuance, the annual fee, plus an amount prorated by month, or part of a month, representing the period during which the certificate was dormant, shall be paid.

Section 9.10.116. Reissuance of Certificates.

The Director of Codes Administration shall have the authority to reissue a Certificate of Qualification without examination, provided such reissuance is requested within two years following expiration. Annual renewal fees for the period during which the certificate was dormant must be paid prior to reissuance. If a certificate is not reissued during this period, the person must begin the application process, including passage of the required examination, anew.

Section 9.10.116.5. Validity of Certificates.

Submitting false information on an application for a Certificate of Qualification renders the certificate, if issued, invalid. No individual may work under such certificate. An individual submitting false information when seeking a certificate shall be barred from obtaining a certificate for two years from the date the certificate is determined by the Director of Codes Administration to be invalid or the information is determined to be false if no certificate is issued. The decision may be appealed to the Building and Fire Codes Board of Appeals.

Section 9.10.117. Certificate Holder's Responsibility.

Chapter 6   License Requirements for States and Cities

(a) General:

All Certificate holders shall be responsible for compliance with the requirements of the Building Code, without limitation, and to the items herein listed.

(1) To present his certificate when requested by any member of Codes Administration.

(2) To faithfully construct without departure from or disregard of approved drawings and specifications.

(3) To obey any order issued under authority of the Building Code.

(4) To pay any fee assessed under authority of the Building Code.

(5) To observe any City ordinances prescribing measures for the safety of workmen and of the public.

(6) To always maintain an active part in the supervision of the workmen under his direction.

(7) To notify the Director of Codes Administration when he leaves the employ of a licensee for whom he is the qualified supervisor.

Section 9.10.118. Suspension or Revocation of Certificate.

(a) Authority:

The Director of Codes Administration may suspend or revoke a certificate issued under the provisions of this Article for any one or more of the following acts or omissions:

(1) Incompetence.

(2) Misuse of the Certificate.

(3) Violation of any provisions of the Building Code.

(4) Failure to comply with any of the Certificate Holder responsibilities as outlined in Section 9.10.117.

(b) Procedure:

When any of the acts or omissions as herein enumerated are committed by a certificate holder and the Director of Codes Administration deems that such certificate shall be suspended or revoked, the action shall be as set forth in Section 9.9.110 for Registered Contractors.

REQUIREMENTS TO OBTAIN A FIREMAN CERTIFICATE OF QUALIFICATION

A. CERTIFICATE OF QUALIFICATION

A Certificate of Qualification is authority to perform certain skills and is issued by the director of codes administration upon successful completion of a written examination. A certificate of qualification is not transferable.

B. FIREMAN CERTIFICATE

A Fireman Certificate shall entitle the holder to operate and maintain boilers carrying less than 100 pounds pressure for the purpose of driving machinery and to operate other steam tanks or steam boilers carrying less than 125 pounds pressure. The certificate is not required for the operation of steam tanks or steam boilers carrying 15 pounds or less.

C. REQUIREMENTS FOR THE CERTIFICATE

An application for the certificate must include: the completed application form, the required letters of reference (as

Chapter 6 License Requirements for States and Cities

stated below), an enclosed $36.00 non-refundable application fee, and a $50.00 fee for an examination which will be given in Kansas City, Missouri, by Block and Associates. (Make checks payable to City Treasurer). The codes administration department must receive the application by the cutoff date shown in the schedule on the reverse side. Please mail application, reference letters, and appropriate fees to the address shown above. The applicant must meet these requirements:

(1) A letter of reference verifying at least one year of experience in power plant operation with boilers and fuel burning equipment. (Excluding material suppliers).

(2) The applicant shall be at least 21 years of age.

(3) The applicant shall be a senior high school graduate or shall have passed the general educational development test. One year of experience may be substituted for each year of high school accreditation.

(4) The applicant shall receive a score of 75% or higher on the examination administered by Block and Associates.

EXCEPTION: The examination and $50.00 Examination fee may be waived if the applicant has a score of 75% or higher on a "standard" Block & Associates examination comparable in content to the examination currently given for the city of Kansas City, Missouri. The applicant must provide verifiable documentation of the examination which is acceptable to the director of codes administration.

D. APPLICANT INFORMATION BOOKLET

When the application is received by codes administration, the applicant will be mailed an information package. The package will include the date, time and location of the next available examination, and a list of required texts and optional texts.

E. APPLICATION REVIEW

The application will be reviewed for compliance with the work experience, education and age requirements. All references will be contacted by the codes administration department for verification of work experience before the applicant can be scheduled for an examination. When all these requirements are met, the applicant will be scheduled for the examination.

F. EXAMINATION PROCEDURES

The examination is open book, forty multiple-choice questions, three hours long. Examination are given on the first Saturday of every month, except holidays.

G. EXAMINATION SCHEDULE

NOTE: Since all experience must be verified before an applicant can be scheduled for an examination, applications should be submitted as early as possible, but in no case later than the following cutoff dates.

Examination Date	Application Cutoff Date
January 9, 1993	December 14, 1992
February 6, 1993	January 11, 1993
March 6, 1993	February 8, 1993
April 3, 1993	March 8, 1993
May 1, 1993	April 5, 1993
June 5, 1993	May 10, 1993
July 10, 1993	June 14, 1993
August 7, 1993	July 12, 1993
September 11, 1993	August 16, 1993

| Chapter 6 | License Requirements for States and Cities |

October 2, 1993.................................September 13, 1993

November 6, 1993................................October 11, 1993

December 4, 1993................................November 8, 1993

INQUIRIES

If you have any questions, please call 816/274-1678.

REQUIREMENTS TO OBTAIN A PLANT OPERATING ENGINEER CERTIFICATE

A. CERTIFICATE OF QUALIFICATION

A Certificate of Qualification is the authority to perform certain skills and is issued by the director of codes administration upon successful completion of a written examination. A Certificate of Qualification is not transferable.

B. PLANT OPERATING ENGINEER CERTIFICATE

A Plant Operating Engineer Certificate shall entitle the holder to operate and maintain the same equipment and accessories as an Operating Engineer but is limited to only designated plant or system of plants with similar equipment.

An Operating Engineer Certificate shall entitle the holder to take charge of and to operate and maintain all steam-generation boilers, steam engines, internal combustion engines, turbines, condensers, compressors, generators, motors, blowers, fuel-burning equipment, refrigeration systems, and all auxiliary apparatus, together with any necessary maintenance of piping used in connection therewith. The certificate is not required for operating the following:

(a) Steam-generating boilers carrying less than 125 pounds pressure.

(b) Boilers carrying less than 100 pounds pressure when used for driving machinery.

(c) Portable boilers less than 10 horsepower in size.

(d) A system containing a Group 1 refrigerant.

(e) A system with a capacity of ten (10) tons or less containing any Group 2 refrigerant.

C. REQUIREMENTS FOR THE CERTIFICATE

An application for the certificate must include: the completed application form, the required letters of reference (as stated below), an enclosed $36.00 non-refundable application fee, and a $50.00 fee for an examination which will be given in Kansas City, Missouri by Block & Associates. (Make checks payable to City Treasurer). The codes administration department must receive the application by the cutoff date shown in the schedule on the reverse side. Please mail application, reference letters and appropriate fees to the address shown above. The applicant must meet these requirements:

(1) A letter of reference verifying three of more years of experience in the Stationary Engineering field, excluding material suppliers. A degree in mechanical engineering from a school accredited by the State of Missouri shall be equivalent to one year of experience.

(2) The applicant shall be at least 21 years of age.

(3) The applicant shall be a senior high school graduate or shall have passed the general education development test. One year of experience may be substituted for each year of high school accreditation.

(4) The applicant shall receive a score of 75% or higher on the examination administered by Block & Associates.

Chapter 6 — License Requirements for States and Cities

EXCEPTION: The examination and $50.00 examination fee may be waived if the applicant has a score of 75% or higher on a "standard" Block & Associates examination comparable in content to the examination currently given for the City of Kansas City, Missouri. The applicant must provide verifiable documentation of the examination which is acceptable to the Director of Codes Administration.

D. APPLICANT INFORMATION BOOKLET

When the application is received by Codes Administration, the applicant will be mailed an information package. The package will include the date, time and location of the next available examination, and a list of required texts and optional texts.

E. APPLICATION REVIEW

The application will be reviewed for compliance with the work experience, education and age requirements. All references will be contacted by the codes administration department for verification of work experience before the applicant can be scheduled for an examination. When all these requirements are met, the applicant will be scheduled for the examination.

F. EXAMINATION PROCEDURES

The examination is open book, forty multiple choice questions, three hours long. Examinations are given on the first saturday of every month, except holidays.

G. EXAMINATION SCHEDULE

NOTE: Since all experience must be verified before an applicant can be scheduled for an examination, applications should be submitted as early as possible, but in no case later than the following cutoff dates.

Examination date	Application cutoff date
January 9, 1993	December 14, 1992
February 6, 1993	January 11, 1993
March 6, 1993	February 8, 1993
April 3, 1993	March 8, 1993
May 1, 1993	April 5, 1993
June 5, 1993	May 10, 1993
July 10, 1993	June 14, 1993
August 7, 1993	July 12, 1993
September 11, 1993	August 16, 1993
October 2, 1993	September 13, 1993
November 6, 1993	October 11, 1993
December 4, 1993	November 8, 1993

H. INQUIRIES

If you have any questions, please call 816/274-1678

Chapter 6 License Requirements for States and Cities

St. Joseph, Missouri
SUMMARY OF QUALIFICATIONS:

1. Exam Location, Address and Phone Number
City Hall
Building Regulations
Room 204
12th & Frederick
St. Joseph, Missouri 64501
816/271-4844

2. Exam Days and Times

2nd Wednesday 4 times per year in January, March, May, July and September at 7 p.m.

3. Examination Cost

$25.00

4. License Cost and Renewal

First Class Operating Engineer.............$50.00

Second Class Steam............................$40.00

Steam Heating Engineer......................$25.00

All licenses expire on December 31st

5. Type of Test

Written and Oral

6. Grades of Licenses

First Class Operating Engineer

Second Class Steam

Steam Heating Engineer

7. Experience and Education Requirements

License Requirements for States and Cities   Chapter 6

First Class Engineer................3 years practical experience

Second Class Steam................1 year practical experience

Steam Heating Engineer..........1 year experience

8. U.S. Citizenship Required?

Not Specified

9. Local Residency Required?

No

10. Recognition of Licenses from other locales

Yes, for experience, with letter from previous employer on company letterhead.

CODES for St. Joseph, Missouri

M-2100.4 Boiler Operator Certificate of Fitness:

M-2100.4.1 Boiler Operator License:

(a) It shall be unlawful for any person to operate a steam boiler of more than ten (10) horsepower or a boiler designed to carry a steam pressure of more than fifteen (15) pounds per square inch, without a license from the building regulations supervisor or other authorized municipal agency. Boilers of less than fifteen (15) pounds per square inch and less than ten (10) boiler horsepower shall be exempted. The licenses for which examinations are given and which are issued are of four (4) kinds, to wit: First Class Operating Engineer's license; Second Class Steam license; Steam Heating license; and Second Class Refrigeration license. All licenses are nontransferable.

(b) Any person who is now operating any equipment which for the first time is subject to this code and for which a license is herein required may, upon payment of the required fee re-

Chapter 6　　　License Requirements for States and Cities

ceive a license to allow such person to continue operating the specific equipment (only) which such person is operating at the time of passage of this ordinance. Such person shall first make application in writing for such license to the building regulations supervisor. Such application shall set forth the information required by Subsection 2100.4.5. When the building regulations supervisor has received all of the information required of the applicant and is satisfied that the information set forth therein is true, then he shall certify such applicant for the appropriate license. This certification shall then be presented to the director of finance; and upon payment of the proper fee as set out in Subsection 2100.4.6, the director of finance shall issue the appropriate license to the applicant.

M-2100.4.2   Equipment Exempt:

Boiler, pressure vessels or other apparatus are exempt from the fee schedule provisions of this ordinance when used under the following conditions:

In buildings owned by the federal government, state government, county government and the city government; not included in code, permit or fee schedule, is railroad locomotives; scientific laboratory equipment; industrial testing equipment except air tanks; or equipment covered under any other section of this code or by another city ordinance or state statute.

M-2100.4.3   Classification of Heating Appliances:

All steam generating boilers, superheaters, and pressure vessels used in conjunction therewith, shall, for the purpose of this section, be divided in the following three (3) classes:

M-2100.4.3.1 Class One (1):

All boilers, pressure vessels and machinery engines subjected to operating pressure of more than fifteen (15) pounds

License Requirements for States and Cities Chapter 6

per square inch, and such equipment either manually or automatically operated must be in charge of a properly licensed First Class Operating Engineer.

M-2100.4.3.2 Class Two (2):

All boilers subjected to operating pressure of fifteen (15) pounds per square inch or less and more than fifty (50) boiler horsepower, either manually or automatically operated, used for heating purposes only, must be in charge of a properly licensed Steam Heating Engineer.

M-2100.4.3.3 Class Three (3):

Lower pressure boilers of fifteen (15) pounds or less and more than ten (10) boiler horsepower with automatic controls. If the boiler is manually fired, then a heating licensed Engineer will be in charge.

M-2100.4.4 ASME Code:

All boilers under First, Second and Third Class shall be required to be installed, repaired and operated according to the ASME Code.

M-2100.4.5 Examination for Licenses:

In order to take the examination for a license required by subsection M-2100.4.1, the prospective licensee shall first make application to the building regulations supervisor. Such application shall set forth the name, place of business or employment and address of the applicant. In the application for a First Class Operating Engineers license, the application shall also contain proof that the applicant has at least three (3) years of practical experience in boiler room or power plant duties or equivalent experience. All applications except those applications made pursuant to Section 2100.4.1(b), shall be accompanied by an

Chapter 6   License Requirements for States and Cities

examination fee in accordance with the fee schedule in Chapter 7, Article VII (Sec. 7-600).

When all of the information required to be included in the application has been received by the building regulations supervisor and the examination fee has been paid, the building regulations supervisor shall examine the applicant. Examinations shall be in writing or may be oral in part and will be administered by the building regulations supervisor or his duly authorized representative. Examinations will be designed to test the applicant's competency and knowledge required to perform work in connection with the operation of the general type and classes of apparatus in general use in the applicant's field. No more than thirty (30) percent of any examination shall be oral. All questions asked, whether written or oral, shall be of equal weight in determining the score of the applicant.

The building regulations supervisor shall make up or provide a prepared standard examination. Prior to the administering of such examination, the building regulations supervisor shall make up and keep on file model answers to all questions asked the applicant during such examinations. All answers will be graded as either right or wrong and will not be graded on a good, fair or poor basis. The building regulations supervisor shall make a complete record of all examination questions asked, the applicant's answers thereto, and model answers to such questions. He shall keep such written record for five (5) years.

In making up the examination questions, model answers thereto, or in grading the examination, the building regulations supervisor may seek the recommendation of the board of Operating Engineers examiners.

<u>**M-2100.4.6 Licensing:**</u>

Upon completion of the examination of the applicant, the board shall grade the examination. If the score of the applicant is equal to or greater than a score published to be a passing score by the building regulations supervisor before the examination is given, and if the building regulations supervisor is satisfied that all the information required to be included in the application is true, then he shall certify such person for the appropriate license. Should an applicant fail to pass an examination for the license applied for, he shall not be eligible for reapplication for another examination for at least three (3) months from the date of his last examination.

The applicant for a license who has received certification for a license shall then present such certification to the director of finance. The director of finance shall then issue to the applicant the proper license upon the payment of the fee for the license, in accordance with the fee schedule in Chapter 7, Article VII (Sec. 7-600).

The score determined to be a passing score by the building regulations supervisor shall be the same for all such examinations given in any one (1) calendar year, and shall be on file in the office of the building regulations supervisor. In determining what score will be a passing score, the building regulations supervisor may first seek the recommendations of the board of Operating Engineers examiners. He shall keep such record for five (5) years.

M-2100.4.7 Appeals:

Whenever an applicant fails the examination he shall have the right to request that the building regulations supervisor re-examine his examination answers. Such request shall be in writing and shall be made within five (5) days from the time the applicant receives notice of his having failed the examina-

Chapter 6 License Requirements for States and Cities

tion. Such re-examination shall take place within ten (10) days of the receipt of such written request. Such re-examination shall be made by the building regulations supervisor and shall be attended by the members of the board of Operating Engineers examiners. The written records of the examination questions asked, the model answers thereto, the answers of the applicant thereto and the score which is a passing score for said examination shall be produced and re-examined by the buildings regulations supervisor at such re-examination. The applicant shall have the right to be present at said re-examination.

The final decision on the re-examination of the applicant's examination answers shall be made by the building regulations supervisor. Prior to making such final decision, the building regulations supervisor may first seek the recommendation of the board of Operating Engineers examiners. The final decision of the building regulations supervisor shall be rendered within one (1) week of the re-examination of the applicant's examination answers. Such decision shall be in writing and shall be kept on record for five (5) years.

M-2100.4.8 First Class Operating Engineer's License:

Any person holding a First Class Operating Engineers license is hereby authorized to operate all steam generated boilers, pressure vessels, superheaters, refrigerating plants and air conditioning refrigeration systems of one hundred (100) tons or more or field installed systems hereinafter mentioned in the boiler and pressure vessel ordinance, provided that to be eligible to hold said First Class Operating Engineer's license said applicant must have at least three (3) years practical experience in boiler room or power plant duties, or the equivalent experience.

M-2100.4.9 Second Class Steam License:

Any person holding a Second Class steam license shall be permitted and is hereby authorized to operate all boilers subject to a pressure of fifteen (15) pounds per square inch or more, provided that to be eligible to hold said Second Class steam license said applicant must have at least one (1) year practical experience in boiler room operation or the equivalent experience.

M-2100.4.10 Steam Heating Operating Engineer's License:

Any person holding a steam heating Operating Engineer's license is authorized to operate all boilers, receptacles or tanks having a pressure of fifteen (15) pounds per square inch or less used for heating purposes only.

M-2100.4.13 Duration, Renewal, Suspension and Revocation of Licenses:

All licenses shall be good for one (1) year with an expiration date of December 31st of each year and shall be renewed upon application of the license holder and payment of renewal fee to the director of finance at any time within thirty (30) days prior to the expiration date thereof. A fee shall be charged for each year's renewal of the above license in accordance with the fee schedule in Chapter 7, Article VII (Sec. 7-600).

The director of public works and transportation shall have the power to suspend for not more than ninety (90) days or revoke any license provided for by this section when he finds that such license holder has willfully violated any provision of this mechanical code. No license shall be suspended or revoked except after a hearing at which both sides may present evidence and be represented by counsel. Ten (10) days written notice of such hearing shall be given the license holder. Such hearing shall be attended by the director of public works and transportation and by the members of the board of Operating

Engineers examiners. The director of public works and transportation shall make the final decision of such suspension or revocation, but prior to reaching such decision, he may first seek the recommendation of the board of Operating Engineers examiners. All decisions shall be rendered within five (5) days of the hearing and shall be in writing.

An ordinance amending Article VII of Chapter 7 of the code of ordinances entitled "Building, Electrical, Plumbing and Mechanical Permit and License Fees"

Be it ordained by the council of the City of St. Joseph, Missouri, as follows:

Section 1. The Article VII of Chapter 7 of the Code of Ordinances of the City of St. Joseph, Missouri, be and hereby is, amended by repealing Article VII and enacting in lieu thereof a new Article VII relating to the same subject to be entitled and read as follows:

Article VII.

Building, Electrical, Plumbing and Mechanical Permit and License Fees

Sec. 7-311. Building Fee Schedule.

(d) Examination and license fees:

(1) With application for examination...............$25.00

(Except as provided in Sec. 2100.4.1(b) of BOCA Mechanical Code)

(2) Licenses

License Requirements for States and Cities — Chapter 6

First Class Operating Engineer's License..........$50.00

Second Class Engineer's License

High Pressure Steam..................................$40.00

Refrigeration...$40.00

Heating Engineer's License...........................$25.00

All licenses are good for one (1) year, and will expire on December 31st of each year.

Examinations shall be given the second Wednesday in the months of January, March, May, July and September of each year.

Chapter 6   License Requirements for States and Cities

St. Louis, Missouri
SUMMARY OF QUALIFICATIONS:

1. Exam Location, Address and Phone Number:

City of St. Louis
Mechanical Section
Room 425
City Hall
1200 Market Street
St. Louis, Missouri 63103
314/622-3375

2. Exam Days and Times

Anytime 8 a.m. to 5 p.m.- Monday through Friday

3. Examination Cost

$10.00

4. License Cost and Renewal

Initial license included in cost of exam.

Renewal-Stationary Engineer Class 1..... $10.00

Renewal-Stationary Engineer Class 2...... $ 7.00

5. Type of Test

The test is five parts - you may take up to 6 months to complete all five parts of the test. When completed you are scheduled to take the oral exam. (Oral exam given by Chief Engineer and 2 examiners)

6. Grades of Licenses

Stationary Engineer - Class 1

Stationary Engineer - Class 2

7. Experience and Education Requirements

License Requirements for States and Cities Chapter 6

See Codes for St. Louis

8. U.S. Citizenship Required?

Yes

9. Local Residency Required?

No

10. Recognition of Licenses from other locales

Not Specified

CODES for St. Louis, Missouri

City of St. Louis

Licensed Engineer Examination Requirements:

(Ordinance 60513 of October 23, 1987)

M-2100.4.8 Qualifications for Stationary Engineer, Class 1:

All applicants for Class 1 Stationary Engineer's licenses shall be a citizen of the United States or shall have made application for such citizenship. The applicant shall be at least twenty one (21) years of age and shall have had at least two (2) years experience as a Class II licensed Engineer under the provisions of the Article or equivalent, or shall be registered with the Missouri State Board of Registration for Architects and Professional Engineers as an Engineer or as an Engineer in training, and shall have been actually employed in the Engineering or research division of a power generating plant in an Engineering capacity for a minimum of twelve (12) months. The applicant shall demonstrate his knowledge, skill, ability and competency to the board to operate boilers of any size or capacity rating which are generating saturated or superheated steam at any pressure in excess of fifteen (15) psig, or hot

Chapter 6 License Requirements for States and Cities

water or any other liquid as defined in this Code, or to operate stationary direct fired natural or manufactured gas turbines and/or engines in excess of five hundred (500) brake horsepower and ammonia compressors in excess of one hundred (100) tons capacity, and to operate associated power plant components and auxiliaries, such as steam turbines, engines, air compressors, ammonia systems, pumps, and feed water heaters, electric generators and other equipment.

M-2100.4.9 Qualifications for Stationary Engineer, Class II:

All applicants for the Class II Stationary Engineer's license shall be a citizen of the United States or shall have made application for such citizenship and shall be at least nineteen (19) years of age. The applicant shall have had at least one (1) year's experience in the operations of steam boilers under the supervision of a Class I or Class II Stationary Engineer under the provisions of this Article, or equivalent or shall have had one (1) year's experience in maintenance work on steam boilers and/or steam engines or steam turbines or direct fired natural or manufactured gas engines in excess of two hundred (200) brake horsepower and/or ammonia compressors in excess of fifty (50) tons capacity or shall be registered with the Missouri State Board of Registration for Architects and Professional Engineers as an Engineer or as an Engineer in training. The applicant shall demonstrate his knowledge, skill, ability and competency to the board to operate boilers which have not more than one thousand five hundred (1,500) square feet of rated heating surface and which are generating saturated or superheated steam in a pressure range of fifteen (15) psig minimum to three hundred (300) psig maximum or hot water or any other liquid as defined in this Code and to operate associated compressors, ammonia compressors, pump and feed water heaters, electric generators and other equipment.

M-2100.4.10 Examination (All Classes) for Stationary Engineers:

The examination for Class II Stationary Engineer license shall be oral. The examination for Class I Stationary Engineer license shall be both oral and written, provided that the applicant shall attain a predetermined percentage as set by the Board of Stationary Engineers in the written examination before the applicant becomes eligible for the oral examination. If the applicant does not pass either the oral or written examination he shall wait ninety (90) days before filing a new application. Any person making application for a certain class and failing to procure it for any cause whatsoever, may be assigned by the Board of Stationary Engineers to the class to which the results of his examination entitle him, and he shall not make any additional application for any class whatsoever until ninety (90) days thereafter.

City of St. Louis, Missouri

Ordinance 60513, October 23, 1987
WHEN A LICENSED ENGINEER IS REQUIRED
M-2100.4.13 OPERATIONS OF BOILERS, 100 TO 1500 SQUARE FOOT OF HEATING SURFACE

Except as provided in M-2100.4.20, any boiler which has not more than 1500 square feet of rated heating surface, and which is rated to generate steam at pressure between fifteen (15) psig and three hundred (300) psig maximum or which is rated to general hot water above 160 psig and 250 degrees F to 300 psig, and associated equipment, shall whenever in operation, be in the charge of an attending Class II or a Class I li-

Chapter 6   License Requirements for States and Cities

censed Stationary Engineer. Square footage shall be determined by the total input to a single header.

M-2100.4.14 PLANT AUXILIARIES:

Any direct fired stationary natural or manufactured gas turbine or engine of two hundred (200) brake horsepower, or more, rated capacity shall be, whenever in operation in the charge of any attending Class I or Class II licensed Stationary Engineer where installed in buildings in the following Use Group Classifications: Use Group I (I-1 and I-2), Use Group R (R-1 and R-2). The above shall apply to all other buildings with the following exceptions: attendance of a Stationary Engineer shall be required only during the times designated as normal working, or occupied hours, by the majority of the building occupants. Any ammonia system totaling fifty (50) tons or more rated capacity shall be, whenever in operation, in the charge of an attending Class I or Class II licensed Stationary Engineer where installed in any and all buildings.

M-2100.4.15 OPERATORS OF BOILERS IN EXCESS OF 1500 SQUARE FEET OF HEATING SURFACE:

Any boiler generating saturated or superheated steam above 212 degrees F in excess of fifteen (15) psig having a rated heating surface in excess of one thousand five hundred (1,500) square feet of hot water or any other liquid as defined in this Code, or any boiler generating saturated or superheated steam or any high temperature liquid above 212 degrees F in excess of three hundred (300) psig, regardless of rated heating surface, and any steam engine or steam turbine, associated with either of the said boilers shall be, whenever in operation in the charge of an attending Class I licensed Stationary Engineer. Square footage shall be determined by the total input to a single header.

M-2100.4.16 PLANT AUXILIARIES:

Any direct fired stationary natural or manufactured gas turbine or engine of five hundred (500) brake horsepower, or more, rated capacity shall be, whenever in operation, in the charge of an attending Class I licensed Stationary Engineer where installed in buildings in the following Use Group classification. Use Group I (I-1 and I-2), Use Group R (R-1 and R-2). The above shall apply to all other buildings with the following exception: attendance for a Stationary Engineer shall be required only during the time designated as normal working, or occupied hours, by the majority of the building occupants. Any ammonia system totaling one hundred (100) tons or more rated capacity shall be, whenever in operation, in the charge of an attending Class I licensed Stationary Engineer where installed in any and all buildings.

M-2100.4.20 CONTINUOUS SUPERVISION BY LICENSED OPERATORS:

The operation of boilers which are generating saturated steam in a pressure range of fifteen (15) psig minimum to one hundred and fifty (150) psig maximum of each of which boilers has not more than one hundred (100) square feet of rated heating surface shall be at all times in the charge of a certified Boiler Operator. Application for a Boiler Operator's certificate is to be made to the Building Commissioner or his duly authorized representative. The Building Commissioner, or his duly authorized representative, upon finding that the applicant is thoroughly familiar with the operational principles which concern the safety and care of the boiler, shall issue to such applicant a certificate of competency as a Boiler Operator. The fee for the examination shall be as listed in Table M-100-B.

Licensed Engineers, City of St. Louis

Chapter 6   License Requirements for States and Cities

An outline of the knowledge which an applicant for Engineer's License must have is set forth below:

BOILERS

The applicant should have a knowledge of the construction of different types of boilers in common use and should be able to answer questions which he might be asked about one particular type of boiler. In order to answer questions satisfactorily, the applicant may find it to his advantage to be able to make a sketch. The applicant should know about different types of construction used on boilers; how flat surfaces are braced and the different types of joints, with a complete understanding of one type of joint commonly used. The applicant should have some knowledge of different types of furnace settings, boiler supports, furnace volumes, and gas passes through the boiler, arrangement of breeching and construction and purpose of the chimney. He should have a knowledge of the different fuels used and should be able to answer questions on the combustion of at least one of the fuels. Some of these questions will pertain to the methods of mixing gases in the combustion chamber so as to get complete combustion. Some questions will pertain to flue gas analysis. Other questions will be relative to the boiler and stack pressure and temperatures. The applicant can expect questions on the elimination of smoke and how this may be accomplished by the use of overfire air. There will also be questions relative to pressure drop through the fuel bed, furnace and boiler. The applicant should know how to determine the heat input to a boiler and how to obtain the heat output.

He should have a knowledge of feed water treatment and why such is desirable and what it accomplishes. He should be able to determine how to blow a boiler down and should know the determining factors for blow down. He should have a knowledge of the heat loss through blow down, of the boiler and the advantage and purpose of using heat exchangers. He should be able to explain to the Board why he would obtain the boiler efficiency. The applicant must have a knowledge of the setting of safety valves according to the A.S.M.E. Code, and should know how to determine the size safety valves to be used on a boiler. His knowledge of water columns, fittings, etc., to the boiler should be determined by what the A.S.M.E. Code has to say.

ENGINES

The applicant should have a general knowledge of the various engines in common use and should know and understand thoroughly the construction and operation of at least one type of engine. He should understand the relationship between engines and boilers in terms of the heat in the steam. He should know the purpose of an indicator and how to interpret the indicator card. He should thoroughly understand how the steam enters the engine and how it can give up heat, causing the engine to do work. He should understand how the flow of steam is controlled by admission valves and should have a knowledge of the different type valves to be found in practice. The applicant must thoroughly understand at least one valve mechanism and be able to convince the Board that he could, if necessary, set the valve with which he is acquainted. The applicant must have a general knowledge of the various governors and must definitely understand completely at least one of the governors and its action. He must know the purpose of the governor and exactly how the governor controls the speed of the engine. The applicant should have a knowledge of the different types of turbines which are found in power plants and should understand the operation of these different types. Where an applicant is not familiar with engine governors he may be called on to explain a turbine governor.

PUMP AND AUXILIARY EQUIPMENT

The applicant must have a general knowledge of the various types of pumps which are used as feed water pumps. He should know how to determine the size of a pump to be used when the boiler size is known and should be able to determine the H.P. of the driving unit under set conditions. Since both plunger type and centrifugal type pumps are ordinarily found in power plants, it is recommended that the applicant know and understand at least one of each of these two types. The applicant should have a knowledge of the various regulation devices which are used on pumps and should thoroughly understand how at least one such device works to control the pump.

He should have a knowledge of feed water regulators and how they are used and for what purpose and should thoroughly understand the construction and operation of at least one. The applicant should understand the effect of temperature in pumping water and should know the determining factors for maximum pump suction. He should have a knowledge of the advantage and purpose of feed water heaters and where the feed water heaters should be located with reference to the pump. He should thoroughly understand the construction of at least one type of feed water heater, preferably the deaerating feed water heater, and should know how connections to the feed water heat should be made and any safety devices that may be used on the heater. The applicant should have a knowledge of reducing valves, steam traps, blow-down tanks and other auxiliary equipment that may be used with the boiler.

REFRIGERATION

The applicant should have a knowledge of the different types of refrigeration systems commonly used and a knowledge of the different refrigerants. He should have a thorough knowledge of the relationship between temperature and pressure for at least one refrigerant. The applicant should understand completely at least one of the systems of refrigeration and should be able to answer any questions pertaining to the cycle of that particular refrigerant. He should understand the safety devices which are used in industrial refrigeration plants and should be able to determine the size of safety valves required. He should understand the complete cycle for one refrigerant and how capacity of the compressor may be affected by a change of suction and head pressure. He should understand the different types of evaporators and condensers which are commonly used and the various types of expansion valves which will be found in refrigeration systems. The applicant should know how to use an indicator and be able to determine the H.P. of a driving unit when the size of the compressor and the properties of the refrigerant are known. The applicant should understand the construction of the compressor efficiency.

AIR COMPRESSORS AND ELECTRICAL EQUIPMENT

The applicant should have a knowledge of the different types of air compressors in general plant use and should understand the different methods of automatic control. He should know the safety requirements on air compressors and how the protection devices work. He should know the problems encountered in the compression and distribution of air.

A knowledge is required of the electrical equipment found in most power plants. The applicant should know the fundamentals of operation of electric motors using direct current, and single phase or three phase alternating current. He should know the safety devices which are used to protect electrical equipment and how to determine if the electrical equipment is adequately protected. He should know the relationship between the electrical and other terms for power. He should know how to determine the size motor to perform a definite duty and he

Chapter 6 — License Requirements for States and Cities

must understand the terms commonly used in discussing electricity.

GENERAL COMMENTS

The examination for Engineer's license requires that sketches of various equipment be made: such sketches must be complete as to convey to the Board of Engineers the knowledge which the applicant has of that particular equipment. The following equipment should be studied with the thought of making sketches on the examination:

A water tube boiler; horizontal return tubular boiler; Corliss engine; duplex pumps; various types of centrifugal pumps; open and closed feed water heaters, and other auxiliary power plant equipment, compression and systems of refrigeration; and an air compressor and system.

RULES & REGULATIONS

Board of Stationary Engineers

City of St. Louis

1. No less than 50% on any part of the written examination.

2. Twenty (20) days for license renewal. Card to be sent and given ten (10) days for renewal.

3. Only one license (class) application at a time.

4. Hold over oral portion of examination for period less than ninety (90) days at discretion of Board.

5. Automatic "zero" for anyone caught conducting themselves improperly or cheating during examination. Oral portion only retaken after failure, not entire exam.

6. Only retake portions of written examination that are failing, not entire exam.

License Requirements for States and Cities   Chapter 6

7. Six (6) months to complete written exam. After notifying and no action taken, file is destroyed after one year.

8. When the signature of a licensed Engineer cannot be obtained because experience was not obtained near St. Louis, the board will consider acceptance of the application based on the quality of the experience of the applicant.

9. No oral exam unless the written is complete and passed.

10. All applicants shall fill out the prescribed application form that is in use presently and shall meet all requirements on said form before acceptance by the Mechanical Supervisor. Any variance or discrepancies by an applicant on said form shall be subject to full review and approval of the Board of Stationary Engineers.

11. Special schooling can be accepted in lieu of work experience as follows:

(a) the curriculum must be approved by the Board.

(b) one year of experience equals 36 - 4 hours classes or 144 totals hours of class time.

12. All questions concerning applicants tests shall be submitted in writing to the Examining Board.

Chapter 6   License Requirements for States and Cities

Montana

State of Montana

SUMMARY OF QUALIFICATIONS:

1. Exam Location, Address and Phone Number

Various testing sites all over the state (some in remote areas)

Helena, Billings, Missoula

Office Location:
Montana Department of Labor & Industry
Safety Bureau
P.O. Box 1728
Helena, Montana 59624-1728
406/444-6401

2. Exam Days and Times

1st and 3rd Friday of each month - at 3 main locations - Helena, Billings, Missoula

3. Examination Cost

Traction Engineer..........................$12.00

Low Pressure Engineer..................$ 8.00

Third Class Engineer.....................$12.00

Second Class Engineer..................$20.00

First Class Engineer......................$30.00

4. License Cost and Renewal

Initial license included in cost of exam.

Renewal of license..........................$4.00

5. Type of Test

License Requirements for States and Cities Chapter 6

All Written

6. Grades of Licenses

Traction Engineer

Low Pressure Engineer

Third Class Engineer

Second Class Engineer

First Class Engineer

7. Experience and Education Requirements

See Codes for Montana

8. U.S. Citizenship Required?

Not specified

9. Local Residency Required?

No

10. Recognition of Licenses from other locales

No

CODES for State of Montana

Title 50

Chapter 74 - Boilers and Steam Engines

Part 3 - Licenses

50-74-301. License required to operate boilers and steam engines.

Chapter 6   **License Requirements for States and Cities**

All boilers and steam engines, except as exempted in 50-74-103, come under the provisions of this chapter, and persons operating same are required to hold the proper Grade of license.

50-74-302. General requirements for licensure.

No person may be granted a license to operate steam or water boilers and steam machinery under the provisions of this chapter who has not met the qualifications for licensing and been found to be competent by examination to perform the duties of an Engineer.

50-74-303. Engineer's license classifications.

(1) Engineers entrusted with the operation, care, and management of steam or water boilers and steam machinery, as specified in 50-74-302, are divided into four classes, namely: First Class Engineers, Second Class Engineers, Third Class Engineers, and Low Pressure Engineers.

(2) Licenses for the operation of steam or water boilers and steam machinery are divided into four classifications in accordance with the following schedule:

(a) First Class Engineers are licensed to operate all classes, pressures, and temperatures of steam and water boilers and steam driven machinery with the exception of traction and hoisting engines.

(b) Second Class Engineers are licensed to operate steam boilers operating not in excess of 250 pounds per square inch gauge saturated steam pressure, water boilers operating not in excess of 375 pounds per square inch gauge pressure and 450 degrees F temperature, and steam driven machinery not to exceed 100 horsepower per unit, with the exception of traction and hoisting engines.

(c) Third Class Engineers are licensed to operate steam boilers operating not in excess of 100 pounds per square inch gauge saturated steam pressure and water boilers operating not in excess of 160 pounds per square inch gauge pressure and 350 degrees F temperature.

(d) Low Pressure Engineers are licensed to operate steam boilers operating not in excess of 15 pounds per square inch gauge pressure and water boilers operating not in excess of 50 pounds per square inch gauge pressure and 250 degrees F temperature.

50-74-304. Requirements for Engineer's license.

Each applicant for an Engineer's license must be physically and mentally capable of performing the required duties and meet the following minimum requirements for the class of Engineer's license for which application is being made:

(1) An applicant for a Low Pressure Engineer's license must:

have at least 3 months' full time experience in the operation of a boiler in this classification, successfully pass a written examination prescribed by the division, have passed his 18th birthday, and be found to be competent to operate a boiler in this classification.

(2) An applicant for a Third Class Engineer's license must:

Chapter 6 License Requirements for States and Cities

have at least 6 months' full time experience in the operation of a boiler in this classification under an Engineer holding a valid Third-Class or higher license, successfully pass a written examination prescribed by the division, have passed his 18th birthday, and be found to be competent to operate a boiler in this classification.

(3) An applicant for a Second Class Engineer's license must:

(a) have at least 2 years' full time experience in the operation of a boiler and steam driven machinery in this classification under an Engineer holding a valid Second Class or First Class license, successfully pass a written examination prescribed by the division, have passed his 18th birthday, and be found to be competent to operate a boiler and steam driven machinery in this classification; or

(b) hold a valid Third Class Engineer's license and have at least 1 year's full time experience in the operation of a boiler and steam driven machinery in this classification under an Engineer holding a valid Second Class or First Class license, successfully pass a written examination prescribed by the division, have passed his 18th birthday, and be found to be competent to operate a boiler and steam driven machinery in this classification.

(4) An applicant for a First-Class Engineer's license must:

(a) have at least 3 years' full-time experience in the operation of a boiler and steam-driven machinery in this classification under an engineer holding a valid First Class license, successfully pass a written examination prescribed by the division, have passed his 18th birthday, and be found to be competent to operate a boiler and steam-driven machinery in this classification;

(b) hold a valid Second Class Engineer's license and have at least 1 year's full-time experience in the operation of a boiler and steam-driven machinery in this classification under an engineer holding a valid First Class license, successfully pass a written examination prescribed by the division, have passed his 18th birthday, and be found to be competent to operate a boiler and steam-driven machinery in this classification; or

(c) hold a valid Third Class Engineer's license and have at least 2 year's full time experience in the operation of a boiler and steam driven machinery in this classification under an Engineer holding a valid First Class license, successfully passed a written examination prescribed by the division, have passed his 18th birthday, and be found to be competent to operate a boiler and steam driven machinery in this classification.

50-74-305. Exceptions to requirements for Engineer's license.

Allowable exceptions or variances to the minimum requirements set out in 50-74-304 are as follows:

(1) An applicant for an Engineer's license in any classifications holding a valid license in that classification from another state having licensing requirements equal to or exceeding the minimum requirements set out in 50-74-304, successfully passing a written examination prescribed by the division, and found to be competent to operate a boiler and steam driven machinery in that classification shall be granted a license in that classification.

(2) Operating experience in a classification accumulated in the United States Military services or the Merchant Marine service satisfactory to the division may be accepted in lieu of

the operating experience required for licensing of Engineers in each of the license classifications.

(3) An applicant having training in the operation of steam or water boilers and steam machinery who has been certified as having satisfactorily completed a prescribed training course from a recognized vocational-technical training school or center or other division approved institution or training program in the classification for which he is applying may, at the discretion of the division, be credited with a maximum of 6 months' experience toward a First, Second, or Third Class Engineer's license.

50-74-306. Traction licenses.

(1) None of the licenses named in 50-74-303 entitle its holder to operate a traction engine.

(2) A person who is entrusted with the care and management of traction engines or boilers on wheels is required to pass an examination testing his competency to operate that class of machinery and procure a traction license.

(3) A traction license does not entitle its holder to operate any other class of steam machinery.

50-74-307. Requirements for Traction licenses.

An applicant for a traction Engineer's license must have at least 6 months' full time experience in the operation of steam traction engines, successfully pass a written examination prescribed by the division, have passed his 18th birthday, and be found to be competent to operate a traction engine.

50-74-308. Waiver of experience requirements for traction licenses.

The division, at its discretion, may waive the experience requirement for operators of traction engines which are maintained and operated as a hobby for the restoration and show purposes of antique equipment.

50-74-309. License fees.

Applicants for Engineer's license shall pay fees according to the class of license for which application is made, as specified in the following schedule.

(1) First Class..................................$30.00

(2) Second Class..............................$20.00

(3) Third Class.................................$12.00

(4) Low Pressure.............................$ 8.00

(5) Traction......................................$12.00

(6) Renewal of License....................$ 4.00

(7) Replacement of Lost Certificate.......$ 2.00

50-74-310. Application to be accompanied by one-half of license fee.

Each application shall be accompanied by a payment equal to 50% of the license fee for which application is being made. The payment shall be forfeited in the event the applicant fails to appear for the examination at the scheduled time or fails to pass the examination.

50-74-311. Waiting period before re-examination permitted.

In case of the failure of any applicant to successfully pass an examination, 45 days must elapse before he can again be examined for license.

50-74-312. Review of license rejection.

(1) If any person who has applied for a license under the provisions of this chapter and has been rejected feels aggrieved, he may, at any time after the lapse of 10 days and within 45 days after the date of his rejection, in writing set forth the causes of his grievance and request a division review. Such request must be addressed to the division and shall be signed by the rejected applicant.

(2) Within 2 days after receiving such request, the division shall notify the applicant in writing that on a certain day, which shall not be less than 5 or more than 30 days after the date the division receives the written request, the division shall review and evaluate the application.

(3) The applicant may appear in person at the review if he so desires. At least 2 days before the day set for the review, the applicant may designate in writing to the division the name of an Engineer holding a valid license of equal or higher Grade with the one applied for, and such Engineer may present himself in behalf of the applicant upon the day and at the hour fixed for the review.

(4) After the review is completed, if the division decides that the applicant is entitled to the license he has applied for, the division shall without delay issue a license accordingly but, if the

division rejects the applicant, it is a final rejection and he must not be granted another examination for the space of 45 days after such last rejection, when he may again apply as provided by 50-74-309 through 50-74-311.

50-74-313. Renewal of licenses.

(1) All certificates of license to Engineers of all classes shall be renewed yearly, except as herein provided.

(2) Any Engineer failing to renew his license as herein provided or within at least 30 days after the date of expiration shall be assessed the fee for the original license of the same Grade before the license will be reissued.

(3) Any Engineer failing to renew his license within 12 months of the date of expiration must reapply for an Engineer's license as required by the provisions of 50-74-303 through 50-74-308.

(4) Any Engineer whose license expired while such Engineer was in the military or naval service of the United States shall have 90 days from the time such Engineer is discharged from such military or naval service within which to renew his license at the renewal fee.

50-74-314. Complaints and revocation of license.

Whenever complaint is made against an Engineer holding a license that he, through negligence, want of skill, or inattention to duty, permitted his boiler to burn or otherwise become in bad condition or that he has been found intoxicated or under the influence of drugs while on duty, it is the duty of the divi-

| Chapter 6 | License Requirements for States and Cities |

sion to make a thorough investigation of the charge and upon satisfactory proof of such charge to revoke the license of the Engineer.

50-74-315. Unlawful to operate boiler or steam engine without license.

(1) It is unlawful for any person in this state to operate a stationary boiler or steam engine or any boiler or steam engine other than engines and boilers exempted by the provisions of 50-74-103 without a license granted under the provisions of this chapter. The owner, renter, or user of any engine or boiler is equally liable for the violation of this section.

(2) A person who operates a boiler or steam engine without first obtaining a license if guilty of a misdemeanor and, upon conviction, shall be fined no less than $50 or more than $100 or be imprisoned in the county jail for any term not to exceed 60 days, or both.

50-74-316. Unlawful to employ unlicensed operator.

It shall be unlawful, except as stated in 50-74-317, for any person, firm, or corporation to employ any person not duly licensed as an Engineer within the meaning of this chapter to run or operate any of the boilers or engines subject to the provisions of this chapter.

50-74-317. When unlicensed person may operate.

(1) In case of accident, sickness, or any unforeseen prevention of the licensed Engineer employed by any owner, renter, or user of an engine or boiler, the owner, renter, or user may for 15 days employ any person of the age of 18 days or over whom he may consider competent to run the engine or boiler.

(2) Although such person so employed may not be the holder of an Engineer's license, he shall have reasonable qualifications acceptable to the division.

(3) The person so employing the unlicensed Engineer shall immediately notify the division.

(4) No owner, renter, or user of boilers or steam machinery shall be allowed to so employ unlicensed Engineers for more than 15 days in any one calendar year.

Chapter 6 — License Requirements for States and Cities

Nebraska

Omaha, Nebraska

SUMMARY OF QUALIFICATIONS:

1. Exam Location, Address and Phone Number

City of Omaha
City-County Building
Permits & Inspections Division
Mechanical Section
1819 Farnam, Room 1110
Omaha, Nebraska 68183
402/444-5370

2. Exam Days and Times

1st Thursday of every month at 9:15 a.m.

3. Examination Cost

First Grade Stationary Engineer..........$40.00

Second Grade Stationary Engineer......$30.00

Third Grade Stationary Engineer.........$20.00

4. License Cost and Renewal

Initial license included in cost of exam

Renewal of license............................$30.00

5. Type of Test

All Written

6. Grades of Licenses

First Grade Stationary Engineer

Second Grade Stationary Engineer

Third Grade Stationary Engineer

License Requirements for States and Cities Chapter 6

7. Experience and Education Requirements

First Grade Stationary Engineer...........5 years experience

Second Grade Stationary Engineer........3 years experience

Third Grade Stationary Engineer..........1 year experience

8. U.S. Citizenship Required?

Not Specified

9. Local Residency Required?

No

10. Recognition of Licenses from other locales

No

EXAMINATION PROCEDURE

STATIONARY ENGINEER - ALL GRADES INCLUDED

1. The examination will be given on the first Thursday of each month only, beginning at 9:15 a.m. Lunch time is 12 noon to 1:00 p.m.

2. Fill out the required application form obtained at the Omaha/Douglas Civic Center, Room 1110 - Permits & Inspections Division, Mechanical Section.

3. All applications must be signed by two (2) citizens of Omaha, one of whom shall be a certified Stationary Engineer of equal grade or higher than the one being applied for.

4. The applicant must be 18 years or older.

5. Third Grade Stationary Engineer must have 1 year experience in the operation of steam boilers.

6. Second Grade Stationary Engineer must have 3 years experience in the operation of steam boilers.

Chapter 6   **License Requirements for States and Cities**

7. First Grade Engineer must have 5 years experience in the operation of steam boilers.

8. The fee for the examination will be paid in Room 1110 on the test day.

9. A work room will be selected prior to the test time for examination.

10. Third Grade examination contains 100 questions. Fee is $20.00.

11. Second Grade examination contains 135 questions. Fee is $30.00.

12. First Grade examination contains 150 questions. Fee is $40.00.

13. Paper and pencils for the examination will be supplied for you.

14. An identification number will be given to you, and it must appear at the top of each sheet of your answers.

15. You will receive one page of questions at a time.

16. Absolutely no use of notes or books and absolutely no conversation with other applicants or you will be disqualified.

17. All mathematical problems must show all work on your answer sheet to get credit.

18. All applicants are recommended the book, "Steam Plant Operation" (4th Edition) by Woodruff and Lammers. It is not necessary to study chapters 8, 9, 10, 11, 12 if you are taking the Third Grade test, as turbines and engines are not involved in this test.

19. All applications must be in this office at least one week before test day, so that the Board can review applicants' qualifications.

20. Results of the examination will be mailed to you, so be sure you have a complete address with zip code on the application.

CODES for Omaha, Nebraska

Ordinance Number 32279

Article III. Boiler and Pressure Vessels

Division 1. Boiler Inspectors

Subdivision B. Deputy Inspectors; Insurance

Division 3. Certificates

Sec. 40-176. Required generally.

No person shall operate, or cause to be operated, any boiler plant unless there shall be in charge of such plant an experienced person who holds a certificate of the proper grade in good standing granted by the board of Engineer examiners as provided in this article, or who holds a temporary permit as provided in Section 40-189.

Sec. 40-177. Exemptions.

The following plants shall be exempt from the provisions of this article:

(a) Heating plants used for heating a private dwelling.

(b) All boilers having a pressure not exceeding fifteen (15) pounds per square inch steam or one hundred sixty (160) pounds per square inch hot water and not exceeding seven hundred fifty (750) square feet of boiler heating surface.

Chapter 6 — License Requirements for States and Cities

(c) All steam plants over (15) pounds pressure under five hundred (500) square feet heating surface.

Sec. 40-178. Continuous operation.

In the event that a boiler plant is operated day and night, the owner or user thereof shall employ certified Engineers, or certified Firemen, as required by this article, during the entire time the plant is in operation.

Sec. 40-179. Preventing compliance.

No owner, agent, lessee or employer shall by means of threats of dismissal nor by any form of compulsion or persuasion attempt to prevent compliance with the provisions of this article by any person amenable thereto.

Sec. 40-180. Grades established.

The board of engineer examiners shall issue grades of certificate based on examinations therefor and upon which the applicant has received passing grades:

(a) First grade Stationary Engineer's certificate.

(b) Second Grade Stationary Engineer's certificate.

(c) Third Grade Stationary Engineer's certificate.

Sec. 40-181. Scope; when required — First Grade certificate.

A First Grade Stationary Engineer's certificate shall entitle the holder to take charge of and operate, any steam, power, refrigeration or compressor plant within the city. At least one (1) such certificate shall be required for each plant having steam prime movers, which in the aggregate amount to one hundred (100) horsepower or more. This examination shall cover the various types of engines, steam generators and auxiliary equip-

License Requirements for States and Cities — Chapter 6

ment, refrigeration and compressor equipment, their construction, operation, maintenance and defects; the use, operation and care of electric generators and motors; and the causes and prevention of smoke.

Sec. 40-182. Same — Second Grade certificate.

A Second Grade Stationary Engineer's certificate shall entitle the holder to take charge of and operate any boiler power plant, having steam prime movers which in the aggregate amount to less than one hundred (100) horsepower. At least one (1) Stationary Engineer's certificate shall be required in each shift for each such plant. The holder of a Second Grade Stationary Engineer's certificate shall be entitled to operate as a shift Engineer under the general supervision of a holder of a First Grade Stationary Engineer's certificate. This examination shall cover hot water boilers and equipment, steam generators and auxiliary equipment, their construction, operation, maintenance and defects; the use, operation and care of electrical generators and motors; and the causes and prevention of smoke.

Sec. 40-183. Same — Third Grade certificate.

A Third Grade Stationary Engineer's certificate shall entitle the holder to take charge of and operate any boiler plant having no steam prime movers. This examination shall cover hot water boilers and equipment, steam generators and their auxiliary equipment, their construction, operation, maintenance and defects; the use, operation and care of electrical motors; and the causes and prevention of smoke.

Sec. 40-184. Reserved.

Sec. 40-185. Same — Responsibility of operator.

Chapter 6   **License Requirements for States and Cities**

(a) It shall be unlawful for any owner, firm or corporation to operate or cause to be operated any engines operated by steam or any turbines operated by steam within the city having an aggregate capacity of one hundred (100) horsepower, or more, unless there be in charge at least one (1) person holding a First Grade Stationary Engineer's certificate; and all other shifts shall be operated by the holder of at least a Second Grade Stationary Engineer's certificate under the supervision of the First Grade Engineer.

(b) Where aggregate capacity of engines operated by steam, and/or turbines operated by steam, is less than one hundred (100) horsepower, there shall be in charge at least one (1) person holding a Second Grade Stationary Engineer's certificate, or a limited certificate at all times the engines or turbines are in operation.

Sec. 40-186. Application.

Every person desiring a certificate required by the provisions of this article shall make his application upon a blank furnished by the permits and inspections division and obtain thereon the endorsement of two (2) citizens of the city, at least one (1) of whom shall be the user of steam or a certified Engineer.

Sec. 40-187. Qualifications of applicants.

(a) Every applicant for a certificate required by the provisions of this article shall be at least eighteen (18) years of age and shall furnish proof of date of birth.

(b) An applicant for a First Grade Stationary Engineer's certificate shall have had at least five (5) years' experience in power and/or heating plants.

(c) An applicant for a Second Grade Stationary Engineer's certificate shall have had at least three (3) years experience in the operation and care of steam boilers under the supervision of a holder of a certificate of like or higher Grade to that applied for or an experience equivalent thereto.

(d) An applicant for a Third Grade Stationary Engineer's certificate shall have had at least one (1) year's experience in the operation and care of boilers under the supervision of the holder of a certificate of like or higher Grade to that applied for or an experience equivalent thereof.

Sec. 40-188. Examination required.

No certificate shall be issued under the provisions of this article until the applicant therefor has passed the examination required for such certificate.

Sec. 40-189. Temporary permits.

The boiler inspector may issue a temporary permit without examination therefor to a person or persons who have applied for a certificate upon proper proof of qualifications and upon recommendation of an intended employer, but such temporary permit shall remain in force only until the applicant has been provided an opportunity to be examined by the Board of Engineer Examiners. All persons to whom such temporary permits are issued shall be subject to and shall comply with all of the provisions of this chapter.

Sec. 40-190. Examination fee.

Along with his application for a certificate required under the provisions of this article, the applicant therefor shall pay the appropriate examination fee as follows:

(a) First Grade Stationary Engineer's certificate.......$40.00

(b) Second Grade Stationary Engineer's certificate..$30.00

(c) Third Grade Stationary Engineer's certificate.....$20.00

Sec. 40-191. Failure to appear for examination.

Any applicant for a certificate required under this article who shall fail to present himself before the Board of Engineer Examiners for examination within sixty (60) days from the date of his application shall forfeit the required fee and his application shall be cancelled unless he has been excused for good cause by the secretary of the Board of Engineers Examiners.

Sec. 40-192. Conduct of examination.

All examinations for a certificate required by the provisions of this article shall be oral, written, or both at the discretion of the Board of Engineer Examiners.

Sec. 40-193. Minimum grade.

A grade of seventy-five (75) percent shall be required to pass any examination required by the provisions of this article.

Sec. 40-194. Record of examination.

A complete stenographic record of all oral examinations given under the provisions of this article shall be made, which, together with the written examinations, shall be kept on file by the secretary.

Sec. 40-195. Reexaminations.

Any applicant for a certificate required by the provisions of this article who shall fail to pass the required examination shall be required to wait at least thirty (30) days after the date of such examination before again making application for the certificate. The applicant shall be required to pay the same fee as for the original examination.

License Requirements for States and Cities   Chapter 6

Sec. 40-196. Duration.

All certificates issued under the provisions of this article, unless revoked for cause, shall be valid until the thirty-first day of December of the year of issuance.

Sec. 40-197. Renewal.

Any certificate issued under the provisions of this article shall be renewed from year to year thereafter on the payment of thirty dollars ($30.00). Such renewal fees shall be paid within thirty (30) days after the date of expiration or the certificates shall become void. If not renewed within one (1) year from the date of expiration, a new application and reexamination shall be required.

Sec. 40-198. Reserved.

Sec. 40-199. Display.

Every holder of a certificate issued under the provisions of this article shall display his certificate in a conspicuous place in the plant where he is employed. The boiler inspector is herewith empowered to suspend for a period not exceeding ninety (90) days, the certificate of any person failing to comply with the foregoing provision.

Sec. 40-200. Temporary assignments to fill vacancies.

The boiler inspector shall have the authority in an emergency, and when the public interests shall require, to make temporary assignments if and when vacancies occur. Such assignments shall be made for a period not exceeding ninety (90) days. Any such appointee shall be the holder of a current certificate and in the judgment of the boiler inspector shall be possessed of proper qualifications for the assignments. Such appointee, if the holder of a Fireman's certificate, shall be exempt from the requirements of Section 40-199.

Sec. 40-201. Leaving boiler unattended.

(a) No boiler shall be left unattended for a period longer than two (2) hours. However, any boiler exempt from the above provisions when, and if, fired by an approved burner or firing device and said boiler is equipped with approved controls which will automatically maintain water level in the boiler and which will extinguish or properly retard the fire if the steam pressure exceeds a predetermined maximum or the water level falls below a predetermined minimum. Such burner, if using gas or oil, shall also be equipped with approved devices, mechanical or electrical, which will automatically prevent an abnormal flow of gas or oil, and which shall completely shut off the flow of gas or oil when combustion ceases. Such burner shall also be equipped with an approved automatic device, so designed, that gas or oil, upon being turned into the combustion chamber, will become ignited substantially immediately or else shut off. In such case, the boiler inspector shall designate on the certificate of boiler inspection the length of time that any such boiler may be left unattended.

(b) Except as hereinbefore provided, no certified Engineer shall leave his post during the operation of the plant unless relieved by a certified man properly qualified. The penalty for any absence, except as hereinbefore provided, shall be the suspension or revocation of the certificate. This shall not include the time the plant is not in operation.

Sec. 40-202. Neglect of duty; suspension of certificate.

(a) Any certified Engineer who shall negligently endanger the life of any person or jeopardize property, by allowing the water to fall below the flues or crown sheet, or otherwise grossly neglects his duty, or who is intoxicated on duty, shall have his certificate suspended indefinitely.

(b) Any certified Engineer or who shall willfully, or through careless firing, cause dense and/or unnecessary smoke or soot to issue from the chimney or smokestack shall have his certificate suspended for a period of not exceeding ninety (90) days.

Sec. 40-203. Reserved.

Sec. 40-204. Board of appeals.

Any person aggrieved by an order, requirement, decision or determination concerning the granting, issuance, denial, revocation or cancellation of any permit or license provided for by any provision of this Code or city ordinance may appeal such order, requirement, decision or determination made by an administrative official, agency, division or department to the administrative appeals board.

Sec. 40-205. Appeals, procedures.

When any person has made application for a Stationary Engineer's and such certificate has been denied or refused by the administrative board of appeals, or when any certificate heretofore or hereafter granted has been revoked or suspended by the boiler inspector, or when any person believes himself otherwise injured or wronged by the administrative board of appeals or the boiler inspector, such person may appeal such action of the Board of Engineer Examiners or the boiler inspector to the board of appeals by filing a written request with the city clerk within three (3) days after receiving notice of such denial, revocation or suspension or other ruling. The board of appeals shall, within three (3) days after filing, hear such appeal and it is hereby authorized to take such action or to make such orders in the premises as in its opinion may be just and proper. All testimony shall be under oath, administered by the chairman of the board of appeals. This section shall not be deemed to deprive or oust the city council of final jurisdiction in any matters covered herein.

Sec. 40-206 - 40-209. Reserved.

Nevada

No State or City License

New Hampshire

No State or City License

New Jersey

State of New Jersey

SUMMARY OF QUALIFICATIONS:

1. Exam Location, Address and Phone Number

Various testing sites all over the state.

They have 21 counties in the state. They use 15 county seats to test at. (combining some of the counties for testing) They sometimes go to the vocational schools in the counties to administer the test if there are enough people.

Office Location:

New Jersey Department of Labor
Office of Boiler and Pressure Vessel Compliance
Station #4, CN 392
Trenton, NJ 08625-0392
609/984-6551

2. Exam Days and Times

Varies per location

3. Examination Cost

$25.00

4. License Cost and Renewal

Initial license included in cost of exam.

Renewal for three years.....$20.00

Renewal for one year.........$10.00

Upgrading your license......$20.00

5. Type of Test

Varies - Multiple choice or Essay

6. Grades of Licenses

Fireman Special

Fireman Low Pressure

Fireman High Pressure

1C Third Class Stationary Engineer

1B Second Class Stationary Engineer

1A First Class Stationary Engineer

7. Experience and Education Requirements

See Codes for New Jersey

8. U.S. Citizenship Required?

Not Specified

9. Local Residency Required?

No

10. Recognition of Licenses from other locales

They recognize experience but you have to take the test

CODES for State of New Jersey

Subchapter 7.

Licensing of Operating Engineers and Firemen

12:90-7.1 Scope of subchapter

This subchapter shall apply to the procedures required to obtain a license as an Operating Engineer or Fireman.

12:90-7.2 Application for licenses

(a) The application shall be typewritten or neatly and legibly printed in ink. Only one application may be submitted at a time.

(b) All applications shall be carefully completed and notarized.

(c) The statements of the applicant shall indicate the actual experience as specified in the eligibility provisions of N.J.A.C. 12:90-7.4 through 7.13. Only pertinent, applicable, lawful and full-time experience shall be listed. This experience shall have been completed within seven years of the filing of application.

(d) Incomplete or improper applications shall not be accepted.

(e) An application for a license shall be made on forms provided by the Office of Boiler and Pressure Compliance. Only one classification or change of Grade may be requested per application.

(f) The statements on the first or original application shall be endorsed by two Engineers each holding a valid blue seal or higher New Jersey license, except as provided in (g) below. These endorsements shall verify the applicant's statement.

(g) Substitution for the signatures of the endorsers of (f) above may be the holder of:

1. A Marine Engineer's license with the holder's experience documented by trip discharges.

2. Another State or city license with the holder's experience documented with a letter from the employer identifying operational experience and equipment;

3. A United States service or Merchant Marine discharge establishing the required Engineering experience; or

Chapter 6 License Requirements for States and Cities

4. Written statements from two former employers or the present employer signed by responsible endorsers and showing required Engineering experience.

(h) The endorsements of (f) above shall not be required for low pressure licenses or special licenses.

(i) No license shall be granted to a person less than 18 years of age.

(j) All correspondence relative to licenses or applications shall be addressed to the Office of Boiler and Pressure Vessel Compliance.

(k) The Office of Boiler and Pressure Vessel Compliance shall be notified of any change of residence. When writing, the license number shall be specified.

(l) The fee for examination for an original license shall be $15.00 and the fee for a raise of Grade or additional classification shall be $10.00.

(m) The fee shall accompany the application.

(n) No annual renewal fee shall be charged for additional classifications on any license.

(o) The fee for application for a license shall be a check or money order made payable to the order of the Office of Boiler and Pressure Vessel Compliance.

(p) No liability shall be assumed by the Office of Boiler and Pressure Compliance for loss in the transmission of the fee.

12:90-7.3 Classifications of licenses for operators

(a) The letters A,B and C shall be used to identify the Grade of the license.

(b) "A" or gold seal shall designate a First-Grade license; "B" or red seal, a Second-Grade license; and "C" or blue seal, a Third-Grade license.

(c) A black seal shall identify a Boiler Operator.

(d) The license stamped on its face, "in-charge" shall identify a Boiler Operator in charge.

(e) Boiler Operator and Firemen classifications shall be identical.

(f) The numbers listed in Table 7.3(f) shall be used to identify the equipment indicated.

Arabic Number - Equipment

- 1.Steam stationary boiler and steam prime mover
- 2.Refrigeration System
- 3.Nuclear Boiler and Prime Mover
- 7.Hoisting Machine and Long Boom Crane
- 8.Steam Portable Boiler and Steam Prime Mover
- 9.Steam Locomotive Crane

(g) Licenses for operators shall be classified as follows:

- 1. Special limited applications;
- 2. Low pressure boiler;
- 3. Operator in charge of low pressure boiler;
- 4. Operator in charge of high pressure boiler;
- 5. 1A, 1B, or 1C steam stationary boiler and steam prime mover;

Chapter 6 **License Requirements for States and Cities**

- 6. 2A, 2B or 2C refrigeration system;
- 7. 3A, 3B or 3C nuclear boiler and prime mover;
- 8. 7C long boom crane;
- 9. 8A, 8B or 8C steam portable boiler and steam prime mover;
- 10. 9A or 9B steam locomotive crane.

12:90-7.4 Eligibility for Boiler Operator's license

To be eligible for a Boiler Operator's license, the applicant shall have had at least three months experience as a helper, apprentice or assistant to a licensed operator of equipment requiring such license.

12:90-7.5 Eligibility for Low Pressure Boiler Operator's license

(a) To be eligible for a Low Pressure Boiler Operator's examination, the applicant shall:

1. Be able to comply with N.J.A.C. 12:90-7.4, or

2. Have had intensive training for 30 full working days in a program established by the Chief Engineer and approved by the Office of Boiler and Pressure Vessel Compliance, prior to the start of the training period. A log shall be established with the licensed operator doing the training, which shall be one-on-one, and the trainee shall have written verification of such training from the chief engineer.

(b) A licensed Fireman may be eligible for examination for an "in-charge" license after three months actual service in the operation of boilers requiring such license.

12:90-7.6 Eligibility for High Pressure Boiler Operator's license

License Requirements for States and Cities — Chapter 6

(a) To be eligible for a High Pressure Boiler Operator in charge examination, the applicant shall:

1. Be able to comply with N.J.A.C. 12:90-7.4, or

2. Have had intensive training for six weeks in a program established by the Chief Engineer and approved by the Office of Boiler and Pressure Vessel Compliance, prior to the start of the training period. A log shall be established with the licensed operator doing the training, which shall be one-on-one, and the trainee shall have written verification of such training from the chief engineer.

(b) If the applicant has had six months experience as a licensed Low Pressure Fireman, the six week period reference in (a) above may be reduced to 30 calendar days.

(c) For special license up to 100 horsepower, an applicant with 30 days experience may be eligible.

12:90-7.7 Eligibility for Third Grade Steam Engineer's license

(a) To be eligible for a Third Grade Steam Engineer's (1-C or 8-C) examination, the applicant shall have a Fireman in-charge high pressure license and shall have had at least six months subsequent experience in the operation of equipment requiring supervision by a Third Grade Engineer, or as an assistant in the operation of equipment requiring a Third Grade license for shift operation.

(b) Experience obtained outside the State of New Jersey may be considered if the applicant has served at least two years as a Boiler Operator of high pressure boilers of over 1,000 horsepower.

12:90-7.8 Eligibility for Third Grade Refrigeration Engineer's license

Chapter 6 License Requirements for States and Cities

(a) To be eligible for a Third Grade Refrigeration Engineer's (2-C) examination, the applicant shall have had at least:

1. Six months experience as an assistant to an operator of a flammable or toxic refrigeration system; or

2. Three months experience as an operator of a flammable or toxic refrigeration system; or

3. Three months experience as an assistant to the operator of a flammable or toxic refrigeration system, provided the applicant has been given intensive training for the period by the licensed operator, and the Chief Engineer verifies such training and experience by letter; or

4. Six months experience as an operator of a nontoxic refrigeration unit of at least 250 tons capacity and three months experience as an assistant to the licensed operator or flammable or toxic refrigeration system; or

5. Six months experience as an operator of a nontoxic refrigeration unit of at least 250 tons capacity and satisfactory proof of completion of sufficient education in the operation of a flammable or toxic refrigeration system in an educational program approved by the Division of Vocational Education of the New Jersey Department of Education.

12:90-7.9 Eligibility for nuclear engineer's license

To be eligible for nuclear Engineer's (3-C) examination, the applicant shall hold certification from the United States Nuclear Regulatory Commission qualifying him to operate nuclear power equipment.

12:90-7.11 Eligibility for Second Grade Engineer's license

To be eligible for a Second Grade Engineer's examination, the applicant shall have a Third Grade license and shall have had at least one year's subsequent practical experience in the operation of equipment requiring supervision by a Second Grade or First Grade Engineer.

12:90-7.12 Eligibility for First Grade Engineer's license

(a) To be eligible for a First Grade Engineer's examination in any classification, the applicant shall have:

(1) A Second Grade license and subsequently served one year as Chief Engineer in a plant requiring supervision by a Second Grade Engineer; or

(2) A Second Grade license with two years subsequent practical experience as an Operating Engineer in a plant requiring supervision by a First Grade Engineer; or

(3) Experience of an equivalent amount for grade or classification from some other jurisdiction.

12:90-7.13 Other eligibility considerations

(a) An applicant for original license, change of classification or raise of Grade may show in writing, as a substitute for a minor portion of the experience listed in N.J.A.C. 12:90-7.4 through 7.12, non-operating experience, such as servicing, maintenance, repair or installation of equipment; or satisfactory proof of completion of sufficient formal education or academic study embracing such equipment.

(b) When an applicant's Operating Engineer experience and training warrants, the Office of Boiler and Pressure Vessel Compliance may determine the classification and Grade of license most suitable.

(c) The Office of Boiler and Pressure Vessel Compliance may consider an applicant's experience of an equivalent

Chapter 6 License Requirements for States and Cities

amount for Grade or classification from some other jurisdiction.

12:90-7.14 Examinations

(a) Examinations shall be held on the first Wednesday of each month at Trenton, and at various other times and places throughout the State when warranted, and shall be conducted by an Examiner.

(b) Applicants shall be notified when and where to appear for the examination.

(c) Failure to appear for the examination shall be considered sufficient cause to void the application, unless a satisfactory explanation is given for failing to appear.

(d) Failure to appear for the examination or to obtain a passing grade shall not entitle the applicant to a refund of any fee.

(e) Examinations for an Engineer's license shall be conducted in a written form and shall consist of as many questions and be of such nature as the office of Boiler and Pressure Compliance shall consider appropriate for the license desired.

(f) Examinations for a Fireman's license shall consist of such questions as the Office of Boiler and Pressure Vessel Compliance shall consider proper and shall pertain to the safe operation of steam and hot water boilers, appliances, auxiliaries and such other equipment covered by the licensing act.

(g) An applicant for up to and including a C Grade license may, upon prior request, be examined through a reader or interpreter accompanying the applicant provided the reader or interpreter is acceptable to the Office of Boiler and Pressure Vessel Compliance.

(h) Questions used in the examination shall not be copied by any applicant or retained by the applicant after examination, or taken from the presence of the authorized agent of the Office of Boiler and Pressure Vessel Compliance during the examination. Violation of this subsection shall be sufficient cause to disqualify the applicant.

12:90-7.15 Granting of license

(a) A license shall be granted and designated Grade A when an average of 80 percent or more is attained on a Grade A examination.

(b) A license shall be granted and designated Grade B when an average of 70 percent or more is attained on a Grade B examination.

(c) A license shall be granted and designated Grade C when an average of 60 percent or more is attained on a Grade C examination.

(d) No license shall be granted on an average of less than 60 percent.

(e) A license issued after examination covering one or more classifications shall be valid in any plant where the class of Engineering is within the scope of the license issued.

(f) Licenses merely bearing the impression of the seal of the Department of Labor shall be issued as special licenses and are limited to the operation of equipment specified on the face thereof. Special licenses may be issued to operators of nonconventional boilers, such as, but not limited to, electric, coil, or waste heat or conventional high pressure boilers of over six to 100 horsepower. These licenses may be transferred to similar equipment when approved following written request by the applicant or employer.

Chapter 6 License Requirements for States and Cities

(g) Duplicate licenses for part-time employment may be issued at the discretion of the Commissioner.

(1) The fee for a duplicate license is $3.00 for one year.

(2) A request by the licensee for a duplicate license shall be accompanied by a letter from the company desiring to employ the licensee.

(3) The duplicate license shall specify the plant where it is to be used and may be transferred when approved following written request.

(4) A Chief Engineer may not request a duplicate part-time license for secondary location employment.

12:90-7.16 Re-examination

(a) The applicant may not be re-examined for a period of at least three months, but may be allowed one re-examination without additional charge, within six months of the original examination. If again unsuccessful, the applicant may request an additional examination, provided that the request is accompanied by a fee of $7.50.

(b) Upon failing the examination for the third time, an applicant who wishes to retake the examination shall wait three months prior to reapplication, at which time a new application form shall be fully completed.

12:90-7.17 Posting of license

(a) All licenses shall be framed and properly posted adjacent to the equipment involved, in the Engineer's office or in the plant office, whichever is suitable.

(b) The license shall be available for examination.

(c) A penalty may be imposed for violation of this section.

12:90-7.18 Suspension or revocation of license

(a) Any license may be suspended or revoked for incompetence, negligence, intoxication, or drug abuse while on duty, or for any other valid reason establishing that the licensee is unfit to hold a license.

(b) Any license or identification card shall be surrendered and immediately revoked if, for any purpose, it is loaned, abandoned or allowed to pass from the personal control of the owner.

(c) All licenses shall expire unless renewed on or before the anniversary month of the original license. Changing of grade or addition of classification shall not change this anniversary date.

(d) A license shall be automatically cancelled on the date of its expiration. Any persons performing the duties of a licensee and holding an expired license shall be subject to the penalty provisions under N.J.S.A. 34:7-6, as is his supervisor and employer.

(e) Any person using fraudulent means to obtain a license shall be subject to prosecution. Any license acquired through such means shall be invalid.

12:90-7.19 Renewal of license

(a) When applying for the renewal of a license it shall be necessary to return only the signed identification card with a fee of $5.00 for a one year renewal or $10.00 for a three year renewal.

(b) A license may be renewed within 60 days prior to the date of its expiration.

(c) An application for a renewal of an expired license shall be approved provided:

Chapter 6   License Requirements for States and Cities

(1) A fee of $7.50 is enclosed for one year of $15.00 for a three year renewal;

(2) The application is made within three years of the expiration date of the expired license; and

(3) All penalties lawfully imposed on the applicant under N.J.S.A. 34:7-6 have been paid.

(d) Application for renewal of a license expired more than three years shall be treated as an original application. All records of the previous license may be destroyed.

(e) An altered, defaced, lost or otherwise mutilated license shall be replaced only after review by the Office of Boiler and Pressure Vessel Compliance. Photostats, photographs or reproduction of a license shall have no status and shall not be recognized. A fee of $5.00 shall be submitted for a replacement license.

12:90-7.20 Employment of unlicensed person

(a) Employers shall immediately request permission from the Office of Boiler and Pressure Vessel Compliance, in writing, if for any reason of emergency it becomes necessary to employ an unlicensed person temporarily for a period not to exceed 15 days explaining fully the circumstances.

(b) Continuation requests for emergency operating permission must be received in the Office of Boiler and Pressure Vessel Compliance prior to the expiration date of the granted permission.

(c) Late requests for continuation are subject to penalties which must be satisfied prior to permission being granted.

(d) The Office of Boiler and Pressure Vessel Compliance shall again be notified when a licensed person is employed, giving the name, address, and license classification, grade and number of such employee.

New Mexico

No State or City License

New York

Buffalo, New York

SUMMARY OF QUALIFICATIONS:

1. Exam Location, Address and Phone Number

City of Buffalo
City Hall
Room 315
Buffalo, New York 14202
716/851-4959

2. Exam Days and Times

Once a month - None given in July and August

3. Examination Cost

$26.25

4. License Cost and Renewal

Initial license..$63.00

Renewal for Chief Engineer......................$42.00

Renewal for First and Second Class..........$31.50

5. Type of Test

All Written (multiple choice & essay questions)

6. Grades of Licenses

Chief Engineer

| Chapter 6 | License Requirements for States and Cities |

First Class Stationary Engineer

Second Class Stationary Engineer

7. Experience and Education Requirements

Chief Engineer - Working as First Class for 3 years

First Class Stationary Engineer - Working as Second Class for 1 year

Second Class Stationary Engineer - 1 year of schooling (Trade School) plus 2 years of experience or 2 years of schooling and 1 year of experience.

8. U.S. Citizenship Required?

Yes, plus 21 years of age.

9. Local Residency Required?

No

10. Recognition of Licenses from other locales

It is prorated - Up to Chief Examiner - They count military experience.

CODES for Buffalo, New York

Buffalo Code

Boilers and Engines; Engineers

94-17 Licensing of Stationary Engineers and Refrigeration Operators required; exceptions.

A. Every person within the limits of the City of Buffalo in charge of or operating any power developing unit or units as defined in this chapter and not specifically excepted herein must hereafter be licensed. The Director shall, within a reasonable time after the filing a written application, notify the applicant to appear at a time and place designated for examination. If such applicant shall, after a written and practical examination, give satisfactory proof of his qualifications as a Stationary Engineer or refrigeration operator, he shall receive from the Director a proper qualifying permit authorizing the Director of Licenses and Permits to issue a license to such applicant as provided by this chapter and authorized by such permit.

B. The provisions of this chapter are not applicable to Engineers of duly incorporated steam railroads while engaged in running or operating the locomotive engines of said railroads and/or any diesel or gas electric driven car used on said railroads, or to persons employed by other public service corporations subject to the jurisdiction of the Public Service Commission of the State of New York while engaged in operating or running any electric-power-generating units in connection with the rendition of public service; or to Engineers duly licensed by the authorities of the United States while engaged in operating or running steam boilers, steam engines or power developing units under the jurisdiction of the United States; or to operators or operation of passenger or freight elevators or escalators permanently installed, except temporary elevators used exclusively for hoisting building materials during construction or alteration; or to gas or diesel engines or operators thereof while such engines are being used for test, repair or emergency purposes; or to heat exchangers used to process industries in which steam is generated during whole or part of the process cycle; or to transit concrete mixers, welding units, overhead traveling cranes, gantry cranes, shop or plant trucks, shop or plant hoists, cupola blowers or any motor vehicles licensed subject to the provisions of the New York State Vehicle and Traffic Law; or to operators or the operation of any boilers being used to furnish temporary heat for any buildings under con-

Chapter 6 License Requirements for States and Cities

struction until the permanent heating system has been completed and final tests made and said heating system and said boilers are accepted by the owner of the buildings and a certificate of acceptance is issued to the heating contractor.

94-18 Citizenship required for issuance of license.

No person shall be granted a license unless he is a citizen or shall have declared his intention to become a citizen of the United States.

94-19 Revocation of license or permit.

A license or permit issued under the provisions of this chapter shall be subject to suspension or revocation for cause by the Director of Licenses and Permits, after a hearing, upon notice to the licensee or permittee. Mental or physical incompetency or negligence in the performance of his duties as an Engineer or refrigeration operator or intoxication while on duty or failure to renew properly such license or a violation of any provisions of this chapter or other pertinent laws or ordinances shall be sufficient cause for revocation or suspension of the license or permit.

94-21 License renewal; scope of authority of engineer or operator.

All licenses must be renewed annually, and no licensed Engineer or refrigeration operator shall maintain or operate more than one (1) power plant as a whole. Nothing herein contained shall be construed to limit the operation of more than one (1) power developing unit by a properly rated Engineer or refrigeration operator where the same are operated or maintained independently of each other in the same building or buildings constituting a power plant as a whole. The required rating of the Engineer shall be determined by the total horsepower of the power developing units, except refrigeration power devel-

License Requirements for States and Cities Chapter 6

oping units, other than steam driven, maintained or operated by him.

94-22 License fees

A. A fee as provided in Chapter 175, Fees, shall be collected by the Director of Licenses and Permits for issuing an original license, expiring one (1) year from date of issuance, authorized by a qualifying permit of the Director. A holder of a license under this chapter who later qualifies for an additional license hereunder shall not be required to again pay such an original fee. Persons licensed as Stationary Engineers or Refrigeration Operators at the time of the effective date of this chapter shall be exempt from payment of the original license fee, except those failing to properly renew the license or who have hereafter suffered a revocation of the license.

B. Yearly renewal fees for licenses shall be as provided in Chapter 175, Fees. Such fees shall be collected for each annual renewal thereof by the Director of Licenses and Permits authorized by a qualifying permit of the Director and the surrender of the previous year's Engineer's or refrigeration operator's license. The Director shall report to the Common Council every three (3) months the number and kind of qualifying permits issued by him.

C. Whenever any person is qualified to receive a license as an Engineer under this chapter and is employed by the City of Buffalo on a part time basis for a period not in excess of ninety (90) days, the payment of fees for the issuance or renewal of such license shall be waived. Such license shall be in full force and effect only during the period said person is employed by the City of Buffalo; and upon termination of said employment, said person shall forthwith surrender to the Director of Licenses and Permits said license if issued without the payment of the fee therefor.

D. There shall be an examination fee as provided in Chapter 175, Fees, which each applicant must pay in order to take an examination for a license to be issued under the provisions of this chapter.

94-23 Classification and qualifications of engineers and operators.

A. All persons licensed under the provisions of this chapter as Engineers or refrigeration operators shall be classified and graded according to the horsepower and pressure of a power developing unit or units, or any combination of such, constituting a power plant of which they shall be found competent, by due written and practical examination, to take charge of or operate, namely, Chief Engineers, First Class Engineers, Second Class Engineers, Special Engineers, Chief Refrigeration Operators, First Class Refrigeration Operators and Second Class Refrigeration Operators.

(1) Chief Engineers shall be Engineers who are found competent to take charge of or operate a power plant of any horsepower and pressure. Every applicant for a license as a Chief Engineer must pass a written and practical examination as to his qualifications and prove to the examiner that he had three (3) years' actual experience as a duly licensed First Class Engineer.

(2) Every applicant for a license as a First Class Engineer must pass a written and practical examination as to his qualifications and prove to the Director that he has been duly licensed as a Second Class Engineer for the period of one (1) year immediately preceding. Persons duly licensed as First Class Engineers shall have authority to take charge of and operate any

power plant or power developing unit or units not exceeding one hundred fifty (150) horsepower and not exceeding two hundred twenty five (225) horsepower of refrigeration power developing units, other than steam driven.

(3) Every applicant for a license as a Second Class Engineer must be least twenty one (21) years of age, have had at least one (1) year's experience in the operation of boilers and/or power plant equipment and, in addition, must have had at least two (2) years of approved mechanical and/or electrical experience, or such applicant must have graduated from a recognized technical school and have had at least one (1) year's satisfactory experience. Persons duly licensed as Second Class Engineers shall have the authority to take charge of and operate any power developing unit or units not exceeding one hundred (100) horsepower and not exceeding one hundred fifty (150) horsepower of refrigeration power developing units, other than steam driven.

(4) A Special Engineer's qualifying permit shall be issued by the Director to the operator, limited to designated premises, of a steam engine, steam boiler or boilers or other power developing unit, not exceeding a total of fifty (50) boiler horsepower where the safety valve or valves are set to operate not over a pressure of one hundred twenty five (125) pounds per square inch. Such qualifying permit shall authorize the Director of Licenses and Permits to issue a Special Engineer's license to such applicant as provided herein and authorized by such permit, upon payment of the proper fee therefor.

B. Any person holding an Engineer's license under the provisions of this chapter may also hold any Grade of refrigeration operator's license for which he can qualify by written and prac-

Chapter 6   License Requirements for States and Cities

tical examination as to his qualifications and experience on refrigeration power developing units.

C. Any person not heretofore required to be licensed and who at the time that this chapter shall become effective has been employed in the operation of or in charge of power developing unit or units within the scope of this chapter shall, upon application, be granted an Engineer's or refrigeration operator's license for a similar type of power developing unit or units of similar brake or rated horsepower to that upon which he is employed at the time this chapter takes effect.

D. In order for any such person not now duly licensed and not excluded by the provisions of this chapter to be granted such a license, it will be necessary for him to prove, to the satisfaction of the Director by documentary evidence in affidavit form, that he has had satisfactory actual experience in the operation of or in charge of a similar type of power developing unit or units of a similar brake or rated horsepower to the unit or units in the operation of or in charge of which he has been employed.

E. Such licenses shall only be granted during a period of ninety (90) days following the effective date of this chapter; except, however, that such licenses granted during such ninety day period may be renewed by the holders thereof from year to year in the same manner as provided for in the issuance of other licenses under this chapter.

F. Engineers duly licensed by the authorities of the United States as Marine Engineers who desire a license to operate power plants within the limits of the City of Buffalo must pass an examination as to their qualifications to operate power plants on land of a grade equal to that of which they hold a license from the government of the United States.

G. Every applicant for an Engineer's or refrigeration operator's license whose experience was gained elsewhere than in the City of Buffalo must prove to the satisfaction of the Director that he has had the same total number of years of experience as is required from applicants in the City of Buffalo.

H. An applicant for a Locomotive Engineer's license as an Engineer on internal combustion locomotive engines must be at least nineteen (19) years of age and have had at least one (1) year's experience in the repair, operation or maintenance of internal combustion engines, and, in addition, he shall present a letter from his employer attesting to his capabilities as a Locomotive Engineer; or he must be at least twenty one (21) years of age and have had at least three (3) years' experience in the repair, operation or maintenance of internal combustion engines, one (1) year of which must have been on internal combustion locomotive engines.

94-24 Exemptions; conditions for unfired pressure vessels.

A. The following shall be exempted from the provisions of this chapter:

(1) Miniature boilers not exceeding sixteen (16) inches inside diameter of the shell, forty two (42) inches overall length of outside to outside of the heads at center, twenty (20) square feet of water heating surface and maximum allowable working pressure of one hundred (100) pounds per square inch.

(2) Unitary or self-contained air conditioning units used for domestic, commercial or industrial purposes, singly or combined. All units over one half horsepower, singly or combined, up to a capacity of sixty (60) horsepower, using ammonia or carbon dioxide refrigerants, or up to a capacity of one hundred (100) horsepower using freon or similar refrigerants shall be equipped with safety devices, as follows:

Chapter 6 — License Requirements for States and Cities

(1) Where water condensers are used: low water cutout and/or High Pressure cut out, and low voltage protection devices.

(2) Air cooled condensers: the same safety devices except low water cutouts.

(3) Additional safety devices providing similar safety protection that the director of Fuel Devices may order.

B. Conditions for unfired pressure vessels. It shall be the duty of every owner or other person using a stationary unfired pressure vessel containing a tank volume of over twenty one (21) cubic feet where the operating pressure is over one hundred (100) pounds to have the same inspected by the Director not less than once in each and every year; provided, however, that any person or owner using such vessel, who shall have had the same inspected and insured by a duly authorized insurance company within six (6) months of the time the Director may offer to inspect the same, which shall be evidenced by a certificate to be filed with the Director within one (1) month after such inspection shall occur or insurance shall attach, shall not be required to comply with the provisions of this subsection.

94-25 Display of license

Engineers or refrigeration operators duly licensed under the provisions of this chapter shall have their licenses suitably framed, under glass, and hung up in a conspicuous place at or near the power developing unit.

94-26 Plants in continuous operation

A. Where a power plant is in service regularly day and night, in the absence of the Chief Engineer, the assistant Engineers must hold First Class Engineer's license if the plant exceeds one hundred fifty (150) horsepower. However, Second Class

Engineers may be employed on the same watch with the Chief Engineer.

B. No power developing unit or units shall be left unattended and without a properly licensed Engineer or refrigeration operator of the required Grade while in operation.

C. One (1) Chief Engineer may be in charge of more than one (1) power plant as a whole on a general property site of the same owner.

94-27 Continuation of existing licenses; renewals.

A. All Engineers or refrigeration operators duly licensed at the time of the effective date of this chapter shall be granted licenses of the same Grade without examination. Each license heretofore issued under the Code of the City of Buffalo for Engineers or Refrigeration Operators shall continue until the expiration date of such license, after which time the provisions of this chapter shall apply.

B. If a license issued under this chapter is not renewed by the licensee within one (1) year of the expiration date thereof, the licensee will be required to take and pass a new examination.

94-29 Licensing of Hoisting Engineers; exemptions

A. No person except as hereinafter provided shall operate in the City of Buffalo any hoisting or portable machines or apparatus in connection with construction, alteration, erection or demolition of buildings in the City of Buffalo without first having obtained the license provided for herein.

B. Any person holding a Chief Engineer's license or a First Class or Second Class Engineer's license provided for by this chapter shall be deemed qualified to operate a hoisting machine in connection with the construction work or erection of

Chapter 6 — License Requirements for States and Cities

buildings and shall not be required to obtain any other license to operate such hoisting machine or apparatus.

C. Any other person desiring to operate in the City of Buffalo any hoisting machine or apparatus in connection with construction work or building operations should make application to the Director of Fuel Devices in writing, duly verified, giving such information as to age, residence, experience and knowledge in hoisting machines and the operation of motors used in running such hoisting machines as the Director shall prescribe. If such application shows the person to be presumptuously capable of operating a hoisting machine or apparatus, the Director may require such tests or further examination as, in his judgment, are reasonably necessary to demonstrate the applicant's ability to operate such hoisting machine with safety to property and the lives and limbs of persons working around or passing by such machines and the motors by which such machines are operated. If upon such examination such applicant is found to possess the necessary qualifications to operate such hoisting machine or apparatus and the engine or motor by which the machine is driven, the Director shall issue to him a qualifying permit authorizing the Director of Licenses and Permits to issue to said applicant a Hoisting Engineer's license authorizing him to operate in the City of Buffalo a hoisting machine, for which license the applicant shall pay the sum as provided in Chapter 175, Fees.

D. Each license, unless sooner revoked, shall expire one (1) year from the date on which it is issued, and for a renewal thereof the licensee shall pay the sum as provided in Chapter 175, Fees. Any misstatement concerning the material facts contained in the application shall be sufficient cause for the revocation of such license.

E. The provisions of this section shall not apply to:

(1) The operation of elevators permanently installed for regular use in any building.

(2) Any hoisting machine operated by muscular power.

(3) Any hoisting machine operated by an engine or motor of five (5) horsepower or less.

(4) Transit concrete mixers, welding units, overhead traveling cranes, gantry cranes, shop or plant trucks, shop or plant hoists or any motor vehicle licensed subject to the New York State Vehicle and Traffic Law.

F. Any person operating any hoisting machine or apparatus in connection with construction work shall at all times carry upon his person while so engaged the license issued to him.

94-30 Replacement of lost licenses or certificates

Where it is provided by this chapter for the issuance of any license or certificate by the Director of Licenses, and any such license or certificate becomes lost or destroyed through no fault of the owner or holder thereof, and a duplicate is desired, said Director of Licenses upon the filing with him of an affidavit reciting the circumstances of such loss or destruction and upon the payment of a fee as provided in Chapter 175, Fees, may issue a duplicate license or certificate in place of the lost or destroyed license or certificate to the person thereto as such owner or holder aforesaid.

94-31 Penalties for offenses.

A. Any person operating in the City of Buffalo a power developing unit, as defined in this chapter, or a hoisting machine or apparatus without first having obtained the license herein required and any person, firm or corporation who or which employs or permits any unlicensed person to operate any power developing unit or hoisting machine or apparatus contrary to the provisions of this chapter shall be liable to a fine or penalty as provided in Chapter 1, Article III, of this Code.

B. Any person who shall violate any other provisions of this chapter shall be liable to a fine or penalty as provided in Chapter 1, Article III, of this Code.

License Requirements for States and Cities   Chapter 6

Mt. Vernon, New York
SUMMARY OF QUALIFICATIONS:

1. Exam Location, Address and Phone Number
City Hall
Roosevelt Square
Room 212
Mt. Vernon, New York 10550
914/699-3983

2. Exam Days and Times
1st and 3rd Tuesday of the month

3. Examination Cost
Free

4. License Cost and Renewal
$25.00 for initial license and renewal

5. Type of Test
All Written, multiple choice, 50 questions

6. Grades of Licenses
Fireman
Second Class Engineer
First Class Engineer

7. Experience and Education Requirements
Not Specified

8. U.S. Citizenship Required?
Yes

9. Local Residency Required?
No

10. Recognition of Licenses from other locales?
For experience requirement only, still must take the exam

Chapter 6   License Requirements for States and Cities

New York City, New York

SUMMARY OF QUALIFICATIONS:

1. Exam Location, Address and Phone Number

New York City Department of Personnel
Skilled Trades and License Division
18 Washington Street
New York City, New York 10004
212/487-6403

2. Exam Days and Times

Twice a year, Dates vary

3. Examination Cost

$170.00

4. License Cost and Renewal

Initial license.............$25.00

Renewal of license....$15.00

5. Type of Test

Written and Oral

6. Grades of Licenses

High Pressure Boiler Operating Engineer

7. Experience and Education Requirements

High Pressure Boiler Operator:

5 years experience operating boilers in New York City or:

if you possess a license from another city then 1 years experience operating boilers in New York City or:

a university degree in engineering and one years experience operating boilers in New York City

License Requirements for States and Cities Chapter 6

8. U.S. Citizenship Required?

No

9. Local Residency Required?

Yes

10. Recognition of Licenses from other locales

Yes, but still need one years experience in New York City operating boilers

Steps in Applying for a License:

1. File application at 2 Washington Street, New York City;

2. Pay $170 application fee;

3. Get written notice in the mail at which public school the examination will be conducted., the examinations are given twice a year;

4. If you pass the written examination, then there is an oral exam;

5. If you pass the oral examination, then there will be an investigation of your application to prove the facts given on the application were correct;

6. If your application passes investigation, then you go to 60 Hudson Street, Building Department License Division, to pick up your license;

7. Pay twenty-five dollars ($25.00) for the original license and $15 for the renewal.

Chapter 6   License Requirements for States and Cities

Niagara Falls, New York

SUMMARY OF QUALIFICATIONS:

1. Exam Location, Address and Phone Number

City of Niagara Falls
City Hall, City Clerk's Office
745 Main St.
P.O. Box 69
Niagara Falls, New York 14302-0069
716/874-9088

2. Exam Days and Times

Monday evening at 6:30 p.m.

3. Examination Cost

$20.00

4. License Cost and Renewal

Fireman-Class C................$15.00

Engineer-Class B...............$20.00

Chief Engineer-Class A.....$25.00

5. Type of Test

All Essay

6. Grades of Licenses

Fireman-Class C

Engineer-Class B

Chief Engineer-Class A

7. Experience and Education Requirements

Fireman-Class C................6 months experience

Engineer-Class B...............2 years as a Fireman

Chief Engineer-Class C.....3 years as an Engineer

8. U.S. Citizenship Required?

Yes

9. Local Residency Required?

No

10. Recognition of Licenses from other locales

No - does allow for waivers of waiting period for license.

Rochester, New York

SUMMARY OF QUALIFICATIONS:

1. Exam Location, Address and Phone Number

City of Rochester
Bureau of Buildings & Zoning
Department of Community Development
City Hall
30 Church Street
Room 121B
Rochester, New York 14614-1290
716/428-6526

2. Exam Days and Times

Every other Monday, starting January 4, 1993, at 6:30 p.m.

3. Examination Cost

$60.00

4. License Cost and Renewal

$60.00 for initial license and renewal

5. Type of Test

Written and Oral

6. Grades of Licenses

3rd Class Engineer

2nd Class Engineer

1st Class Engineer

Chief Engineer

7. Experience and Education Requirements

See Codes for Rochester below

8. U.S. Citizenship Required?

Not Specified

9. Local Residency Required?

Not Specified

10. Recognition of Licenses from other locales

Not Specified

CODES for Rochester, New York

Stationary Engineers & Refrigeration Operators

Licensing Ordinance

Chapter 103

From the Code of the City of Rochester

Section 103-3. License required; responsibilities of licensees.

A. General requirement. Whenever a boiler is in use and the premises are in productive operation, a licensed stationary engineer shall be on the premises and shall be responsible for the operation of such boiler.

B. No person other than a licensed stationary engineer or refrigeration operator shall operate equipment as defined herein. No person who is a licensed stationary engineer or refrigeration operator may have charge of a plant in excess of that which is authorized by the scope of his license. No person whose license has been revoked, canceled or suspended may operate equipment, as defined herein, during such period that the license is revoked, canceled or suspended.

C. No person may employ or permit any person who is not a licensed stationary engineer or a refrigeration operator to be in charge of or operate equipment as defined herein, or any per-

Chapter 6	License Requirements for States and Cities

son who is a licensed stationary engineer or refrigeration operator to be in charge of or operate equipment, as defined herein, in excess of that which is authorized by the scope of his license.

D. If a licensed stationary engineer of the proper class has charge of equipment as defined herein, a licensed stationary engineer of a lower classification may operate such equipment under his responsibility, but this privilege does not apply to watchmen. If a licensed refrigeration operator of the proper classification has charge of equipment as defined herein, a refrigeration operator of a lower classification may operate such equipment under the former's responsibility.

E. No person shall have charge of or operate any combination of a steam plant and a refrigeration plant, as defined in this chapter, without being duly licensed as both a Stationary Engineer and a Refrigeration Operator.

Section 103-4. Classes of Stationary Engineers licenses.

There shall be the following classes of Stationary Engineers' Licenses, Third Class being the lowest and Chief Engineer being the highest, with other classes being of the degree as respectively set forth:

A. Third Class Engineers:

shall be entitled to take charge of and operate any steam plant not to exceed one hundred (100) boiler horsepower. No person shall be permitted to take an examination for this license unless he has had at least one (1) year's practical experience acceptable to the Board or has such experience as may be required of an applicant for a certificate of competence as a boiler Operating Engineer issued by the Industrial Commissioner of the State of New York.

B. Second Class Engineers:

shall be entitled to take charge of and operate any steam plant not to exceed an aggregate of five hundred (500) boiler horsepower. No person shall be permitted to take an examination for this license unless he has held a Third Class Stationary Engineers License and has had at least one (1) additional year's practical experience acceptable to the Board.

C. First Class Engineers:

shall be entitled to take charge of and operate any steam plant not to exceed an aggregate of one thousand five hundred (1500) boiler horsepower. No person shall be permitted to take an examination for this license unless he has held a Second Class Stationary Engineers License and has had at least one (1) additional year's practical experience acceptable to the Board.

D. Chief Engineers:

Shall be entitled to take charge of and operate any steam plant of any horsepower. No person shall be permitted to take an examination for this license unless he has held a First Class Stationary Engineers License and has had at least two (2) additional year's practical experience acceptable to the Board.

Section 103-5. Custodian and watchmen licenses.

There shall be the following custodian and watchmen licenses:

A. Custodians shall be entitled to take charge of and operate a steam boiler or boilers for heating purposes only in an apartment type building, where boiler or boilers in aggregate exceed fifty (50) horsepower, provided that such building is not used for the accommodation of transient residents.

B. Watchmen shall be entitled to take charge of and operate a steam boiler or boilers at an establishment when the estab-

Chapter 6   **License Requirements for States and Cities**

lishment in which the boilers are situated is not in productive operation.

Section 103-7. Application for licenses; fees.

A. Every applicant for a Stationary Engineer's or Refrigeration Operator's license or a Custodian's or Watchman's license shall make application for a license on forms as prescribed by the Board. The applicant shall state his name, address and experience in operating steam plants or refrigeration plants sufficient to qualify for the class and type of license applied for.

B. Fees.

(1) The applicant shall pay the following application fees at the time the application is made at the Permit office of the Department of Community Development:

(a) Application fee for Stationary Engineers and Refrigeration Operators: Sixty ($60).

(b) Application fee for custodians and watchmen; sixty dollars ($60).

(2) Said fee is forfeited in the case of failure of the examination or failure to appear for the examination.

Section 103-8. Qualification for license.

A. An applicant for examination shall be required to meet the following prerequisites:

(1) Applicants must submit a completed application form.

(2) Applicants must satisfy the Board as to their experience.

(3) Applicants must be at least eighteen (18) years of age

(4) Applicants must be able to read, write, and speak in a manner sufficient to ensure proper and safe operation of all equipment covered by the type of license applied for.

B. The requirements of this section may be waived by the Board on a temporary basis if warranted.

Section 103-9. Examination of applicants.

Every applicant shall submit to an examination by the Board. Examinations shall be oral or written, or a combination thereof, as the Board may direct. In the event that an applicant fails an examination, a minimum period of ninety (90) days shall elapse before he becomes eligible for reexamination.

Section 103-10. Issuance of licenses.

A. The Board shall issue a license to each person deemed competent by it after such examination, of the kind and class for which the examination is held, which license shall be countersigned by the Commissioner.

B. The Board shall file in the office of the Commissioner all records made by it of examinations, a statement of the results thereof, the name and address of every person to whom a license is issued, the kind and class of license and the date of issue thereof, of which a record shall be kept by the Commissioner.

C. Upon payment of fees prescribed herein, a license shall be issued to any person who meets the requirements contained herein and who passes the examination prescribed by the Board. The Board may also give an examination to an applicant who, not meeting the requirements contained herein, presents to the Board a valid license or its equivalent of another properly constituted authority or satisfies the Board as to experience obtained. It shall be discretionary with the board as to the classification of examination such person shall be allowed to undergo.

103-11. License fees.

A. The following fees shall be paid by the applicant for the type of license as indicated or a renewal of said license:

Stationary Engineers	Refrigeration Operators	License/Renewal Fees
3rd Class	Grade R4	$60.00
2nd Class	Grade R3	$60.00
1st Class	Grade R2	$60.00
Chief Engineer	Grade R1	$60.00
Custodian/ Watchman		$60.00

B. All fees for licenses or renewal of licenses are payable to the City Treasurer. All fees are payable at the Permit Office of the Department of Community Development.

C. An applicant who applies for both types of licenses must pay an application fee for each type of license. After being granted both licenses, he must pay the license fees for each license. When renewing both types of licenses, an applicant must pay the renewal fee for each license.

103-12. Payments and disbursements.

Payments of any fee required by this chapter and any other sums shall be made to the order of the City Treasurer. All necessary expenses incurred by the Board shall be paid by the City Treasurer upon order to him, and certified by the Secretary of the Board.

103-13. Exclusions.

Nothing in this chapter shall apply to:

A. Workmen acting under the direct supervision of a licensed stationary engineer or refrigeration operator.

B. The operator of a high pressure boiler operated by the State of New York or a political subdivision thereof, provided that the operator has qualified in accordance with the New York State Civil Service Law and the rules promulgated thereunder.

C. Steam or refrigerating plants, as defined herein, under the jurisdiction of the Interstate Commerce Commission.

D. Steam or refrigerating plants, as defined herein, located in establishments operated by a department of the Federal Government.

E. Boilers used in residential buildings where the aggregate boiler horsepower does not exceed fifty (50) horsepower and operates at less than fifteen (15) p.s.i.g. (pounds per square inch gauge).

F. Boilers used in any other establishment which do not exceed forty (40) aggregate boiler horsepower.

G. Refrigeration equipment located in an establishment which is not in productive operation.

H. Refrigeration equipment which circulates a Group One refrigerant.

I. Refrigeration equipment which circulates a Group Two refrigerant, or Group Three refrigerant and has an aggregate capacity of seventy-five (75) tons or less.

J. No refrigeration unit of twenty (20) tons or less shall be taken into consideration in the calculation of aggregate tonnage of a refrigeration plant.

103-14. Licenses to be displayed.

Every Stationary Engineer's or Refrigeration Operator's license shall at all times be exposed to view where the licensee is normally engaged at work.

103-15. Expiration and renewal of licenses.

A. All licenses issued shall expire on the 31st day of December of the first year in which they are issued and shall be renewed annually between October 1 and December 31. In the event that an applicant for renewal applies for renewal later than December 31, the fee for such renewal shall be one and one-half (1 1/2) times the regular fee.

B. A license which is not renewed by March 31 shall be canceled without notice and may be reinstated by examination only.

103-16. Revocation of licenses.

Licenses issued are subject to revocation or suspension at any time by a majority vote of the Board members. Any person may bring charges against a licensee. The Board shall establish a time and place to hear charges and may suspend a license for a period not to exceed thirty (30) days pending such hearing. After the hearing, the Board may suspend or revoke a license upon a finding that the licensee has failed to comply with any state or local law, ordinance, rules or regulation relating to the conduct of the business for which the license is issued.

103-17. Penalties for offenses.

A violation of this chapter shall be punishable in the manner as set for in 68-15 of the Municipal Code the same as if specifically set forth herein.

103-18. Holders of Third Class Licenses.

Persons holding a valid Third Class License as a stationary engineer on the effective date of this chapter shall be deemed to have met the requirements of a new applicant for said license, and such licenses may be renewed or upgraded in accordance with the terms of this chapter.

103-19. Severability.

Should any section, paragraph, sentence, clause or phrase in this chapter be declared unconstitutional or invalid for any reason, the remainder of the chapter shall not be affected, and to this end the provisions of this chapter are declared to be severable.

The Rochester City Council recently approved legislation amending the City Code by revising Chapter 103, the Stationary Engineers and Refrigeration Operators Licensing Ordinance. The following revisions became effective on July 1, 1992.

Stationary Engineers

Class	License/Renewal Fees
Custodian/Watchman	$60.00
Third Class Stationary Engineer	$60.00
Second Class Stationary Engineer	$60.00
First Class Stationary Engineer	$60.00
Chief Engineer	$60.00

Chapter 6　　　License Requirements for States and Cities

There will be an application fee for all of the above exams of $60.00.

Suggested Study outline for 3rd Class Stationary Engineers

An Applicant for this grade of license must be able to discuss intelligently:

1. His own plant's procedure for loss of water.
2. His own plant's equipment.
3. How, why, and when to blow down a boiler.
4. How to blow down a water column and gauge glass.
5. The location and purpose of a fusible plug.
6. The difference between a water tube and fire tube boiler.
7. An HRT boiler.
8. The replacement of a water gauge
9. Different methods of supplying water to a boiler.
10. The locations and purpose of all hand valves, check valves, etc.
11. All phases of safety, including causes of explosions and other damage to a boiler.
12. How a steam trap works.
13. The most important gauge in a boiler room.

He must be able to define:

1. A boiler

License Requirements for States and Cities   Chapter 6

2. Steam, saturated and superheated
3. BTU
4. Boiler Horsepower
5. Draft

Suggested Study Outline for Second Class Stationary Engineers

An applicant for this class license must know everything expected of a 3rd Class Engineer, plus the following topics:

1. Tensile strength
2. Efficiency of riveted joints
3. Compound gauge
4. Why factor of safety is changed
5. Boiler bursting pressure, given working pressure and factor of safety
6. What superheating does to steam
7. Hydrostatic test
8. Measurements of water, including weight of a column of water
9. Types of steam turbines
10. Feedwater regulators
11. Difference between a tube and a flue
12. Latent heating of vaporization at different temperatures
13. Converting to centigrade scale

Chapter 6 — License Requirements for States and Cities

14. Principle of synchronous motor
15. Efficiency of different types of boilers
16. Actual evaporation of a boiler
17. Restriction on size of HRT boilers
18. Different types of gear pumps, rotary pumps, and centrifugal pumps
19. How centrifugal pump is rated
20. Conditioning of water boiler
21. Combustion of coal
22. Proper carbon dioxide in flue gasses
23. BTU content in coal
24. Absolute zero
25. Conversion of vacuum to absolute pressure
26. Types of gauges
27. Specific gravity
28. Construction of HRT boilers
29. Deaerator heater
30. Equivalent evaporation
31. Oil in boiler water
32. Types of modern boilers
33. Economizers
34. What are Orsat tests for in flue gasses

35. The evaporation and parts of a duplex pump

36. Elastic limit

37. Riveted joints

38. Stays

39. Describe a water column and where it is connected to boiler

Note: This list is drawn up by the Board of Examiners of Stationary Engineers. The bulk of the examination will be taken from this list.

Suggested Outline for 1st Class Stationary Engineers

An Applicant for this class license must know everything expected of a 2nd Class and 3rd Class Engineer. He must also be prepared to work out the following problems:

1. Given the necessary information, calculate tensile strength, efficiency of joints, thickness of plate, factor of safety and diameter of shell by using the basic working pressure formula,

2. Calculate the indicated horsepower, efficiency, or brake horsepower of an engine, given the necessary information.

3. Calculate the developed horsepower of a reciprocating or centrifugal water pump.

In addition to problems involving specific formula, there is a need to be prepared to work out general problems involving latent heat and sensible heat, developed boiler horsepower, weight of water, capacity of pumps, differential pressures in reciprocating pumps, efficiency of boilers, equivalent and actual evaporation of boilers, etc.

N.B. This is not meant to be an exhaustive list of suitable topics to be covered in an exam. The exam for 1st Class Engineer does not test the applicant's ability to memorize formulas. It is designed to demonstrate his ability to employ basic facts with which he should be familiar. The exams are prepared with this thought in mind. The applicant should so prepare himself.

White Plains, New York

SUMMARY OF QUALIFICATIONS:

1. Exam Location, Address and Phone Number

Exam Location:
Fire Station #2
21st Avenue
White Plains, NY 10601
914/422-6377

Location of Class:
Tonawanda Senior High School
Hinds & Fletcher
White Plains, NY 10601

Office Location:
Public Safety Building
77 S. Lexington Ave.
White Plains, NY 10601

2. Exam Days and Times

No set time - every four to five months - usually afternoon - minimum 20 people

3. Examination Cost

$10.00 application fee (bring picture photo)

License Requirements for States and Cities   Chapter 6

4. License Cost and Renewal

Initial license................$20.00

Renewal of license.......$20.00

5. Type of Test

All Written - forty questions - short answer

6. Grades of Licenses

Chief Engineer - Unlimited

First Class Stationary Engineer - over 150 hp

Second Class Stationary Engineer - up to 150 hp

7. Experience and Education Requirements

Chief...............Three years experience with 1st Class license

First Class.......Two years experience as 2nd Class engineer

Second Class...Three years experience around boilers with competent engineer

8. U.S. Citizenship Required?

Yes

9. Local Residency Required?

Yes, must live or work in White Plains

10. Recognition of Licenses from other locales

Recognize for experience

North Carolina

No State or City License

North Dakota

No State or City License

Chapter 6 License Requirements for States and Cities

Ohio

State of Ohio
SUMMARY OF QUALIFICATIONS:

1. Exam Location, Address and Phone Number
1st & 3rd Monday (If holiday -Tuesday) at 8 a.m.

Cincinnati, Toledo, Akron, Youngstown

2nd & 4th Monday

Columbus

Cambridge

4th Tuesday

Cleveland

800/686-1524

Office Location:
State of Ohio
Department of Industrial Relations
2323 W. Fifth Avenue
P.O. Box 825
Columbus, OH 43216
614/644-2248

2. Exam Days and Times

See above

3. Examination Cost

$50.00

4. License Cost and Renewal

Initial license............$35.00

Renewal of license....$35.00

5. Type of Test

All Essay

6. Grades of Licenses

Low Pressure Boiler Operator

High Pressure Boiler Operator

First Class Steam Engineer

Second Class Steam Engineer

Third Class Steam Engineer

7. Experience and Education Requirements

See Codes for Ohio below

8. U.S. Citizenship Required?

Not Specified

9. Local Residency Required?

No

10. Recognition of Licenses from other locales

Recognize military experience.

CODES for State of Ohio

4101:11-1-01. Applications

(A) Examinations shall be given by appointment only and each applicant's request for examination must be received fourteen days prior to taking such examination. Each initial request shall be accompanied by a certified statement of experience on the form provided by the division of examiners of Steam Engineers. In lieu of this form, a license from another state or political subdivision thereof, Marine license or a D.D.

Chapter 6   License Requirements for States and Cities

214 will be accepted with proper classification and service time.

(B) If the request for examinations is approved the applicant will be examined within sixty days of date request is received. The applicant will be notified at least five days prior to assigned examination date and a copy of such notice sent to the Chief of examiners of Steam Engineers.

(C) Each applicant shall furnish two personal facial photographs taken not more than six months prior to the date of the applicant's examinations, one to be attached to the application at the time of examination: the second to be attached to the original license when issued.

(D) An applicant when appearing for the examination shall make application in ink on a blank furnished by the examiner and shall make affidavit attesting that each and every statement in this application is true and shall acknowledge that any false statement included therein constitutes grounds for revocation of license.

(E) An applicant may be examined in a district other than that in which the applicant is employed or resides, provided the applicant secures from the Chief examiner of Steam Engineers permission to take such examination.

4101:11-1-02. Experience for Low-Pressure Boiler Operators.

No person shall be permitted to take an examination for Low-Pressure Boiler Operator's license, unless:

(A) The applicant has reached the age of eighteen years.

(B) The applicant must have one of the following experience qualifications:

1. 600 hours practical experience as a Steam Engineer, Oiler, Boiler Operator, Boiler Operator's helper, or boiler repair person experienced with duties that pertain to steam boiler operation.

2. 300 hours practical experience as a Steam Engineer, Oiler, Boiler Operator, Boiler Operator's helper, or boiler repair person experienced with duties that pertain to steam boiler operation and 300 hours of boiler construction and/or boiler maintenance experience.

3. 300 hours practical experience as a Steam Engineer, Oiler, Boiler Operator, Boiler Operator's helper, or boiler repair person experienced with duties that pertain to steam boiler operation and the successful completion of a thirty-six hour course in an approved program teaching steam boiler operation and boiler maintenance.

4101:11-1-03. Experience for High-Pressure Boiler Operators.

No person shall be permitted to take an examination for high-pressure Boiler Operator's license, unless:

(A) The applicant has reached the age of eighteen years.

(B) The applicant must have one of the following experience qualifications:

1. 1200 hours practical experience as a Steam Engineer, Oiler, Boiler Operator, Boiler Operator's helper, or boiler repair person experience with duties that pertain to steam boiler operation.

2. 600 hours practical experience as a Steam Engineer, Oiler, Boiler Operator, Boiler Operator's helper or boiler repair person experience with duties that pertain to steam boiler op-

Chapter 6   License Requirements for States and Cities

eration and 600 hours of boiler construction and/or boiler maintenance experience.

3. 600 hours practical experience as a Steam Engineer, Oiler, Boiler Operator, Boiler Operator's helper, or boiler repair person experienced with duties that pertain to steam boiler operation and the successful completion of a sixty hour course in an approved program teaching steam boiler operation and boiler maintenance.

4101:11-1-04. Experience for Stationary Steam Engineers.

No person shall be permitted to take an examination for stationary Steam Engineer's license, unless:

(A) The applicant has reached the age of eighteen years.

(B) The applicant must have one of the following experience qualifications:

1. 1800 hours practical experience as a Steam Engineer, Oiler, Boiler Operator, Boiler Operator's helper, or boiler repair person experienced with duties that pertain to the operation of steam reciprocating engine, turbine or boiler.

2. 900 hours practical experience as a Steam Engineer, oiler, Boiler Operator, Boiler Operator's helper, or boiler repair person experienced with duties that pertain to the operation of steam reciprocating engine, turbine or boiler and 900 hours of steam engine, turbine or boiler construction and/or maintenance experience.

3. 900 hours practical experience as a Steam Engineer, Oiler, Boiler Operator, Boiler Operator's helper, or boiler repair person experienced with duties that pertain to the operation of steam reciprocating engine, turbine or boiler and the successful completion of a ninety-six hour course in an ap-

proved program teaching Steam Engineering and power plant maintenance.

4101:11-1-05. Experience and Approved Course.

(A) The experience for all examinations in rules 4101.11-1-02.03 and 04 must be obtained on equipment with a manufacturer's rating of at least thirty horsepower.

(B) An approved course shall be a course approved by the state board of education, division of vocational education.

(C) The curriculum of the course and the instructor shall meet minimum requirements of the state board of education, division of vocational education.

(D) The request for approval of the course or instructor must be by written application to the state board of education, division of vocational education stating course curriculum and instructor's qualification.

4101:11-1-06. Fee.

The application for examination under rules 4101:11-1-02,03 and 04 must be accompanied by the fee required by section 4739.14 of the Revised Code regardless of the applicants' success or failure on examination, this fee is non-refundable.

4101:11-1-07. Grade Required to Qualify for Licenses.

Except as provided in rule 4101:11-1-11, a license shall be issued within thirty days after the completion of the examination to any applicant who receives a score of at least seventy percent on the questions submitted at the examination.

4101:11-1-08. Examination Procedures.

All examinations shall be in writing by the applicant and be in the English language.

Chapter 6 License Requirements for States and Cities

(A) Applicants shall not be permitted to converse with other applicants during examination. No information references shall be used during an examination except those supplied by the Chief examiner of Steam Engineers.

(B) Examiners of Steam Engineers may use the common slide valve engine and duplex pump models in examination at their home offices. If used, a record shall be made upon the application showing whether or not the applicant understands the principle of valve setting.

4101:11-1-09. Scope of Examinations.

(A) The examination for Engineer's license shall consist of a list of fifty practical questions on the construction and operation of stationary steam boilers and stationary steam engines, steam pumps, and on the subject of hydraulics.

(B) The examination for Boiler Operator's license shall consist of a list of forty practical questions on the construction and operation of stationary steam boilers, steam pumps, and hydraulics.

(C) The examination for a Low-Pressure Boiler Operator's license shall consist of a list of thirty practical questions pertaining to the operation of stationary steam boilers of less than fifteen pounds pressure and their appurtenances.

(D) An examiner shall not submit to an applicant for license any questions which have not been supplied by the Chief examiner of Steam Engineers. The Chief examiner of Steam Engineers shall with the assistance of other designated by the Chief review all test questions and update as needed every five years or less.

4101:11-1-11. Classifications of Licenses

License Requirements for States and Cities — Chapter 6

(A) A Third-Class Engineer's license shall be granted to an applicant who qualifies for and submits to an Engineer's examination and receives a percentage of seventy or more.

(B) A Second-Class Engineer's license shall be granted to an applicant who qualifies for and submits to an Engineer's examination and receives a percentage of seventy-five or more, provided such applicant has had 3600 hours practical experience as an actual operating stationary Steam Engineer.

(C) A First-Class Engineer's license shall be granted to an applicant who qualifies for and submits to an Engineer's examination and receives a percentage of eighty-five or more, provided such applicant has had 5400 hours practical experience as an actual operating stationary Steam Engineer.

(D) A Boiler Operator's license shall be granted to an applicant who qualifies for and submits to a Boiler Operator's examination and receives a percentage of seventy or more.

(E) A low-pressure Boiler Operator's license shall be granted to an applicant who qualifies for and submits to a Low-Pressure Boiler Operator's examination and receives a percentage of seventy or more.

4101:11-1-15. Revocations.

(A) If after an examination, it develops that a license may have been obtained through fraud, such procedure may be followed as provided by Ohio laws in the revocation of license for cause.

(B) Provisions of Ohio laws pertaining to license revocation shall be followed in any action seeking to revoke the license of any person.

4101:11-1-16. License Renewal

| Chapter 6 | License Requirements for States and Cities |

(A) A license that is not renewed within sixty days after expiration, shall be expired permanently, unless good and sufficient reason is given for the failure to have it renewed with the time specified.

(B) A license shall not be renewed without the renewal fee prescribed by section 4739.14 of the Revised Code. The renewal of a license shall be made through the examiner of the district in which the license holder is employer or resides, or the central office, if so notified.

Section 4739.05 (1048). License Application; Rules.

Each person who desires to act as a Steam Engineer shall make application to the district examiner of Steam Engineers for a license, upon a blank furnished him, and shall pass an examination in the construction and operation of steam boilers, steam engines, and steam pumps, and in the subject of hydraulics. Examination questions shall pertain only to the practical operation of stationary steam engines, stationary steam boilers, and steam pumps. The examination shall be conducted under the rules and regulations adopted by the Chief examiner which shall be uniform throughout the state. The district examiners, assistant chief examiner, and Chief examiner may administer all oaths or affirmations to any applicant whenever the same is made necessary by the rules and regulations adopted by the Chief examiner.

Section 4739.06 (1049). License and Revocation.

If, upon examination, the applicant is found proficient in the subject mentioned in section 4739.05 of the Revised Code a license shall be granted him to have charge of and operate station steam boilers and engines of the horsepower required by section 4739.04 of the Revised Code, for one year from the date on which it is issued. Upon written charges, after notice

License REQUIREMENTS for States and Cities   Chapter 6

and hearing, the district examiner may revoke the license of a person guilty of fraud in obtaining such license, or who becomes insane, or is addicted to the liquor or drug habits to such a degree as to render him unfit to discharge the duties of a Steam Engineer.

Common pleas jurisdiction. 2305.01.

Section 4739.07 (1050) Renewals.

Upon application, the person to whom a license is issued under section 4739.01 to 4739.10 inclusive, of the Revised Code, shall be entitled to a renewal thereof annually, unless the district examiner for a cause named in section 4739.06 of the Revised Code and upon notice and hearing shall refuse such renewal.

Section 4739.08 (1053) Must Exhibit License.

No Steam Engineer shall neglect or refuse to exhibit his license under glass in a conspicuous place in his engine room

Penalty. 4739.99 (A)

Section 4739.09 (1056) Prohibition.

No Engineer or owner, or user, or agent of an owner or user of a steam boiler or engine shall violate sections 4739.01 to 4739.10, inclusive, of the Revised Code.

Penalty 4739.99 (B)

(Amended June 14, 1957)

Section 4739.10 (1057) Exceptions.

Section 4739.01 to 4739.10, inclusive, of the Revised Code do not apply to boilers and engines under the jurisdiction of the United States, or to locomotive boilers and engines nor to the owners or users thereof.

Chapter 6 — License Requirements for States and Cities

Section 4739.11 (1058-5) Exemptions.

Sections 4739.04 of the Revised Code, insofar as it has relation to the operation and having charge of stationary steam boilers, shall not apply to persons holding license issued under sections 4739.11 to 4739.15, inclusive, of the Revised Code.

Section 4739.12 (1058-1) Licenses.

Any person who desires to operate or have charge of a stationary steam boiler of more than thirty horsepower, except boilers which are in charge of a licensed Engineer, shall make application to the district examiner of Steam Engineers for a license upon a blank furnished by the examiner, and shall pass an examination on the construction and operation of steam boilers, steam pumps, and hydraulics under such rules and regulations as may be adopted by the Chief examiner of Steam Engineers, which rules and regulations and standard of examination shall be uniform throughout the state. Examination questions shall pertain only the practical operation of stationary steam boilers and steam pumps. If the applicant is found proficient in said subjects, a license shall be granted him to have charge of and to operate stationary steam boilers of more than thirty horsepower. Such license shall continue in force for one year from the date the same is issued, and upon application to the district examiner may be renewed annually without being required to submit to another examination. The district examiner may, on written charges, after notice and hearing, revoke the license of any person guilty of fraud in passing the examination, or who, for any cause has become unfit to operate or have charge of stationary steam boilers. Any person dissatisfied with the action of any district examiner in refusing or revoking a license or renewal thereof, may appeal in accordance with sections 119.01 to 119.13 inclusive, of the Revised Code.

License Requirements for States and Cities   **Chapter 6**

Section 4739.13 (1058-1a) Application for License to Operate Stationary Steam boilers; Examination; License; Renewal; Revocation; Appeal.

Any person who desires to operate or have charge of a stationary steam boiler of more than thirty horsepower, carrying a pressure of no more than fifteen pounds per square inch, except a person licensed under section 4739.05 or 4739.12 of the Revised Code, shall make application to the district examiner of Steam Engineers for a license upon a blank furnished by the examiner, and shall pass an examination on the construction and operation of steam boilers and appurtenances, under such rules and regulations as may be adopted by the Chief Examiner of Steam Engineers, which rules and regulations and standards of examination shall be uniform throughout the state and shall be in accordance with sections 119.01 to 119.13, inclusive, of the Revised Code. Examination questions shall pertain only to the practical operation of stationary steam boilers carrying less than fifteen pounds of pressure and appurtenances.

If the applicant is found proficient in said subjects a license shall be granted him to have charge of and to operate stationary steam boilers of more than thirty horsepower and operating at less than fifteen pounds pressure. Such license shall continue in force for one year from the date the same is issued, and renewed annually without the holder being required to submit to another examination. The district examiner may, on written charges, after notice and hearing, revoke the license of any person guilty of fraud in passing the examination, or who, for any cause has become unfit to operate or have charge of stationary steam boilers. Any person dissatisfied with the action of any district examiner in refusing or revoking a license or renewal thereof, may appeal to the Chief Examiner who shall re-

Chapter 6 — License Requirements for States and Cities

view the proceedings of the district Examiner and decide the merits of the appeal.

Section 4739.14 (1058-1b) Fees.

The fee for examination of applicants for licenses under Section 4739.01 to 4739.15, inclusive, of the Revised Code, shall be fifty dollars to be paid at the time of examination. The fee for each original or renewal of a license issued under section 4739.01 to 4739.15, inclusive, of the Revised Code, shall be twenty-five dollars. Upon renewal, the licensee shall be furnished a certificate of renewal. All fees provided for in this section shall be paid by cash, money order, or certified check to the division of examiner of Steam Engineers which shall transmit the same to the treasurer of state to the credit of the general revenue fund.

License Requirements for States and Cities　　　Chapter 6

Oklahoma

Oklahoma City, Oklahoma
SUMMARY OF QUALIFICATIONS:

1. Exam Location, Address and Phone Number
City Hall
200 North Walker
Oklahoma City, Oklahoma 73111
405/297-2381

2. Exam Days and Times
At least 25 people must sign up before test will be given

3. Examination Cost
$10.00

4. License Cost and Renewal
$4.00 for initial license and renewal

5. Type of Test
All Written, 100 multiple choice questions

Can have an oral exam if requested

6. Grades of Licenses
Class 2 Low Pressure Boiler Operator

Class 1 High Pressure Boiler Operator

7. Experience and Education Requirements
See Codes for Oklahoma City below

8. U.S. Citizenship Required?
Yes

9. Local Residency Required?
No

Chapter 6   License Requirements for States and Cities

10. Recognition of Licenses from other locales

Recognized for experience, must take exam

CODES for Oklahoma City, Oklahoma

Division 3. Boiler Operator Qualifications and Examination

Sec. 29-186. Application and examination for Boiler Operator's license.

Any person desiring to engage in or work at the trade of operating boiler systems either as a Class One or Class Two Operator shall apply to the Board of Boiler Engineer Examiners and shall at such time and place as the Board of Boiler Engineer Examiners may designate, first pass such examination as to his qualifications as may be directed by the Board.

Sec. 29-187. Examination fee.

The applicant shall pay to the City a non-refundable fee for the examination. The amount of such fee shall be as established by ordinance; copies of such ordinance are on file in the Office of the City Clerk.

Sec. 29-188. Contents of examination.

The examination may be whole or in part in writing, and shall be of such a nature as to show that the applicant has practical experience, sufficient knowledge, training and ability to properly carry on or engage in or work at the trade of operating boiler systems in the class for which he makes application. No applicant shall be considered unless applicant is at least 18 years of age, a citizen of the United States, or has declared his intention of becoming such, is of good character and reputation and these qualifications when answered in any application shall be vouched for in writing by at least three resident taxpayers or may be verified under oath by the applicant.

Sec. 29-189. Passing grade on examination; issuance of certificate.

(a) If any applicant shall make a grade of at least 75 percent with 100 percent as perfect in the examination, he shall be issued the appropriate license.

(b) Applicants applying to take the Class One Boiler Operator examination shall possess a current Class Two Boiler Operator license and show satisfactory proof that he/she has had experience on boiler systems of a type and size which indicates to the Boiler Examining Board that the applicant is competent. Approved technical schools with a specialization in boiler skills can be considered as experience.

Secs. 29-190 - 29-194. Reserved.

Division 4. Boiler Operator License

Sec. 29-195. License required.

No person shall operate or assist in operating any boiler or other steam generating apparatus under pressure in excess of 30 horsepower, or work at or engage in the trade of Class One or Class Two Boiler Operator, without having a valid Class One or Class Two Boiler Operator's license.

Sec. 29-196. Application.

A person who has passed the Boiler Operator examination for his class and received such certificate from the Board of Boiler Engineer Examiners, or who otherwise qualified according to the provisions of this article, may make application for a license to the Supervisor of Licenses.

Sec. 29-197. Fee.

A person making application for a license or license renewal pursuant to the provisions of this article shall first pay to the city the fee established by ordinance. Copies of such ordinance are on file in the office of the City Clerk.

Sec. 29-198. Term; renewal.

The term of the license shall be for one year from the date of issuance. Licenses may be renewed annually upon the payment of the required fee without further examination. However, any person who fails to renew their license within ninety days of the date of expiration shall be required to pay an additional fee for renewal. No license may be renewed if it has been expired for more than 12 months.

Secs. 29-199 - 29-205. Reserved.

Division 5. Boiler Operator Regulations.

Sec. 29-206. Authority to operate equipment.

(a) First Class Boiler Operator's license. Any person holding a First Class Boiler Operator's license is hereby authorized to operate all steam generating boiler, pressure vessels and superheaters.

(b) Second Class Boiler Operator's license. Any person holding a Second Class Boiler Operator's license is hereby authorized to operate and maintain heating boilers subject to a pressure not to exceed fifteen (15) psig pounds steam, one hundred (100) horsepower or one hundred twenty-five (125) pounds psig water safe working pressure and all pumps and apparatus in connection therewith.

Division 8. License Suspension or Revocation.

Sec. 29-235. Suspension or revocation for cause.

A license issued pursuant to this article may be suspended or revoked for cause.

(1) Action by Chief Mechanical Inspector. The Chief Mechanical Inspector, or his assistants, may suspend a license if he finds that the licensee has failed to perform his work in accordance with the requirements of the mechanical code and such violation or violations are such that property and the public health safety and welfare are in imminent peril. Notice of such action shall be given to the licensee by certified mail, return receipt requested. The notice will contain information on how such action may be appealed. The licensee shall remain suspended until a hearing is held before the Mechanical Code Review and Appeals Commission. Such hearing shall be held in conformance with the provisions of this chapter.

Chapter 6 License Requirements for States and Cities

Tulsa, Oklahoma
SUMMARY OF QUALIFICATIONS:

1. Exam Location, Address and Phone Number

City Hall
Permit Center, Room 101
200 Civic Center Drive
Tulsa, Oklahoma 74103
918/596-9679
918/596-1841
Location of test will be announced in letter to applicant.

2. Exam Days and Times

3rd Tuesday of every month from 5:30 p.m. to 8:30 p.m.

3. Examination Cost

$35.00

4. License Cost and Renewal

$20.00 for initial license and renewal

5. Type of Test

All Written (From Block & Associates)

6. Grades of Licenses

First Class Engineer

First Class Limited Engineer

Third Class Engineer

Steam Specialist Engineer

7. Experience and Education Requirements

See Codes for Tulsa below

8. U.S. Citizenship Required?

Yes

9. Local Residency Required?

No

10. Recognition of Licenses from other locales

Up to Mechanical Appeals Board

CODES for Tulsa, Oklahoma

TITLE 59 - Mechanical Code

Chapter 2 - Stationary Engineers

Section 200. EXAMINATION AND LICENSING OF STATIONARY ENGINEERS

A. Examination of Stationary Engineers; Qualifications.

The Mechanical Examiners and Appeals Board shall administer written examinations to all eligible Stationary Engineer applicants. Examinations shall be appropriate to the license sought and shall reflect the knowledge and experience required to perform the work of the particular class. Each applicant shall pay the sum of twenty five dollars ($25.00) to the City of Tulsa for such examination. Applicants for the following classes of licenses shall have the minimum experience designated for each class of license as follows:

1. First Class Engineers License.

All applicants for a First Class Engineer's License shall:

(a) Have at least three (3) years experience in the operation or assisting in the operation of boilers exceeding 15 psig steam or 160 psig water pressure or 250 degrees water temperature and refrigeration or air conditioning units exceeding 100 tons, or,

(b) Shall be a currently licensed Mechanical Journeyman who has worked for a Mechanical Contractor licensed with the City of Tulsa for at least three (3) consecutive years, installing the equipment described in A.1.a. herein, or,

(c) Shall have a First Class Limited license and have served two (2) or more years operating or assisting in the operation of the equipment described in Paragraph 2 herein.

2. First Class Limited Engineer's License.

All applicants for a First Class Limited Engineer's License shall:

Have at least one (1) year's experience in the operation or assisting in the operation of low or high pressure boilers and at least three (3) years experience in operation or assisting in the operation of refrigeration or air conditioning units exceeding 100 tons. A maximum credit of one (1) year for refrigeration or air conditioning experience shall be given for experience in the operation of heating and air conditioning equipment using steam and chilled water furnished from a remote plant.

3. Third Class Engineer's License.

All applicants for a Third Class Engineer's License shall:

Have at least one (1) year's experience in operation or assisting in the operation of low or high pressure boilers.

4. Steam Special Engineer's License.

All applicants for a Steam Special Engineer's License shall:

Have at least three (3) years experience in the operation or assisting in the operation of boilers exceeding 15 psig steam or 30 psig water pressure.

5. Refrigeration Special Engineer's License.

All applicants for a Refrigeration Special Engineer's License shall:

Have at least three (3) years experience in the operation of or assisting in the operation of a refrigeration or air conditioning unit exceeding 100 tons.

All experience required for any of the aforementioned classes of license defined in this section and so claimed by an applicant for such class of license shall be proven by a notarized certification from former and/or present employers concerning the examinee's experience and qualifications. Only experience on the equipment itemized in Section 201 of this Code shall be considered.

B. Issuance of Certificates of Competency.

Stationary Engineer applicants receiving a grade of seventy five percent (75%) or higher on a required written examination shall be issued a Certificate of Competency appropriate to the examination administered by the Mechanical Examiners and Appeals Board. All Certificates of Competency shall show the date of passage of the examination, or if issued without examination as herein provided, the reason therefor. Such certificates shall be consecutively numbered and the office of Mechanical Inspections shall keep a record of all certificates issued.
Should a certificate be lost or destroyed, a duplicate may be obtained under a new registration number by submitting a signed, written request and fee of Ten Dollars ($10.00). Old numbers will automatically become invalid.

C. Issuance and Renewal of Stationary Engineer Licenses.

Every holder of a current Stationary Engineer Certificate of Competency shall be issued an annual license appropriate to such certificate by the Mechanical Examiners and Appeals

Chapter 6   License Requirements for States and Cities

Board upon payment of a license fee of Twenty Dollars ($20.00).

D. Annual License Registration.

All Stationary Engineer licenses shall expire on December 31 and shall be renewed annually. Failure to renew within thirty (30) days of expiration will subject the licensee to a penalty of Two Dollars ($2.00) per month thereafter until renewal. When any license has expired for one (1) year or more, the licensee shall be examined as to his qualifications by the Board before the license is renewed.

E. Stationary Engineer License Classifications Established and Work for Which Qualified Defined.

Stationary Engineer licenses shall be divided into the following categories and issued to qualified applicants.

1. First Class Engineer's License.

Any person holding a First Class Engineer's License is qualified to operate and maintain all steam generating boilers, pressure vessels, superheaters, refrigeration plants and air conditioning units of unlimited tonnage, horsepower and pressures, and all pumps and apparatus in conjunction therewith.

2. First Class Limited Engineer's License.

Any person holding a First Class Limited license is qualified to operate and maintain all boilers with an operating temperature of less than 250 degrees F. and subject to a pressure not exceeding 15 psig steam or 160 psig water and refrigeration and air conditioning units of unlimited tonnage including all pumps and apparatus in conjunction therewith.

3. Third Class Engineer's License.

License Requirements for States and Cities Chapter 6

Any person holding a Third Class Engineer's license is qualified to operate an maintain heating boilers with an operating temperature of less than 250 degrees F. and subject to a pressure not to exceed 15 psig steam or 160 psig water and all pumps and apparatus in connection therewith.

4. Steam Special License.

Any person holding a Steam Special License may operate and maintain all boilers, engines, pumps and apparatus in conjunction therewith of unlimited size and pressure.

5. Refrigeration Special License.

Any person holding a Refrigeration Special License may operate and maintain refrigeration plants and units or unlimited tonnage or horsepower.

F. Display of Certificates and Licenses Required.

Licensees shall keep Certificates of Competency and current licenses displayed at all times in the place where they are employed, so that same may be readily seen. In case a copy should be used, it must show the address where the original is displayed.

G. Revocation and Suspension of Stationary Engineer Licenses and Certificates of Competency.

Subject to the procedures governing the Mechanical Examiners and Appeals Board, Stationary Engineer Certificates of Competency or Stationary Engineer Licenses may be revoked or suspended by the Board for incompetence, gross carelessness in the maintenance or operation of equipment, or intoxication while on duty.

H. Appeals.

Any person aggrieved by any decision of the Board shall have a right of appeal to the City Council in accordance with the procedures in Chapter 1, herewith.

APPLICATION AND INFORMATION FOR TESTING FOR STATIONARY ENGINEERS- REFERENCE LIST - PLEASE READ CAREFULLY

MECHANICAL EXAMINERS & APPEALS BOARD

Information for License Applicants

(1) Applications may be obtained from the Public Works Permits and Licensing, Room 101, City Hall. Notarized "Employer's Certification of Examinee's Experience Qualifications" form, which is attached to this application, is required to prove experience. If this form cannot be secured, you may present notarized letters in person to the Mechanical Board for their approval.

(2) Your application must be turned in to the Chief Mechanical Inspector's office (before) the third (3) Tuesday of the month preceding the month in which the examination is desired.

(3) It is your responsibility to call the Mechanical Inspector's office after 11:00 a.m. following the third (3) Thursday to see if your application was accepted or rejected. You also will receive a letter from this office stating the following:

Approval of your application

Date of testing

Time of testing

Place of testing

Test ordered according to your application

If you application is rejected, the letter will state the reason.

(4) The required test fee must be paid at the time you submit your application. This fee is refundable by request if your application is rejected.

Testing Fees:

All Engineers Examination fees are......................$25.00

(5) Please read the attached detailed information concerning the filling out of the application.

TESTING INFORMATION:

(1) Tests are given the third (3) Tuesday of each month at 8:30 p.m. Location of test will be announced in the letter to applicant.

(2) Test scores will be available approximately three weeks from taking the examination. For test scores, call 596-9660. After passing this test, your Certification of Competency will be mailed to the address on your application.. Also a letter will accompany your certificate giving you information concerning the required fee to activate your license.

(3) If you fail your test, you may be retested by paying another fee of the same amount. A new application is not necessary.

THE FOLLOWING SCHEDULE IS FOLLOWED FOR RETAKING THE TEST

First Failure...............................2 months waiting period

Second Failure..........................4 months waiting period

All failures thereafter................4 months waiting period

Chapter 6   License Requirements for States and Cities

Only a total of three times per year beginning with month of first test.

(4) Failure to appear for examination at scheduled time will result in a forfeiture of the examination fee.

PREPARATION FOR TESTING

(1) A reference list is available, on request, at the Mechanical Inspector's office for each specific examination.

(2) All examinations are open book so you may bring references. No notes.

(3) #2 pencils are to be brought by the student. Make sure these are sharpened. A pencil sharpener is not available.

(4) Calculators: The use of calculators will be permitted only if the machine is non-electric, producing no tape, quiet, memory on calculator must be clear when entering or leaving classroom.

(5) Everything else needed will be supplied by the examiners.

EXAMINATION RULES AND INFORMATION

A. References:

(1) Each applicant must provide their own references; no material of any sort may be shared by applicants.

(2) Only books and references will be permitted in the examination room. All briefcases, books, etc., may be inspected by the proctors.

(3) No notes of any sort will be allowed.

B. Scratch paper:

(1) No paper of any sort is to be brought to the examination. Use the back of the pages in the examination for scratch paper.

C. Note taking:

(1). No part of the examination may be copied or reproduced in any manner, or removed from the examination room. If notes are made in references, the proctor will pick up the references and turn them in to the examining board.

D. Talking:

(1) No talking will be permitted in the examination room during the examination, except by the proctor.

E. Smoking:

(1) No smoking will be permitted in the examination room.

F. Applicant Number:

(1) Do not put your name on the exam, use the assigned applicant number only.

VIOLATION OF ANY OF THE ABOVE RULES WILL DISQUALIFY THE APPLICANT AND NO GRADE WILL BE ISSUED. LEGAL AND DISCIPLINARY ACTION CAN BE TAKEN BY THE EXAMINATION BOARD.

G. Questions:

(1) No technical questions of any sort will be answered by the proctor. If you do not understand a questions on the examination, do your best to answer it.

(2) Answer all questions. There is no penalty for guessing.

H. Materials:

(1) Bring all references, a #2 pencil, an eraser, a calculator, and personal identification documents (one with your picture).

Oregon

No State or City License

Pennsylvania

Erie, Pennsylvania

SUMMARY OF QUALIFICATIONS:

1. Exam Location, Address and Phone Number
City of Erie
City Hall, Room 507
626 State St.
Erie, Pennsylvania 16501
814/870-1253

2. Exam Days and Times

Twice a year - in April and November on the Saturday following the 2nd Wednesday of the month.

3. Examination Cost

Class 1 - Chief Stationary Engineer...............$50.00

Class 2 - Stationary Engineer........................$40.00

Water Tender...$30.00

4. License Cost and Renewal

Initial license included in cost of examination.

Renewal fee for Chief Engineer....................$50.00

Renewal fee for Class 2 Engineer.................$40.00

Renewal fee for Water Tender......................$30.00

5. Type of Test

All Essay

License Requirements for States and Cities   Chapter 6

6. Grades of Licenses

Class 1 - Chief Stationary Engineer

Class 2 - Stationary Engineer

Water Tender

7. Experience and Education Requirements

Class 1 - Chief Stationary Engineer - 2 years practical experience as 2nd Class (Technical School certificate in field related counts as 1 year practical experience)

Class 2 - Stationary Engineer - 2 years practical experience as Water Tender

Water Tender - 1 year practical experience (2 licensed Engineers must sign application and it must be notarized)

8. U.S. Citizenship Required?

No

9. Local Residency Required?

No

10. Recognition of Licenses from other locales

No

CODES for Erie, Pennsylvania
319.03 RULES AND REGULATIONS.

(a) These regulations shall apply to:

(1) All boilers within the city having a safety valve setting of fifteen pounds per square inch or above or normal operating aggregate of 1500 square feet of heating surface or more, except those located in private dwellings.

Chapter 6   License Requirements for States and Cities

(2) Steam driven machinery within the City having a normal operating aggregate of 50 HP or above.

(3) Refrigeration machinery within the City having a normal operating aggregate of 50 HP or above and using ammonia refrigerant.

(4) Every owner or lessor of any steam boiler in theaters, motion picture theaters, hospitals, schools, hotels, motels, organized homes or institutions within the city not requiring a licensed operator shall cause the boiler to be inspected annually in accordance with State law and a certificate of pressure vessel operation shall be made available to the Chief of the Bureau of Engineer Licenses when requested.

(b) Licenses shall be granted in three classes as follows:

(1) Class 1 - Chief Stationary Engineer.

This class entitles licensee to have charge of or operate any steam boilers, steam driven machinery and refrigeration machinery. To qualify, an applicant must hold a Class 2 - Stationary Engineer's license for a period of two years.

(2) Class 2 - Stationary Engineer

This class entitles the licensee to have charge of or operate steam driven machinery up to and including 200 HP, boiler or boilers up to and including in the aggregate of 4000 square feet of heating surface, and refrigeration machinery. He may also operate any size boiler or boilers or steam driven machinery under a Chief Stationary Engineer in charge. To qualify, an applicant must hold a Class 3 - Water Tender's license for a period of two years.

(3) Class 3 - Water Tender

License Requirements for States and Cities Chapter 6

This class entitles licensee to have charge of or operate steam driven machinery up to and including 100 HP, boiler or boilers up to and including in the aggregate of 2,000 square feet of heating surface, and refrigeration machinery. He may also operate any size boiler or boilers under the supervision of a duly classified licensed Engineer. Any operation which is not presently in compliance with this provision will have a maximum of five years from effective date of this article (May 8, 1974) to meet the requirements.

To qualify an applicant must have experience in and around steam generating equipment and auxiliaries for a period of one year. A degree held by the applicant from any recognized technical school shall be equivalent to one year's experience for each class of license. It shall be the responsibility of a licensee when on duty to thoroughly examine all equipment and any defects or sign of neglect shall be reported to the Chief of the Bureau of Engineer Licenses.

319.04 GENERAL REQUIREMENTS.

(a) All applicants must be twenty one years of age and have a high school education or its equivalent. He must pass a written examination for each class of license.

(b) All licenses shall be framed under glass, in a conspicuous place where it can be readily seen.

(c) Any licensee shall assist the inspector in his examination of any equipment covered under this section. He shall point out any defects known to him to the inspector.

(d) Where a licensed operator is required by this article, such operator must have his post of duty within the building housing the boilers, steam driven machinery, or refrigeration machinery, and is further required to be present when the machinery is in operation.

(e) Every owner or lessee, or agent of the owner or lessee of any steam boiler, steam driven machinery or refrigeration machinery embraced within the provisions of this article, or any appliances connected therewith, or any person acting for such owner, lessee or agent, is hereby forbidden to delegate or transfer in any manner whatsoever, the responsibility or liability for the management, operation or maintenance in good condition and repair of any such steam boiler, steam driven machinery or refrigeration machinery, or appliances connected therewith to any person or persons other than a duly classified licensed Engineer in charge thereof; provided that in case of the absence from duty of any such licensed Engineer on account of illness or other unavoidable cause, an unlicensed person may take the place of the Engineer for a period of not exceeding ten days; provided further that the Chief of the Bureau of Engineer Licenses shall be notified and approve of any such temporary change at the time the change is made.

(f) Any person desiring authority to perform the duties outlined under this article shall apply, in writing, to the Chief of the Bureau of Engineer Licenses of the City. The Board of License Examiners shall review all applications to insure eligibility requirements are met. The board shall also review the applicant's character, habit of life and experience to determine that all those components substantiate the belief that the applicant is a suitable and safe person to be entrusted with the powers and duties required.

319.05 FEES AND FINES.

(a) Application for license shall include a non-refundable fee of thirty dollars ($30.00) for Water Tenders and Portable Engineers, forty dollars ($40.00) for Second and Third Class Engineers, and fifty dollars ($50.00) for Chief Engineers and Refrigeration Engineers, payable to the City Treasurer.

(b) All class licenses shall be renewed annually within ten days of expiration and shall include a fee of thirty dollars ($30.00) for Water Tenders and Portable Engineers, forty dollars ($40.00) for Second and Third Class Engineers, and fifty dollars ($50.00) for Chief Engineers and Refrigeration Engineers, payable to the City Treasurer.

(c) (Editor's Note: This subsection was repealed by Ordinance 35-1983, passed July 13, 1983.)

(d) Anyone entering the active duty in the armed services shall have his license remain in effect until such time of discharge without any renewal fees. Such license shall be renewed within six months after discharge and requires the payment of the annual renewal fee.

(e) Any person violating any provisions of this article shall on conviction thereof, be liable to a fine of not less than one hundred dollars ($100.00) nor more than two hundred dollars ($200.00) and costs of prosecution to be collected by summary proceedings before any duly authorized magistrate and in default of payment of any such fine and cost may be committed to Erie County Jail for a period not to exceed thirty days. In addition, such person shall be liable to a suspension of revocation of his license as the Board of License Examiners shall determine.

Chapter 6　　　License Requirements for States and Cities

Philadelphia, Pennsylvania
SUMMARY OF QUALIFICATIONS:

1. Exam Location, Address and Phone Number

Municipal Services Building
1401 JFK Blvd, Room 580
Philadelphia, Pennsylvania 19102
215/686-2494

2. Exam Days and Times

Engineer - Grade A - 2nd Tuesday of each month at 1 p.m.

Engineer C & D - Fireman - 1st Tuesday of odd numbered months at 1 p.m.

3. Examination Cost

Engineer A................$25.00

Engineer C................$25.00

Fireman D................$25.00

Test Fee....................$75.00

4. License Cost and Renewal

For initial license and renewal

Engineer A $25.00

Engineer C $25.00

Fireman D $25.00

5. Type of Test

All Written

6. Grades of Licenses

Engineer - Grade A

License Requirements for States and Cities   Chapter 6

Engineer - Grade C - Hoisting

Engineer - Grade D - Fireman

7. Experience and Education Requirements

Two years experience

8. U.S. Citizenship Required?

Not Specified

9. Local Residency Required?

Not Specified

10. Recognition of Licenses from other locales

Not Specified

GENERAL INFORMATION

Applicants requiring occupational licenses must pass a qualifying examination. Please submit two (2) recent photographs (taken within the last two years) along with application. Photographs should be approximately 1 1/2 inches squared. After application is approved, you be notified when and where to appear for testing. Application must be received at least one week prior to the scheduled examination. Please do not pay the license fee with application for exam. License fee is due upon notification of passing grade.

SCHEDULE FOR OCCUPATIONAL LICENSING EXAMINATION

All examinations begin at 1:00 P.M. in room 580 Municipal Services Building, 1401 J.F.K. Boulevard.

Engineer Grade A on 2nd Tuesday of each month

Chapter 6 — License Requirements for States and Cities

Engineer Grade B, C, and D on the 1st Tuesday of odd number months.

Make check or money order payable to the City of Philadelphia (application fee only)

Mail to:
Department of Licenses and Inspections
License Issuance Section
PSC MSB, 1401 JFK Blvd.
Philadelphia, PA 19102
Phone number 215/686-2494 for any questions
(215) 686-2418
(215) 686-2489

LICENSE TYPE

(B) Engineer Exam (7314, 7315, 7316, & 7317)

Grade A- Operate steam boiler, stationary and refrigeration Engineer

Grade B-Refrigeration Engineer

Grade C-Portable and Stationary Engineering (hoisting)

Grade D-Fireman

Applicant must be at least 21 years of age and have a minimum two (2) years experience as an assistant Engineer or helper, and is recommended by two (2) Philadelphia licensed Engineers.

(1) State qualifying experience

(2) Obtain employer vouchers attesting to good moral character and competence to perform duties of license applied for.

(3) Obtain recommendations from two (2) Philadelphia licensed Engineers. Must include license number.

(4) Military Personnel recently separated from duty may submit "DD214" in lieu of employer voucher.

Note:

This section does not apply to:

(1) Boilers, engines or machinery in dwelling houses used exclusively for residential purposes.

(2) Boilers, engines or machinery under the supervision of federal authorities or used in connection with interstate commerce.

(3) Boilers carrying less than 15 pounds of pressure.

(4) Boilers carrying more than 15 pounds of pressure but not exceeding 30 HP (1 boiler horsepower is equal to the evaporation of 34.5 pounds of water at 212 F.

The Expiration Date and License Number will be completed by the Department. To arrive at TOTAL FEES, add up the fees for all examinations being applied for and enter total. Receipt Number will be filled in by Department. Ordinarily no receipt will be sent to applicant. Your cancelled check is your receipt, or the agent who sells money orders supplies a receipt for the purchase of the money order. Please do not mail cash. Cash payments cannot be accepted.

If re-examination, check box and fill in date of most recent exam for license applied for. Two photographs are required. Check box indicating that photographs are attached. No additional photographs required for re-examination if less than two years has lapsed since photographs were submitted.

APPLICATION FEES

Engineer A $25.00

Engineer B $25.00

Engineer C $25.00

Engineer D $25.00

LICENSE FEES

Engineer A $25.00

Engineer B $25.00

Engineer C $25.00

Engineer D $25.00

Pittsburgh, Pennsylvania
SUMMARY OF QUALIFICATIONS:
1. Exam Location, Address and Phone Number
City of Pittsburgh
City Office Building
1600 W. Carson St.
Pittsburgh, Pennsylvania 15219
412/255-2181

2. Exam Days and Times
2nd Wednesday of January, March, May, July, September and November

From 1 p.m. to 3 p.m.

3. Examination Cost
$50.00

4. License Cost and Renewal
Initial license................$50.00

Renewal of license.......$50.00

5. Type of Test
Multiple Choice - Open Book

6. Grades of Licenses
Boiler Fireman

Stationary Engineer

7. Experience and Education Requirements
Boiler Fireman ....2 years practical experience

Stationary Engineer...................2 years practical experience

8. U.S. Citizenship Required?

Chapter 6   License Requirements for States and Cities

No
9. Local Residency Required?
No
10. Recognition of Licenses from other locales
No

CODES for Pittsburgh, Pennsylvania

Chapter 745 - Stationary Engineers (Also covers Boiler Fireman)

745.01 License Required: Exceptions

No person shall control or operate pipes, boilers, stationary or locomotive engines, or any other containers, tanks or vessels under pressure of water, liquid, gas or steam without first obtaining a license from the License Officer. However, the following are excepted from licensing requirements:

(a) Pipes, containers or vessels used in the transportation of water, liquid, gas or steam:

(b) Hot water tanks used for domestic service as defined in the Mechanical Code;

(c) Portable compressor operators; and

(d) All vessels defined as Low Pressure in Mechanical Code or any vessel of fifteen (15) pounds per square inch or less of pressure.

745.02 Written Examination

(a) A written examination shall be given by the License Officer to insure all applicants comprehend the operation of pressure vessels for the safety of life and property, and a license shall only be issued to the applicant who receives a passing grade. Examinations and licenses may be issued under different classifications as determined by the License Officer.

(b) The fee for such examination shall be fifty ($50.00) dollars which, if passed, shall be credited to the license fee.

745.03 Qualifications of Applicant

In addition to the qualifications enumerated in Section 701.04 for all applicants for licenses an applicant for a Stationary Engineer license, or any classification thereof, shall:

(a) Be eighteen (18) years of age and able to read and speak the English language.

(b) Possess two (2) years practical experience as evidenced by a certification thereof by a licensed Stationary Engineer; and

(c) Pass a test in accordance with Section 745.02.

754.04 Only Licenses to Operate Equipment

No owner, lessee or agent thereof, of any pipes, containers, boilers, stationary or locomotive engines, tanks or other vessels within the scope of this chapter, or any person acting for such owner, lessee or agent shall permit any unlicensed person to operate or control such equipment. No unlicensed person shall operate or control such equipment. The licensed person in charge of or operating the aforesaid equipment must be located on the same premises where said equipment is located during such operation or control.

745.05 Suspension or Revocation

The License Officer, upon investigation shall suspend the license of any licensee who is incompetent, has been guilty of negligence, has endangered life or property or willfully vio-

Chapter 6	License Requirements for States and Cities

lated any provision of this chapter. Revocation shall be as required in Section 701.14(b).

745.06 License Fee

The fee for an initial Stationary Engineer license shall be one hundred ($100.00) dollars per year and the fee for a renewal thereof shall be fifty ($50.00) dollars if renewed within thirty (30) days of expiration, otherwise the initial fee shall be paid without re-examination.

FIREMAN'S TEST INFORMATION

City of Pittsburgh
Department of Public Safety
Bureau of Building Inspection

Description of Examination

The boiler fireman's examination is a 25 question, open book, multiple choice examination with a two hour time limit. Only bound, copyrighted materials are permitted. Mini-calculators, silent, hand held, battery operated, non programmable, without paper tape are permitted. No notes, no scratch paper, and no loose papers of any kind are permitted. The only scratch paper allowed will be provided with the examination booklet, and that scratch paper must be turned in with the examination.

Schedules and Procedures

The examination is scheduled to be given six (6) times a year, in January, March, May, July, September and November. It is given on the second Wednesday of the month, starting at 1:00 p.m. and ending at 3:00 p.m. You must arrive at the test site not later than 12:30 p.m. to allow time for registration, and examination instructions. If you arrive late, you will be permitted to take the examination but the instructions will not be repeated and no additional time will be allowed. The examination ends at 3:00 p.m.

Fees, Application Date Deadline and Admission

The fee for examination is $50.00. A check or money order, payable to: Treasurer, City of Pittsburgh, must accompany the application.

The completed application and fee must be received by the City thirty (30) days prior to the examination date. If for any reason the application is not acceptable, your fee and application will be returned with an explanation. If your application is accepted, you will be sent an admission letter which will be sent to you approximately two weeks before the examination date. You must save this letter. It will be required for admission to the test.

You will be required to show this admission letter and a photo I.D. to receive your examination. Only those with the admission letter and a photo I.D. will be given the examination.

If, after being accepted for an examination, you are not able to take the examination on the date scheduled, you must return your admission letter to City within seven (7) days of the examination date. Your examination fee will be credited to the next administration of the exam and a new admission letter will be sent. Failure to notify the City will result in forfeiture of the test fee. In no case will the test fee be refunded; it can only be applied to a future examination.

Examination Results

Results will be reported in writing to the examinee only and will be mailed within 30 days. Results will not be given over the telephone. Test results will be reported as "P" for passing or the numerical grade for failure.

If an examinee is disqualified during the test, their examination will not be scored and the fee will not be refunded.

Taking the Examination

Each candidate will be given an exam booklet paper and an answer sheet.

Each question in the booklet has one and only one best right answer. In answering questions, select the one response which best answers the questions and mark the corresponding letter on the answer sheet. If more than one response is marked, no credit will be given for the question.

All marks other than your chosen answers must be erased completely.

All questions have equal value. Do not spend too much time on any one question. Each questions will be scored right for a correct answer and wrong for an incorrect or blank answer or omission.

Your score on the examination will be based on the number of correct answers you give; therefore, it is to your advantage to answer every question. There is no penalty for guessing.

Regulations at the Examination

The following regulations and procedures will be observed at the examination:

- The examination will be held only on the dates and at the times scheduled. Be prompt.
- Admission letter and photo identification required for admission.

- Bring three or four sharpened soft (No. 2) pencils and a good eraser. No pencils or erasers will be furnished. You will not be permitted to use ballpoint pens, colored pencils or felt pens.
- You should bring a watch. You will not be permitted to continue a test beyond the established time limit for any purpose. Although time is not intended to be a factor, you must keep track, you must not allow yourself to become bogged down on any one question.
- All materials that you have at the examination are subject to inspection on arrival and prior to leaving the examination.
- If you must leave the room during the examination, you must ask permission as only one examinee will be permitted to leave at a time.
- Writing on or in reference materials is prohibited.
- Examinees may not share reference materials.
- Examination materials may not be removed from the room.

Dismissals

Any examinee who gives or receives assistance during the examination will be required to turn in his or her materials immediately and leave the room. In addition, the examinee shall be dismissed for any of the following:

- Creating a disturbance.
- Attempting to remove materials or notes from the room.

Chapter 6   License Requirements for States and Cities

- Having a copy of any examination question (which shall be confiscated), unauthorized scratch paper, photo copies or other unauthorized items.
- Writing in reference materials.
- Copying questions.

License Requirements for States and Cities   Chapter 6

Rhode Island

Providence, Rhode Island

SUMMARY OF QUALIFICATIONS:

1. Exam Location, Address and Phone Number

City of Providence
City Building
190 Dyer Street
Providence, Rhode Island 02903-1712
401/421-7740 ext. 367

2. Exam Days and Times

Tuesday & Thursday at 9 a.m. every week (2 hrs)

3. Examination Cost

Stationary Engineer..........................$15.00

Boiler Operator................................$12.00

4. License Cost and Renewal

Initial license included in cost of exam

Renewal-Stationary Engineer................................$10.00

Renewal-Boiler Operator..$ 8.00

Renewal: Apprentice Fireman (good 6 months)....$ 5.00

5. Type of Test

Choice of Written, Oral or Practical Test

6. Grades of Licenses

Stationary Engineer

Boiler Operator

Apprentice Fireman - No Exam

Chapter 6　　　　License Requirements for States and Cities

7. Experience and Education Requirements

See Codes for Providence below

8. U.S. Citizenship Required?

No

9. Local Residency Required?

No

10. Recognition of Licenses from other locales

No

CODES for Providence, Rhode Island

Sec. 1105.0 - Maintenance and Operation

Sec. 1105.1 - License (Certificate of Fitness).

It shall be unlawful for any person to operate a steam boiler of more than thirty (30) horsepower without first obtaining a license from the Director in accordance with the following provisions:

Sec. 1105.2 - Stationary Engine Operator and Boiler Operator.

To be eligible to receive a license as a stationary engine operator and/or a Boiler Operator, the applicant must be at least twenty one (21) years old. He must be of good character and reputation and the license shall be granted according to his competence, and shall be classified as follows:

.21 - Stationary Engine Operator.

To be eligible for examination for a license to have charge of and operate any steam plant, steam engine, steam boiler, diesel plant and/or refrigerating machinery, a person must have been employed as a Boiler Operator for five (5) years, or assistant to a duly licensed operator for at least three (3) years preceding

License Requirements for States and Cities   Chapter 6

the date of application, or he must have graduated as a mechanical Engineer from a recognized Engineering school and must have been employed one (1) year in connection with the operation of a steam or diesel plant after receiving a permit (no fee) from the Director.

.22 - Boiler Operator.

This classification includes Firemen and Water Tenders. Any person to whom this license is issued is allowed to have charge of and operate any boiler or boilers not over one hundred (100) Boiler H.P. To be eligible for examination for this license a person must have obtained a permit (issued free by the Director) and worked for six (6) months as an apprentice or helper under the direct supervision of a duly licensed stationary engine operator or Boiler Operator.

Sec. 1105.3 - Applications.

Applications for licenses shall be on forms prescribed and furnished by the Director. Applications must be properly notarized.

Sec. 1105.4 - Examinations.

Examinations may be written, oral or practical as determined by the Director. The purpose is to determine the qualifications and ability of the applicant. If the candidate fails to pass the examination he may apply for re-examination at the expiration of six (6) months and may be re-examined once without payment of an additional fee. Additional subsequent examinations may be granted upon payment of the regular fee. Fees must accompany application and are not returnable.

Sec. 1105.5 - Authority to Issue Licenses.

The Director is hereby authorized and empowered to examine and license Engineers and Boiler Operators other than

Chapter 6 License Requirements for States and Cities

hoisting Engineers, under such suitable rules and regulations as he shall make. He shall cause all rules and regulations so made by him to be issued in printed form. He shall employ a Chief examiner to carry out the provisions of this Code under the supervision of the Chief Inspector of Mechanical Equipment and Installations. Such examiner shall be a practical Steam Engineer of not less than five (5) years' experience.

.51 - Application for License.

Each person who desires to act as a stationary engine or Boiler Operator in the City of Providence, shall make application to said Director upon a form to be furnished by said Director and shall pass an examination in the construction and operation of steam boilers, steam pumps, diesel engines, refrigerating equipment, boiler accessories and appliances and in the subject of hydraulics. The examination shall be conducted under rules and regulations to be provided and issued by the Director and he may administer oaths or affirmations to anyone whenever the same is made necessary by rules and regulations adopted.

.52 - Issuance of License.

If, upon examination, the applicant for a stationary engine or Boiler Operator's license is found proficient in such subjects as may be required by the Director, a license shall be granted him to be in direct charge of and operate such equipment for one (1) year from the date of which such license is issued.

.53 - Revocation of Licenses.

Upon written charges and after notice and hearing, the Director may revoke or suspend the license of a person guilty of fraud in obtaining such license, or who for any reason has become unfit to discharge the duties of a stationary engine or

Boiler Operator, subject to a right of appeal as set forth in Section 128.0.

.54 - Renewals Upon Application.

Upon application, the person to whom a license is issued, under this ordinance, shall be entitled to renewal thereof annually, unless the Director for a cause named in the preceding section and upon notice and hearing shall refuse such renewal. Such refusal shall be subject to a right of appeals as set forth in Section 128.0.

.55 - Fees.

Each applicant for an examination for a license as a stationary engine or Boiler Operator shall pay to the Director at the time of application a fee and for each renewal of such license a fee. (See Schedule)

.56 - Posting of License.

Each stationary Engineer or Boiler Operator shall exhibit his license under glass in a conspicuous place in his engine or boiler room, and for each neglect or refusal to comply with the provisions of this section shall be fined not to exceed Ten ($10.00) dollars.

Sec. 1105.6 - Exemptions

The provisions set forth herein shall not apply to railroad boilers and engines when used as such, nor to hoisting engines or to the owners or users thereof or boilers that are used for the purpose of tempory heat on any building during construction, alteration or repairs.

Sec. 1105.7 - Licenses issued prior to the effective date of this code.

Chapter 6 License Requirements for States and Cities

Any person who holds a license which is in effect on the passage of this Code, and any person who is employed as Stationary Engine Operator or Fireman or Boilerman prior to the passage of this ordinance, shall be granted a license as provided herein, without examination, upon application, and the payment of the regular examination fee, if such application is made within six (6) months after the passage of this ordinance.

Schedule

Stationary Engineer.............$15.00.....Renewal-$10.00

Boiler Operator....................$12.00.....Renewal-$ 8.00

Refrigeration Machine Opr..$12.00.....Renewal-$ 8.00

Apprentice Fireman-$ 5.00.....Good only six (6) months.

South Carolina

No State or City License

South Dakota

No State or City License

License Requirements for States and Cities　　　Chapter 6

Tennessee

Memphis and Shelby County, Tennessee

SUMMARY OF QUALIFICATIONS:

1. Exam Location, Address and Phone Number
Construction Code Enforcement
6465 Mullins Station Road
Memphis, Tennessee 38134
901/385-5177

2. Exam Days and Times
1st Wednesday of March, June, September, and December.

3. Examination Cost
$50.00

4. License Cost and Renewal
Initial license included in cost of exam.
Yearly renewal............ $50.00

5. Type of Test
All written, open book, all multiple choice questions, three hour time limit.

6. Grades of Licenses
Third Class Steam Operating Engineer (up to 15 Horsepower)
First Class Steam Operating Engineer (unlimited)

7. Experience and Education Requirements
Third Class Steam Operating Engineer - Working under a licensed engineer for 2-3 years.
First Class Steam Operating Engineer - Working under a First Class Engineer for 2-3 years

8. U.S. Citizenship Required?
Yes, and 18 years of age

9. Local Residency required
No

10. Recognition of Licenses from other locales
Yes, this will allow you to take the First Class Engineer examination

Texas

Houston, Texas

SUMMARY OF QUALIFICATIONS:

1. Exam Location, Address and Phone Number

Exam Location

Department of Planning and Development
Mechanical Section
2nd Floor
1801 Main St.
Houston, Texas
713/754-0255

Mailing Address

Department of Planning and Development
Mechanical Section
P.O. Box 1562
Houston, Texas 77251-1562

2. Exam Days and Times

2nd and 4th Thursdays each month except holidays

Application must be in by 12:00 noon on test day

Application is good for 30 days

Exam is from 6:00 pm to 8:00 pm

3. Examination Cost

$20.00

4. License Cost and Renewal

$20.00 - Initial License and Renewal

Type of Test

All Written, Multiple Choice, Short Essay, True/False, Math

6. Grades of Licenses

First Grade Stationary Engineer.........Unlimited

Second Grade Stationary Engineer.....6,696,000 BTU input

Third Grade Stationary Engineer........1,674,000 BTU input

7. Experience and Education Requirements

Third Grade Stationary Engineer - Two years practical experience or a graduation certificate from an accredited Engineering school with at least six months practical experience

Second Grade Stationary Engineer - Three years practical experience or a graduation certificate from an accredited Engineering school with at least one years practical experience

First Grade Stationary Engineer - Five years practical experience, or a graduation certificate from an accredited Engineering school with at least two years experience, or a United States Department of Labor diploma showing completion of a three year apprenticeship program with one year experience as a Second Grade or two years experience as a Third Grade Stationary Engineer.

8. U.S. Citizenship Required?

No

9. Local Residency Required?

No

10. Recognition of Licenses from other locales

Not Specified

CODES for Houston, Texas

Uniform Mechanical Code

Chapter 6 — License Requirements for States and Cities

1988 Edition

City of Houston Amendments

License

Sec. 204. (a) Stationary Engineer's License.

Persons desiring to secure a stationary engineer's license shall apply to the Boiler Board for an application blank and shall pay to the building official the fee hereinafter required.

Licenses so granted shall be graded into three grades:

1. First Grade Stationary Engineer's license:

Shall entitle the licensee to have direct charge of, operate or supervise a stationary boiler or boilers of any size.

2. Second Grade Stationary Engineer's license:

Shall entitle the licensee to have direct charge of, operate and supervise any boiler or boilers having an aggregate amount of heat input not to exceed 6,696,000 Btu per hour and to act as assistant or watch engineer under the charge and supervision of the holder of a first grade stationary engineer's license in any steam plant.

3. Third Grade Stationary Engineers license:

Shall entitle the licensee to have charge of, operate or supervise any boiler or boilers having an aggregate amount of heat input not to exceed 1,674,000 Btu per hour and to act as assistant or watch engineer under the care and supervision of the holder of a second grade stationary engineer's license of any boiler or boilers having an aggregate amount of heat input not to exceed 6,696,000 Btu per hour.

An applicant for a First Grade Stationary Engineer's license:

Shall present to the Board service letters, or certified copies of same, showing that such applicant has had either the following specified experience, or the specified combination of experience and education:

(i) at least five years of practical experience as a stationary engineer, oiler, water tender, fireman, or mixed service of five years in such capacity or capacities, provided, however, three years of such five-year period shall have been serviced as a certified stationary engineer; or

(ii) a graduation certificate from an accredited engineering school with at least two years of practical experience as a stationary engineer, oiler, water tender, fireman, or mixed service of two years in such capacity or capacities; or

(iii) a United States Department of Labor diploma showing the applicant finished a full three-year course as an apprentice stationary engineer and one year of practical experience as a second grade stationary engineer or two years of practical experience as a third grade stationary engineer.

Applicants for Second Grade Stationary Engineer's license:

Shall present to the Board service letters or certified copies of same, showing that they have at least three years of practical expereience as a stationary engineer, oiler, water tender, fireman, or mixed service of three years in the above capacities, provided, however, that one of the three years shall have been served as a certified engineer; or a graduation certificate from an accredited engineering school, with at least one year practical experience as a stationary engineer, oiler, water tender, fireman, or mixed service of one year in above capacities.

Applicants for a Third Grade Stationary Engineer's license:

Chapter 6 — License Requirements for States and Cities

Shall present to the Board service letters or certified copies of same, showing that they have had at least two years of practical experience as a stationary engineer, oiler, water tender, boiler repairman, fireman, or mixed service of two years in above capacities, or a graduation certificate from an accredited engineering school with at least six months of practical experience as listed above.

No person may take an examination for a stationary engineer's license unless he has submitted the service letters, certificates and diplomas to the Board as required by this section.

Applicants will be required to correctly answer at least 70 percent of the questions composing the examination and will be required to attain at least a grade of 70 percent in examination in order to qualify for a stationary engineer's license of any grade. All questions and answers will be written in the English language.

When an applicant for a stationary engineer's license shall fail to satisfactorily pass an examination, he shall forfeit the fee as paid to the City of Houston and shall not be re-examined for the grade in which the applicant failed, or examined for a higher grade within a period of less than 90 days.

Each applicant shall pay the following examination fee for each and every such examination an applicant applied for:

First Grade Stationary Engineer.............$20.00

Second Grade Stationary Engineer.........$20.00

Third Grade Stationary Engineer............$20.00

Said fee is to be paid to the building official at the time such application is filed. All service letters or certified copies of same presented by anyone taking such examination shall be filed with said application. All applicants shall be examined by the Board not later than 30 days after the filing of completed application and payment of the fee, but no application shall be filed after 12:00 noon on examination days.

Applicants who have successfully passed the examination shall pay a $20.00 license fee to the city prior to the issuance of the license. The license issued under the provisions of this chapter shall expire on December 31 of the year of issuance, unless such licenses have been previously suspended or revoked. Thereafter, such licenses may be renewed annually pursuant to the provisions set forth below. The receipt for payment of a license renewal fee shall be displayed with the license and failure to do so shall constitute a ground for the suspension or revocation of such license.

(b) License Renewals.

License renewals shall be granted without re-examination upon payment of a fee of $20.00, provided such fee is paid within 30 days after the expiration date of the license and not thereafter. When an application for renewal is filed more than 30 days after the expiration date of the license, the fee for such renewal shall be $25.00 during the first year after the expiration date of the license and $15.00 for each additional year or part of year thereafter. When the annual license renewal fee has not been paid for a period of five consecutive years, the license shall not be renewed until the applicant has successfully passed a re-examination.

Each certificate or license issued under the terms and provisions of this ordinance shall be signed by not less than two members of the Board, the secretary of the Board, and the building official, who shall sign such certificate of license after said certificate has been signed by at least two members of the

Chapter 6   **License Requirements for States and Cities**

Board and all fees as herein provided for have been fully paid to the building official.

(c) Validity, Renewal and Replacement of License.

When the holder of a license is examined by the Board and granted a license in a higher grade, the higher grade license shall not be issued until the license of the lower grade is surrendered and all required fees paid to the building official.

When a license becomes lost or destroyed, the Board, at its own discretion, may grant a new license in the same grade, provided proof of such loss or destruction is presented to the satisfaction of the Board. The fee for such replacement license shall be $10.00. If the proof of such loss or destruction is not satisfactory to the Board, re-examination in the same grade shall be required and the fee for such re-examination shall be as provided in the first paragraph of (b) in this section.

Upon expiration, said license may be renewed as provided in the second paragraph of (b) in this section.

(d) Reciprocity.

A person who has been issued a marine engineer's license by the United State Coast Guard, which license is currently in full force and effect, shall be qualified for examination by the Board for a City of Houston stationary engineer's license, of equal or lower grade , provided the fee set forth in Section (a) has been paid.

A person who has been issued a stationary engineer's or a steam engineer's license by a state municipality or governmental agency, which license is currently in full force and effect, may be qualified for examination by the Board for a City of Houston stationary engineer's license, within the discretion of the Board, provided the Board shall determine the grade of

City of Houston license to which such applicant may be entitled, and provided further the holder of such license presented proof to the satisfaction of the Board that said license was granted as a result of boiler operating experience and passing grade on a written examination on the operation, maintenance and repair of boilers and boiler accessories and safety rules for the boilers.

Neither a license issued by a foreign government nor a graduation certificate from a foreign school, college or university, nor a service letter from an employer in a foreign country shall qualify the holder thereof to be examined by the Board for a City of Houston stationary engineer's license of any grade unless sufficient evidence of the validity of such license or certificate of service letter is presented to the satisfaction of the Board. Upon examination of the evidence of validity presented, the Board, at its sole discretion, shall designate the grade in which the applicant may be examined, if such evidence is found by the Board to be satisfactory.

(e) Expiration of License.

All licenses issued for stationary engineers which were in effect immediately preceding the adoption of this code by city council shall expire on the 31st day of December of the year in which this code is adopted. Any such license may be renewed as though it had been originally issued pursuant to this code.

Third Grade Stationary Engineer License Examination Guide

Applicants for Third Grade Stationary Engineer's License shall present to the Board service letters or certified copies of the same, showing that they have had at least two years of practical experience as a Stationary Engineer, Oiler, Water Tender,

Chapter 6 License Requirements for States and Cities

Boiler Repairman, Fireman, or mixed service of two years in above capacities, or a graduation certificate from an accredited Engineering school with at least six months of practical experience as listed above.

No person shall take the examination for a stationary Engineer's license unless he has submitted the service letters and/or certificates and/or diplomas to the Board as required by this section.

Proof of Experience

1. Service letters submitted as proof of experience must be on the letterhead of the company for which the applicant is (or was) employed. The letter must be recent and signed by a responsible member of the firm or corporation for which the applicant is (or was) employed.

2. Service letters and/or documents must clearly state the length of time in service complete with dates indicating employment "from" and "to." Service letters should also indicate the type and size of boilers and the use of those boilers operated during that period of service. Applicants who have gained their experience within the state of Texas, but not within the city limits of Houston, should present along with their service letters a copy of the State of Texas Boiler Certificate - as issued for each boiler by the Texas Department of Labor and Standards - for each boiler operated by the applicant.

3. Service letters must be originals or certified photocopies.

4. Applicants who obtained boiler experience in the U.S. Navy must submit "Evaluation Reports" indicating the length of service, classification and assigned duties of the applicant during that time period.

5. Maritime and other service documents will be reviewed by the Boiler Licensing and Review Board on a case by case basis.

6. Neither a license issued by a foreign government nor a graduation certificate from a foreign school, college or university, nor a service letter from an employer in a foreign country shall qualify the holder thereof to be examined by the Board for a Stationary Engineer License of any Grade unless sufficient evidence of the validity of such license, diploma or service letter is presented to the satisfaction of the Board.

Application Process

1. An application form will be provided by the City at the Mechanical Section Office located at 1801 Main Street. Upon completion of the form, review and acceptance of proof of experience letters or documents, the applicant shall pay an examination fee of $20.00.

2. The application must be submitted no later than 12:00 noon the day of the examination.

3. The application is valid for thirty (30) days. The applicant may take any one examination within the thirty day period following the date of the application.

4. If an examinee takes an examination and does not achieve a passing score, that applicant must wait (90) days before taking another examination.

5. There are no provisions for refund of the examination fee

Examination Dates and Time Period

Stationary Engineer examinations are held twice per month on the 2nd floor at 1801 Main Street between the hours of 6:00 pm - 8:00 pm. Examination dates are posted in the Mechanical

Chapter 6   License Requirements for States and Cities

Section Office. At the time of application you may choose a date (within a thirty day period following the date of application) to take the examination.

Sample Examination Topics

Note: This is not a comprehensive list. It is a sample of topics which may appear on the examination. Questions pertaining to other boiler code, construction, and operation topics may appear on the examination.

Water Glass

Water Column

Safety Valve

Steam Gauge and Siphon

Feed Water Piping

Boiler Tests

Condensation

Non-Return Valves

Pump

Continuous Blowdown

Corrosion

Pitting

Breeching

Priming

Foaming

Celsius

Blowback

Factor of Safety

License Limitations

Draft: induced/forced/etc.

Thickness of Boiler Tubes

Container Capacity

Absolute Pressure

Manholes

Static Head

Heating Surface

Temperature Conversions

BTU

Horsepower

Force Calculations

Safe Working Pressure

FIRST AND SECOND GRADE STATIONARY ENGINEER LICENSE EXAMINATION GUIDE

Qualifications for First and Second Grade Examination

Sec. 204(a) An applicant for FIRST GRADE Stationary Engineer's License shall present to the Board service letters or certified copies of same, showing that the applicant has had either the following specified experience, or the specified combination of experience and education:

- (i) at least five years experience as a stationary Engineer, oiler, water tender, Fireman, or mixed service of five years in such capacity or capacities, provided however, three years of such five year period shall have been served as a certified Stationary Engineer or
- (ii) a graduation certificate from an accredited Engineering school with at least two years of practical experience as a stationary Engineer, oiler, water tender, Fireman, or mixed service of two years in such capacity or capacities or
- (iii) a United States Department of Labor diploma showing the applicant finished a full three year course as an apprentice Stationary Engineer and one year of practical experience as a Second Grade Stationary Engineer or two years of practical experience as a Third Grade Stationary Engineer.

Sec. 204 (a): An applicant for a SECOND GRADE Stationary Engineer's License shall present to the Board service letters or certified copies of same, showing that they have at least THREE YEARS of practical experience as a stationary Engineer, oiler, water tender, Fireman, or mixed service of three years in the above capacities, or a graduation certificate from an accredited Engineering school with at least ONE YEAR practical experience as a stationary engineer, oiler, water tender, fireman, or mixed service of one year in above capacities.

Sample Examination Topics

Note: This is not a comprehensive list. It is a sample of topics which may appear on the examination. Questions pertaining to other boiler code, construction, and operation topics may appear on the examination.

License Requirements for States and Cities — Chapter 6

- Water Glass
- Water Column
- Safety Valve
- Steam Gauge and Siphon
- Feed Water Piping
- Staybolts
- Boiler Tests
- Condensation
- Non-Return Valves
- Pumps
- Continuous Blowdown
- Dry Pipe
- Superheaters
- Corrosion
- Pitting
- Breeching
- Priming
- Foaming
- Celsius
- Blowback
- Factor of Safety
- License Limitations

Chapter 6   License Requirements for States and Cities

Draft: induced/forced/etc.

Superheated Steam

Thickness of Boiler Tubes

Container Capacity

Absolute Pressure

Manholes

Static Head

Heating Surface

Safe Working Pressure

Temperature Conversions

BTU

Horsepower

Try-Cocks

Force Calculations

Safe Working Pressure

This is a closed book test. Do not bring study materials or notes into the examination area.

Utah

Salt Lake City, Utah

SUMMARY OF QUALIFICATIONS:

1. Exam Location, Address and Phone Number

Salt Lake City Corporation
Permits & Licensing Dept.
City & County Building

License Requirements for States and Cities — Chapter 6

451 South State Street
Room 215
Salt Lake City, Utah 84111
801/535-7752

2. Exam Days and Times

Monday through Friday

3. Examination Cost

Steam Engineer - Unlimited...................................$30.00

Steam Engineer - Limited..$25.00

Boiler Operator - High Pressure.............................$25.00

4. License Cost and Renewal

Initial license included with cost of exam.

Duplicate License..$10.00

Renewal of License prior to expiration....................$15.00

Renewal of license to one (1) year after expiration...$25.00

5. Type of Test

All Written - multiple choice.

6. Grades of Licenses

Steam Engineer-Unlimited

Steam Engineer-Limited

Boiler Operator-High Pressure

7. Experience and Education Requirements

Steam Engineer-Unlimited...5 years experience or education

Steam Engineer-Limited.......4 years experience or education

Boiler Operator-High Pressure-1 year experience

Chapter 6   **License Requirements for States and Cities**

8. U.S. Citizenship Required?

Not Specified

9. Local Residency Required?

No

10. Recognition of Licenses from other locales

Yes - for experience

CODES for Salt Lake City, Utah

Sec. 5-4-8. Boiler operation without license unlawful.

It shall be unlawful to operate, have control of, manage, or take charge of any portion of a steam or hot water plant, absorption system, or appliance connected therewith while working under pressure unless such plant, system, or appliance is being operated by a Steam Engineer or Boiler Operator who has been duly licensed for such operation.

Sec. 5-4-9 Application for Boiler Operator examination.

All persons desiring to be licensed to perform the duties of Engineer and/or Boiler Operator of stationary or portable boilers shall make application to the Division of Building and Housing Services on forms to be furnished by said division after payment of the fees, as hereinafter provided.

Sec. 5-4-10. Classes of Boiler Operator licenses.

The license required for this title shall be divided into two (2) classes for Engineers and one (1) class for Boiler Operators as follows:

(1) Steam Engineer - Unlimited

(2) Steam Engineer - Limited

(3) Boiler Operator - High Pressure

Sec. 5-4-11. Qualifications of a Steam Engineer - Unlimited.

An applicant for a Steam Engineer's license - unlimited, must be an Engineer, Oiler, or Boiler Operator having at least five (5) years actual operating experience in the management, control, or operation of High Pressure steam boilers or steam plants of 300 horsepower (10,200,000 BTU) or more. Up to four (4) years will be credited an applicant having previous equivalent training course in Mechanical Engineering at an accredited Engineering school.

Sec. 5-4-12. Steam Engineer - Limited.

An applicant for a Steam Engineer license - limited, must be an Engineer, Oiler, or Boiler Operator having at least four (4) years actual operating experience in the management, control, or operation of High Pressure steam boilers, or steam plants of not less than 100 horsepower (3,400,000 BTU). Up to four (4) years will be credited an applicant having previous equivalent training course in Mechanical Engineering at an accredited Engineering school.

Sec. 5-4-13. Boiler Operator - High Pressure.

An applicant for a High Pressure Boiler Operator's license must have at least one (1) year's actual operating experience under the supervision of a licensed Steam Engineer or Boiler Operator in the management, control, and operation of a high pressure steam boiler above fifteen (15) pounds operating pressure or a hot water boiler above thirty (30) pounds pressure of 160 degrees F.

Sec. 5-4-14. Exception for Boiler Operator experience requirements.

If an applicant for a High Pressure Boiler Operator's license has not had actual operating experience in the management,

Chapter 6 — License Requirements for States and Cities

control, or operation of boilers, a provisional license may be granted by the director for not more than one (1) year period, providing the applicant demonstrates sufficient knowledge and ability to safely and properly operate any boiler to which he may be assigned. A provisional Boiler Operator's license is subject to review of qualifications at any time at the discretion of the Building Official.

Sec. 5-4-15. Fees for examination and license.

The fees for Steam Engineer or Boiler Operator examination and license shall be as follows:

Steam Engineer - Unlimited.....................................$30.00

Steam Engineer - Limited..$25.00

Boiler Operator - High Pressure................................$25.00

Duplicate License...$10.00

Renewal of License prior to expiration......................$15.00

Renewal of License to one (1) year after expiration...$25.00

Late renewal fee $25.00 first year, $30.00 second year late, $35.00 third year late. Examination required thereafter.

Sec. 5-4-16. License granted upon passage of examination.

If said applicant, after examination by the Building Official, is found to have the requisite knowledge of mechanical equipment and experience to safely operate same and minimize fire, explosion, asphyxiation hazards, and smoke nuisance, said applicant shall be granted the appropriate license.

Sec. 5-4-17. License Boiler Operator certificate.

Each person who has successfully passed the required examination shall be issued a license certificate of the proper Grade.

License Requirements for States and Cities   Chapter 6

Sec. 5-4-18. Renewal of Boiler Operator license.

Any person who holds a license issued under the provisions of this title may have the same renewed annually without examination provided that said person pays the annual renewal fee as stated in section 5-4-15, above, and presents a receipt for the same from the City Treasurer. A new examination must be taken by any applicant who fails to renew his license within three (3) years from the expiration of said license. The expiration date of each license issued shall be the June 30 next following the date of issuance. All records of licenses and examination may be removed from Salt Lake City files after three (3) years from expiration date of license and destroyed.

Sec. 5-4-19. Powers granted by Boiler Operator license.

Steam Engineer - Unlimited. Persons holding a Steam Engineer - Unlimited license may take charge of and operate any steam or hot water plant.

Sec. 5-4-20. Steam Engineer - Limited.

Persons holding Steam Engineer, Limited license may take charge of and operate any steam or hot water plant not exceeding 300 horsepower (10,200,000 BTU).

Sec. 5-4-21. High Pressure Boiler Operator.

Persons holding High Pressure Boiler Operator license shall have the right to operate any steam or hot water boiler not exceeding 75 horsepower (2,555,000 BTU). Any High Pressure Boiler Operator operating high pressure boilers greater than 75 horsepower (2,555,000 BTU) shall be under the supervision of a qualified Steam Engineer.

Sec. 5-4-22. Exception for emergency operation.

In all cases of temporary disability from accident, sickness, or other cause, a Steam Engineer with a lower Grade of license, or without a license, may operate boilers and machinery of an Engineer of a high Grade, having charge of the plant for a period of not to exceed thirty (30) days, provided that the employer of the Steam Engineer so disabled shall notify the Building Official as soon as possible after the occurrence of such disability and an application is submitted for a provisional license pursuant to the provisions of Section 5-4-14 of this chapter.

Sec. 5-4-23. Duty of Building Official.

Each steam boiler plant and each hot water boiler plant shall have a person in charge who is in possession of a proper and valid license, as provided by this chapter, when said boiler plant is working under pressure. The Building Official shall have authority to order discontinuance of boiler operation for non-compliance of the provisions of this code.

Sec. 5-4-24. Duty of Boiler Operator licenses.

Each person licensed under this chapter shall display his license certificate in a conspicuous place in the boiler room at all times. It shall be the duty of each person licensed under this title to operate the boiler under his charge in a safe and proper manner in accordance with the provisions of the Uniform Mechanical Code as adopted by these revised ordinances and with standards approved by the Building Official. Copies of said approved standards shall be available for public inspections at the office of the Salt Lake City Division of Building and Housing Services.

Sec. 5-4-25. Reports required of Boiler Operator licenses.

It shall be the duty of every licensed Steam Engineer and Boiler Operator to report to the Building Official any defect, or any accident in any steam boiler, hot water boiler, stack, stoker, burners, furnace, smoke prevention device, absorption system, or appurtenances belonging thereto, any fire hazard

License Requirements for States and Cities   Chapter 6

coming under his care, or any condition of the plant in violation of the Uniform Mechanical Code, as adopted by Salt Lake City ordinances or contrary to the standards adopted by Section 5-4-24 of this chapter.

Sec. 5-4-26. Suspension, revocation, and reinstatement of Boiler Operator license.

The Building Official shall have the power to suspend or revoke the license of a Steam Engineer or Boiler Operator for permitting water to get too low in the boiler; for carrying a higher pressure of steam than allowed; for allowing or permitting a fire hazard to exist in any boiler room after being duly notified by the authority having jurisdiction over such hazard; for persistent violation of the air pollution code; for an unnecessary absence from his post of duty; for the excessive use of intoxicating liquors or other neglect or capacity; provided, however, that no license shall be suspended or revoked without first giving an accused person an opportunity to be heard in his own defense. When a license of an Engineer or Boiler Operator, shall be revoked, no license shall be issued to such person for ninety (90) days thereafter; and for any subsequent revocation, no license shall be issued to said person.

Vermont

No State or City License

Virginia

No State or City License

Chapter 6 License Requirements for States and Cities

Washington

Seattle, Washington

SUMMARY OF QUALIFICATIONS:

1. Exam Location, Address and Phone Number

Seattle Municipal Building
600 Fourth Avenue
Room 102
Seattle, Washington 98134
206/386-1298

2. Exam Days and Times

Written Test - Tuesday and Thursday - by appointment only.

Oral Test -   Friday only

3. Examination Cost

$20.00

4. License Cost and Renewal

Initial license.................$45.00

Renewal of license.......$45.00

5. Type of Test

Written and Oral

6. Grades of Licenses

Grade 1 Steam Engineer

Grade 2 Steam Engineer

Grade 3 Steam Engineer

Grade 4 Boiler Fireman

Grade 5 Boiler Fireman

License Requirements for States and Cities — Chapter 6

Small Power Boiler Fireman

7. Experience and Education Requirements

See Codes for Seattle below

8. U.S. Citizenship Required?

No

9. Local Residency Required?

No

10. Recognition of Licenses from other locales

Yes

DEPARTMENT OF LICENSES AND CONSUMER AFFAIRS

We have prepared this information sheet to help you prepare for your license exam. Now that you have decided to become a licensed Boiler Fireman or Engineer, you probably have many questions. This information sheet will provide some answers.

YOUR EXAM:

Our goal is to ensure the safe operation of boilers and steam engines in the City of Seattle. To do this, we want you to be familiar with the Seattle Steam Engineers' and Boiler Firemen's Law and understand the safe maintenance, care, and operation of steam equipment. By taking our license exam, you will be able to demonstrate your knowledge and understanding of steam engineering.

THE WRITTEN TEST:

The written test is designed to test your knowledge of equipment maintenance, care, and operation. The questions will be

Chapter 6 License Requirements for States and Cities

straightforward and can be answered in a few short sentences, or in some cases with a simple sketch. You may take the written test Tuesdays and Thursdays by appointment only, in Room 102, Seattle Municipal Building, 600 Fourth Avenue. After you complete the written test, you will be scheduled for an oral examination. Fridays are reserved for oral tests. The telephone number for information and appointments is 386-1298.

THE ORAL TEST:

The oral test is designed to further examine you knowledge of steam equipment operation, care and maintenance. You will have an opportunity to discuss the questions you answered in the written test as well as other aspects of steam plant operation and maintenance.

The people who will interview you have been especially selected for their knowledge of steam engineering and their ability to interview and assess applicants. They will not try to trick or embarrass you; rather, they want to find our just what you know so that you can get the highest license for which you are qualified. Please come to your interview at the scheduled time. Since exam time is limited, if you cannot make your appointment, please call 386-1298 and arrange for another time. We can then assign your appointment time to someone else.

CODES for Seattle, Washington

Steam Engineer and Boiler Fireman License Law

6.204.090 Steam Engineers and Boiler Firemen.

The annual fee for a Steam Engineers' and Boiler Firemen's license shall be: Steam Engineers and Boiler Firemen (all grades) $35.00.

6.230.010 Scope.

License Requirements for States and Cities Chapter 6

The regulation and licensing of Steam Engineers and Boiler Firemen in this chapter and the regulations relating to the operation of boilers and steam engines as defined in this chapter provide the means for ensuring safe operation of such boilers and steam engines.

6.230.020 Application of other provisions.

The licenses, endorsements and certificates provided for in this chapter are subject to the general provisions of the new Seattle license code set forth in Chapter 6.202 as now or hereafter amended. In the event of a conflict between the provisions of Chapter 6.202 and this chapter, the provisions of this chapter shall control.

6.230.040 License required - Expiration.

It is unlawful to have charge of, or operate or permit anyone to have charge of, or operate, any boiler or steam engine without a license, endorsement, or certificate to do so issued by the Director under this chapter. All licenses, endorsements, and certificates shall expire at midnight on the thirtieth day of September of each year, and shall not be transferred or assigned. All renewals shall specify the same Grade and be subject to such conditions or limitations as may be provided under the license to be renewed. Renewal of a license which has been expired for more than one (1) year requires the holder to submit a new application and to be re-examined.

6.230.050. Exemptions from license requirements.

A steam engineer's or boiler fireman's license shall not be required of any person in charge of, or operating, the following:

A. Any boiler or steam engine subject to federal regulations;

B. Heating boilers in group R1 occupancies of less than six (6) units, group R3 occupancies and group M occupancies as

Chapter 6 **License Requirements for States and Cities**

defined in the Seattle Building Code as now or hereafter amended;

C. Low-pressure hot water, low-pressure steam and hotwater supply boiler plants having inputs of less than two million five hundred thousand (2,500,000) BTU per hour:

D. Any boiler having an input of less than one hundred thousand (100,000) BTU per hour and a maximum pressure of one hundred pounds per square inch (100) psi or less;

E. Water heaters,

6.230.060 Grades of licenses.

A. The Grades of Steam Engineers' and Boiler Firemen's licenses shall be as follows:

Grade One.......... (1) Steam Engineer

Grade Two.......... (II) Steam Engineer

Grade Three........(III) Steam Engineer

Grade Four..........(IV) Boiler Fireman

Grade Five...........(V) Boiler Fireman

Grade Small Power.....Boiler Fireman

B. The scope of each grade of license as related to the type of equipment and capacity subject to any limitations or conditions imposed pursuant to SMC Section 6.230.100 shall be as set forth in the following table:

License Requirements for States and Cities — Chapter 6

Maximum Capacity Allowable For Grades of License

Grade of License	Steam Engine (brake horsepower)	Boilers (BTU/hr. input total)	Small Power Boiler
Grade I (a) Steam Engineer	Unlimited	Unlimited	Unlimited
Grade II (a) Steam Engineer	1,500	300,000,000	(f)
Grade III (a) Steam Engineer	250	50,000,000	(f)
Grade IV (b) Steam Engineer	0	20,000,000	(b)
Grade V (c) Steam Engineer	0	5,000,000	(d)
Small Power Boiler Fireman	0	0	Less than 350,000 BTU/hr output

Notes to table:

Chapter 6 **License Requirements for States and Cities**

(a) A boiler supervisor endorsement on a Steam Engineer license shall permit the licensee to supervise automatic boilers up to the combined capacity of each individual boiler plant permitted by his/her Steam Engineer license.

(b) A Grade IV boiler Fireman may operate a battery of not more than two (2) boilers with a combined capacity not greater than twenty million (20,000,000) BTU per hour total input; except, that when he/she is the head Fireman on duty and under the direct supervision of a licensed Steam Engineer hereunder, he/she may operate boilers with greater capacity but not to exceed the capacity permitted by the license of such supervising Engineer.

(c) A Grade V boiler Fireman may operate one (1) low-pressure hot-water heating boiler or one (1) low-pressure steam-heating boiler and may also operate one (1) electric small power boiler limited to one hundred pounds per square inch (100 psi) and not exceeding two hundred kilowatts (200 kw) per hour input.

(d) A small power endorsement on a Grade V boiler Fireman license shall permit the licensee to operate one (1) small power boiler.

(e) For license determination purposes, BTU per hour input ratings of a boiler shall be computed equal to burner input.

(f) Grades II and III Steam Engineer may operate any number of small power boilers provided that the combined capacity of all boilers operated does not exceed the allowable limits as stated on his/her license.

6.230.070 Issuance of licenses.

Persons desiring a license or certificate described in Section 6.230.060 Shall make written application to the Department of

License Requirements for States and Cities Chapter 6

Licenses and Consumer Affairs on the forms provided by the Department. Such application shall include the applicant's full name and address, and if the applicant is an employee, the name of his/her employer. Applications shall be accompanied by a receipt showing payment to the City Treasurer of the required examination fee as provided under Section 6.230.090.

  A. Applicants for a Steam Engineer's license, Grade I, II, or III shall show by competent evidence one (1) of the following:

  1. That he/she has been employed at least three (3) years in a position directly responsible for the care and operation of boilers or steam engines, or in the design or supervision of boilers, boiler systems, boiler firing and automatic control and safety systems, or under the direct supervision of a licensed Steam Engineer, Grade I, II or III; or

  2. That he/she has at least three (3) years of practical experience as a machinist apprentice in a steam engine works together with one (1) year of employment in the direct care and operation of boilers and steam engines; or

  3. That he/she has graduated from a recognized school of technology and has had at least one year of employment in the direct care and operation of boilers and steam engines.

  Completion of a Boiler Fireman's course approved by the Department of Licenses and Consumer Affairs shall be the equivalent of one (1) year of practical experience under subsections 1 and 2 above, however, each applicant will be entitled to only one such credit.

  B. Any licensed Grade I, II, or III Steam Engineer may apply for a boiler supervisor endorsement. Such applicant shall show by competent evidence that he/she has been employed at least three (3) years in one (1) of the following:

1. In a position directly responsible for the care or operation of boilers, or steam engines.

2. In the design or supervision of boilers, boiler systems, boiler firing, and automatic control and safety systems.

3. In the direct supervision of a licensed Grade I, II or III Steam Engineer.

C. Any professional Engineer licensed in the State of Washington pursuant to RCW Chapter 18.43 may apply for a boiler supervisor certificate. A boiler supervisor certificate shall permit the holder to supervise, but not operate, automatic boilers without any limitation as to boiler equipment or capacity thereof. Such applicant must meet the requirements of Section 6.230.070 B.

D. Applicants for a Grade IV Boiler Fireman license shall show by competent evidence one (1) of the following:

1. One (1) year of practical experience in the care and operation of a boiler, or

2. Completion of an in-service training course in the fundamentals of boiler operation as approved by the Department of Licenses and Consumer Affairs which shall include at least forty (40) hours of classroom work together with eighty (80) hours of on-site training or instruction relating to the care and operation of boilers under the direct supervision of a Steam Engineer with a license of Grade I, II or III, without limitations.

E. Applicants for a Grade III Steam Engineer License limited to hoist and portable boilers, shall show by competent evidence, one (1) of the following:

1. Three (3) years of practical experience in the care and operation of boilers and steam engines; or

2. Completion of an in-service training course on the fundamentals of boiler operation, as approved by the Department of Licenses and Consumer Affairs which shall include fifty-five (55) hours of classroom work, together with one hundred twenty (120) hours of work relating to the care and operation of a minimum of two separately located hoist and portable boilers, under the direct supervision of a Steam Engineer with a license of Grade I, II or III.

F. All persons applying for a license or certificate under this Chapter shall be examined by the Department of Licenses and Consumer Affairs according to the provisions of Section 6.230.100. Upon determination by the Department of Licenses and Consumer Affairs that the applicant has passed the applicable examination and is otherwise qualified under this chapter, including payment by the applicant of the license fee, the Director of Licenses and Consumer Affairs shall issue the license, certificate or in the case of application for a boiler supervisor endorsement, endorse the words "boiler supervisor" on the applicant's Steam Engineer license.

6.230.080 Special license.

Any person having been employed at least two years as a licensed Steam Engineer or Boiler Fireman operating any boiler plant the capacity of which is enlarged or changed beyond the limits of his/her license may apply to the Department of Licenses and Consumer Affairs for a special license with the limits extended to apply only to such plant. The Director of Licenses and Consumer Affairs, upon receipt of such application, shall forward the application to the Director of the Department of Construction and Land Use who shall make an investigation of the changed boiler plant conditions together with such examination of the applicant as may be necessary to determine whether the applicant is qualified under the provisions of this chapter to operate such enlarged or changed boiler plant. When such investigation and examination reveals that the applicant is qualified to operate such plant in its changed condition, the Director of Licenses and Consumer Affairs shall certify approval of the application and issue such special license.

6.230.090. Examination fees for licenses.

The examination fee for a license or certificate described in this chapter shall be Twenty dollars ($20.00).

6.230.100. Department of License and Consumer Affairs.

A. In connection with the regulation and licensing of Steam Engineers and Boiler Firemen, the Department of Licenses and Consumer Affairs is authorized to perform the following:

1. Provide qualifying examinations for persons applying for Steam Engineer or Boiler Fireman licenses under this chapter. Such examinations shall be practical in their character and shall relate to those matters which will fairly test the capacity, skill, experience, and habits of sobriety of each person examined to safely operate and properly care for a boiler and/or steam engine, within the scope of the license sought.

2. Provide qualifying examinations for persons applying for a boiler supervisor endorsement or a boiler supervisor certificate as described in this chapter. Such examination shall be practical in character and shall relate to those matters which will fairly test the applicant's capacity, skill, experience, and habits of sobriety to safely use, operate, and maintain boilers and automatic boilers under applicable city and state regulations.

3. When approving any license under this chapter, the Director of Licenses and Consumer Affairs may impose stated conditions or limitations to such license restricting the licensee to

the operation and maintenance of particular equipment at a stated location, or to the operation and maintenance of a certain class of boilers or steam engines, or to specified permitted services in connection with the operation and maintenance of boilers and steam engines. Such restrictions shall be based upon the applicant's qualifications under this chapter and be reasonably related to the protection of the public in the safe operation and maintenance of boilers and steam engines.

B. The Department of Licenses and Consumer Affairs may require affidavits regarding an applicant's character, training, experience and record, and such other supporting credentials as may be necessary to determine his/her fitness.

C. The Department of Licenses and Consumer Affairs shall refuse to certify the applicant for a Steam Engineer's or Boiler Fireman's license if the result of the examination is such that the Department of Licenses and Consumer Affairs determines he/she has not sufficient knowledge of, and experience in, the care or operation of boilers or steam engines, or if the applicant is found to be mentally or otherwise unfit to safely operate boilers or steam engines. The action of the Department of Licenses and Consumer Affairs shall be final.

D. It shall be sufficient cause to refuse an original Steam Engineer's or Boiler Fireman's license, or any renewal thereof, if the applicant, through neglect or incompetency while in charge of a boiler or steam engine, has caused serious damage to property or has endangered the lives of others.

6.230.110. Licenses to be posted or carried.

Every licensed Steam Engineer or Boiler Fireman on duty shall display his/her license in a conspicuous place in the room wherein the boiler or steam engine is located, and such shall be effective only for the operation of the plant where it is displayed. If the licensee is in charge of, or is operating, a portable boiler or steam engine and the posting of his/her license is not practicable, such license shall be carried on his/her person, and on demand he/she shall exhibit same. A boiler supervisor shall display a legible copy of his/her license or certificate in the boiler room of each boiler that he/she supervises.

6.230.120 Notice of place of employment.

Every licensed Steam Engineer or Boiler Fireman shall keep the Boiler Operations Inspector for the City informed of any change of place of employment. Notice shall be given within twenty-four (24) hours after leaving and/or accepting a position. Such notice shall be in writing, addressed to the Boiler Operation's Inspector, 1st Floor, Seattle Municipal Building, Seattle, Washington, giving the licensee's name, number and Grade of license and the name and address of the plant of last employment and of new employment.

6.230.130 Reporting of defective boilers.

A. Every licensed Steam Engineer or Boiler Fireman before operating any boiler shall first examine the boiler permit issued for such boiler or boilers to see that the permit is in force, and if the permit has expired he/she shall notify his/her employer. If the permit has been expired for more than ninety (90) days, he/she shall notify his/her employer and then the City Boiler Inspector of the date of expiration. He/she shall note the pressure allowed by the permit and shall test the operation of the boiler and its control and safety devices for proper operation.

B. Whenever the Steam Engineer or boiler Fireman believes any part of a boiler or steam engine to be in defective or potentially unsafe condition, he/she shall report the fact to his/her employer in writing. If immediate corrective action is not taken, he/she shall report such defective or potentially unsafe conditions to the City Boiler Inspector.

C. The City Boiler Inspector shall thereupon investigate the same, and report any lack of proper care on the part of any licensed person to the employer and the Department of Licenses and Consumer Affairs. The Department of Licenses and Consumer Affairs shall record the facts on the records of the licensee.

D. The Steam Engineer or Boiler Fireman in charge of any boiler or steam engine shall report to his/her employer and to the City Boiler Inspector any damage or injury to any such boiler or steam engine under his/her charge or care which affects the safe operation of such boiler or steam engine. Failure to make such reports to his/her employer and the City Boiler Inspector shall be sufficient cause for the suspension or revocation of the license of the person in charge.

E. It shall be the duty of every licensed Steam Engineer and Boiler Fireman to report serious negligence in the care of boilers and steam engines to his/her employer and the Boiler Operations Inspector.

6.230.140 Duties of Steam Engineers and Boiler Firemen.

A licensed Steam Engineer and Boiler Fireman shall perform the following duties in connection with his/her operation and maintenance of boilers and steam engines;

A. Test the operation of the boiler and its control and safety devices periodically on a routine basis in accordance with nationally recognized standards and/or boiler and control manufacturer's written recommendations;

B. Maintain and operate the equipment in a safe manner and according to nationally recognized standards such as those rec-

| Chapter 6 | License Requirements for States and Cities |

ommended by the American Society of Mechanical Engineers for boilers and as adopted by the City Steam License Advisory Board. Such standards shall be filed with the City Comptroller;

C. Prepare and maintain a boiler log book and record, at least daily, such pertinent boiler readings and data as may be required by the Boiler Operations Inspector and/or the senior license holder or other authorized person in charge of the boiler operation.

5TH GRADE - STUDY OUTLINE FOR SMALL BOILER FIREMAN

I. Describe the boiler you are now firing including how it is fired, the type of heat source (natural gas, fuel oil, electric power, or a combination), type of flame failure device protection, type of main boiler outlet valve, how vented, location and type of safety valve, and how water is heated in the boiler (fire tubes, water coils, etc.) and the bottom type blow or drain.

II. Discuss the effect of low water in a steam generating boiler fired by fuel oil or natural gas.

III. Discuss the effect of low water in an electric element heated boiler.

IV. Discuss what you will do if you observe overpressure on the pressure gage.

V. Discuss what you will do if you observe no water level showing in the gage glass and you do not know how long the water has been out of sight.

VI. Describe how to test your flame failure device for proper operation.

License Requirements for States and Cities　　　　**Chapter 6**

VII. Describe how you start the boiler from checking the boiler, water level, water supply, vent, purging, feedwater pump, etc. When and how should you test the safety valve?

VIII. Discuss means for preventing corrosion inside the boiler and in firebox, scale in the boiler, the buildup of dissolved and non-dissolved solids in the boiler water and control of alkalinity in the boiler.

IX. How do you prevent and extinguish paper or wood, oil and electric fires.

X. Discuss the explosion potential and possible damage that could result from a boiler explosion. Note that one pound of steam requires over 1600 times as much space as one pound of water at atmospheric pressure.

XI. Discuss proper care and maintenance of boiler controls, water level control, low water cut-out, and safety valve.

XII. State who issues boiler permits and operators license and list information shown on the boiler permit.

XIII. Knowledge of the City of Seattle "Steam Engineer and Boiler Fireman License Law." Section 6.204.090 and Chapter 6.230 of the Seattle Municipal Code.

1. Definitions

2. License required.

3. Maximum capacity allowed for Grade V.

4. License to be posted.

5. Notice of place of employment.

6. Reporting defective boilers.

Chapter 6   License Requirements for States and Cities

7. Duties of boiler fireman.

8. Observation and inspection of boiler.

9. Enforcement

10. Posting of regulation

4TH GRADE - STUDY OUTLINE FOR BEGINNING STEAM FIREMAN

I. Types of valves found on and around boilers. Describe and tell where they are used.

1. Gate Valves. Rising Stem. Rising Valve.

2. Globe Valves. Rising Stem. Rising Valve.

3. Non-Return Valves.

4. Check Valves.

5. Needle Valves.

6. Pressure Reducing Valve. (PRV)

7. Balance Valves

8. Plug Valves. Pet Cocks.

9. Quick Closing Valves. "Guillotine like"

10. Seatless Blowdown Valves.

11. Ball Valves.

12. Motorized Valves.

13. Solenoid Operated Valves.

License Requirements for States and Cities — Chapter 6

II. Neatly sketch and label a fire tube boiler showing at least the following parts.

1. Shell and or drum.
2. Fire Tubes.
3. Stay Tubes.
4. Furnace.
5. Stays. Through and Diagonal.
6. Dampers.
7. Baffles in some cases.
8. Breeching.
9. Stack Connection.
10. Safety Valve.
11. Boiler Stop Valve.
12. Header Stop Valve.
13. Feedwater Connection. Check Valve. Proper Use of Globe Valve.
14. Vent Valve.
15. Gage Glass With Valve Connection.
16. Water Column.
17. Try Cock.
18. Gage Glass Drains.
19. Water Column Drain.
20. Bottom Boiler Blowdown Connection.

Chapter 6   License Requirements for States and Cities

21. Surface Blow. Continuous Blow.

22. Steam Gauge and Test Gauge Connection with Pig Tail.

23. High Pressure Cut Out with Manual Reset.

24. High pressure Cut Out - Recycling.

25. Modulating Load Control.

26. Water Level Control. Choice of Float, Thermostatic, or Vapor Pressure Types.

27. Flame Failure Device.

III. Describe circulation in a steaming boiler to include the following:

1. The effect of the water level when a normal load is applied to include what is actually taking place with the water and steam generation in the boiler. Take the description step by step from the application of the load, the resultant drop in boiler pressure, the immediate reaction in the boiler and the follow-up action of the combustion controls and feedwater controls to meet the new load demands.

2. Consider the same steps when the load is dropped.

3. Describe the danger caused by loading a steaming boiler too rapidly.

4. Describe the action in the steaming boiler if the feedwater and fuel were shut off immediately - say a power outage or a complete securing of the system.

NOTE: In the discussion describe what is actually physically taking place in the water and steam in the boiler in both the liquid and steam bubble forming parts of the boiler.

IV. Discuss the indications of faulty gage glass readings. Tell how the "water bounce in the gage glass" indicates possible pluggage. Give step by step procedure to properly clear connections to the gage glass. Discuss what will happen if the top valve to the gage glass is closed. Describe proper procedure to replace a gage glass on a steaming boiler to include safety precautions.

V. Define an interlock and a relay and a sample of their use in boiler controls.

VI. List the steps to follow in putting a boiler on the line starting with an empty boiler.

VII. Discuss gage pressure, vacuum pressure, furnace draft pressure, and absolute pressure measurements. What are the terms or units and what is their relationship to atmospheric pressure.

VIII. Describe a water level control to include how the level is determined and how this information gets from the level device to the point of use. The following lists some devices.

1. Float (McDonnel Miller)

2. Thermostatic (Copes)

3. Vapor Pressure (Bailey)

4. Differential Pressure (DP Cell)

IX. Describe the operation of the following flame failure indicating devices:

1. Lead Sulfide Cell. (Fireye)

2. Flame Rod. (Rectification by Flame)

3. Ultra Violet Flame. (UV Cell)

Chapter 6   **License Requirements for States and Cities**

X. Define the following heat energy terms.

1. Sensible heat. How measured.

2. Latent Heat. How measured and relationship to pressure.

3. British Thermal Unit. (BTU)

4. Therms.

XI. Give examples of the following heat transfer items.

1. Radiation

2. Convection

3. Conduction

XII. Steam Characteristics.

1. Boiling point temperature. Discuss the effect of pressure.

2. Steam volume per lb. Discuss the effect of pressure and a comparison with a pound of water under the same conditions.

3. Note the change in BTU's per pound of steam at different pressures.

4. Describe superheated steam and discuss reasons for superheating steam.

5. Define steam quality.

XIII. Fuel commonly used in Seattle.

1. Fuel oil grades and BTU's per lb.

2. Natural Gas. BTU's per cu. ft.

IX. Furnace draft and combustion air. Discuss how it is obtained.

License Requirements for States and Cities — Chapter 6

1. Stack
2. Induced draft fan
3. Forced draft fan
4. Dampers
5. Balanced draft

X. Furnace Pressure.

1. Describe means for measuring.
2. Means for controlling.
3. Discuss a negative pressure and a positive pressure furnace.

XI. Describe natural gas and fuel oil supply systems.

1. Oxygen in atmosphere by volume is 21% and nitrogen is 79%.
2. Oxygen is active ingredient which combines with carbon (C), hydrogen (H_2), and sulphur (S) to release heat.
3. The reason for maintaining proper fuel air ratio to obtain efficient combustion. The effect of too little air and too much excess air.
4. Products of combustion. Carbon monoxide (CO); Carbon dioxide (CO_2); water (H_2O); sulphur dioxide (SO_2); sulphur trioxide (SO_3); nitrogen oxides (NOX); water (H_2O).
5. Good combustion requirements.

Time-temperature-turbulence

6. Means of determining good combustion.

Stack observation. O_2 meter. CO_2 meter. Flame observation. Ringleman chart.

XIII. Means of ignition.

1. Hand torch
2. High voltage electric spark
3. Pilot gas

XIV. Proper boiler and furnace maintenance.

1. Periodic cleaning of water side.

A. Proper feedwater treatment.

B. Bottom blow.

C. Steam drum continuous blow.

2. Fire side.

A. Brick and baffle maintenance.

B. Thorough tube cleaning.

C. Proper soot blower adjustment.

XV. Proper safety procedures to follow:

1. Describe action in case of boiler overpressure.

2. Discuss action to take in case of low water.

3. Discuss action to take to reduce the chance of a furnace explosion.

4. Discuss safety valve testing for the boiler.

XVI. Control system using changes in boiler pressure.

1. On and off control.

2. Modulating Control.

A. Modulating device for sensing boiler pressure changes.

B. Modulating motor controlled by modulating device.

The motor controls the fuel feed and air damper according to demand by means of cams and levers.

XVII. Discuss the location and use of safety and limiting devices on the boiler/furnace.

1. Manual reset overpressure sensing device.

2. Recycling overpressure control (On-Off Control)

3. Modulating boiler pressure sensing load for combustion control.

4. Low water cutout - recycling.

5. Low water cutout - manual reset.

6. Hi-gas pressure shut down with manual reset.

7. Low-gas pressure shut down with manual reset.

8. Blower proving sensor.

9. Furnace pressure sensing for positive furnace pressure.

10. Low fuel oil temperature on some furnaces.

11. Low fuel oil pressure.

12. Flame failure.

13. Programmed purge.

XVIII. Describe a boiler feedwater cycle including the following:

Chapter 6   License Requirements for States and Cities

1. The hot well and hot well pumps.

2. The deaerating feedwater heaters.

3. The centrifugal pump and include proper starting procedure proper hook-up and discuss the necessity of a recirculation line.

4. The basic principle of a duplex and simplex feedwater pump.

XIX. Miscellaneous.

1. Discuss water hammer and ways that it can be caused.

2. Name different types of steam and water traps and where used.

XX. Information on boiler permit and office of issue.

1. Working pressure of the boiler.

2. Location or address of boiler.

3. Grade of operator license required to operate the boiler.

4. Kind of boiler.

5. Boiler Number.

XXI. Knowledge of the City of Seattle "Steam Engineer and Boiler Fireman License Law." Section 6.204.090 and Chapter 6.230 of the Seattle Municipal Code.

1. Definitions.

2. License required and expiration.

3. Exemptions from license requirements.

4. Grades of licenses.

5. Maximum capacity allowable for grades of licenses.

6. Issuance of licenses. Method and requirements.

7. Special license.

8. Department of licenses and consumer affairs.

9. Licenses to be posted or carried.

10. Notice of place of employment.

11. Reporting of defective boilers.

12. Duties of steam engineers and boiler fireman.

13. Observation and inspection of boilers.

14. Posting of regulations.

ec, 5/89

opeiu8 afl/cio

3RD GRADE - STUDY OUTLINE FOR INTERMEDIATE STEAM ENGINEER

NOTE: THE APPLICANT SHOULD BE INFORMED OF THE INFORMATION RECOMMENDED FOR THE BEGINNING STEAM FIREMAN AS WELL AS THE FOLLOWING:

I. Sketch a fire tube boiler. See study outline for beginning steam fireman.

II. Sketch a complete heavy grade fuel oil burning system including the following:

1. Tank.

Chapter 6 — License Requirements for States and Cities

   2. Vent.

   3. Tank Heater.

   4. Return Line.

   5. Full Flow Relief Valves.

   6. Heater After Fuel Oil Pumps.

   7. Fuel Control Valves.

   8. Mechanical Atomizing Burners.

   9. Steam Atomizing Burners.

      A. Constant Differential Pressure Valve.

   10. Temperature Interlock.

   11. Pressure Interlock.

III. Sketch complete natural gas fuel firing system to include the following:

   1. Gas Pressure Reducing Valve With Vent.

   2. Hi-Gas Pressure Sensing and Interlock.

   3. Low-Gas Pressure Sensing and Interlock.

   4. Gas Quick Closing Valves.

   5. Vent Line and Valve.

   6. Cock or Plug Valves.

   7. Gas Burner.

   8. Flame Failure Device.

   9. Gas Fuel Control Valve.

Seattle, Washington

License Requirements for States and Cities — Chapter 6

10. Gas Control With Valve Closed Position Proving Micro Switch.

IV. Sketch a feedwater system for two or more boilers from hot well or source of supply to boiler showing types of protective valves and means of water level control in the boilers.

V. Describe complete operation of the following steam traps:

1. The inverted bucket trap and explain how the size of the discharge orifice is related to the trap; input or line pressure.

2. Impulse Trap.

3. Bucket Trap.

4. Disc Trap.

5. Thermostatic Trap.

6. Float Operated Trap.

VI. Discuss the difference between alternating current, direct current, and the phases of alternating current used in the boiler room. Include the voltages involved with the controls, motors, and ignition spark igniters. Discuss overcurrent or electric short protection.

VII. Describe generally accepted current feedwater treatment procedures for control of scale forming water hardness, reduce oxidation or rusting, control of acidity or alkalinity, control of the concentration of dissolved and undissolved solids in the boiler water, helping undissolved material to stay in suspension, and a means to prevent corrosion in the heating system.

VIII. Discuss means to improve the quality and purity of steam. This includes dry pipes, dryers, scrubbers, and baffles in the steam drum.

Chapter 6   License Requirements for States and Cities

IX. Describe a positive displacement and a centrifugal pump and where each may be found in the boiler room.

X. Describe procedure to protect water side and fireside when securing a boiler for an extended period of time.

XI. Discuss the differences between fire tube and a water tube boiler and describe briefly an example of each. Include construction details, means of support, superheaters, air preheaters, and economizers.

XII. Explain the differences between a "pop" safety valve, a relief valve, and a rupture disk to include the reason for the "popping" action of the pop safety valve and the blow down reseating of the same type of valve.

XIII. Describe how an injector pump works including possible malfunctions and where pumps are used.

XIV. Describe the importance of furnace draft and means of obtaining a balanced draft. Include the difference between a negative and positive pressure furnace.

XV. Describe a deaerating feedwater heater to include location with respect to the feedwater pump, need for venting, operation, how fed, how water level is controlled and reason for having a vacuum breaker.

XVI. Discuss necessity for soot blower, how operated, how installed, and proper alignment.

XVII. Discuss Class A, B, and C fire extinguisher and types of fires that the extinguisher are used for:

XVIII. Describe the combustion process and what is meant by fuel air ratio to include what a good minimum excess air percentage by volume is for burning natural gas, fuel oil and coal. Include the means used to determine the amount of excess air.

XIX. Describe procedures for boiler/furnace maintenance to include cleaning inside and outside, testing for leaks, examination inside for corrosion, inspection for weakness, feedwater valves and connections, surface blow, dry pipes, baffles, water column lines, safety valves, vent valve connection, gage connection, cleaning fireside of tubes and tube sheets, fireside corrosion, pitting, grooving, fire cracks, bulging, bagging, blisters, etc.

XX. Discuss proper start-up of auxiliary steam turbines and describe governing, overspeed trip, types of bearings, lubrication, thrust bearings, governing when used to drive a feedwater pump, etc. when used to drive a centrifugal feedwater pump to a battery of boilers and in parallel with other feedwater pumps, discuss how the turbine driving centrifugal feedwater pump should be piped, how the pump is brought on the line and precautions that must be taken when loading the pumps and running them in parallel.

XXI. Discuss running continuously running air compressors to include starting, means of unloading, cooling requirements, water traps, intake filters, air relief valves, and safety alarm devices.

XXII. Discuss steam tables including a comparison of steam and water density at different pressure up to "critical pressure" and what takes place at that point. Discuss the difference in volume between one pound of water and one pound of steam from atmospheric pressure on up. Know the temperature and pressure at critical pressure. Discuss the enthalpy of water, the enthalpy (latent heat) of evaporation, and the temperature of saturated steam at various gage pressures from atmospheric

Chapter 6   **License Requirements for States and Cities**

pressure to critical pressure. Explain how these characteristics effect the circulation and generation of steam in the boiler up to and past critical pressure. Discuss superheat, describing what it is and why it is used in a power plant. Describe what happens to the heat stored in water at the boiling point temperature when the boiler pressure is reduced by an additional load being placed on the boiler and there is a natural lowering of boiling point temperature. Explain how this and the lower pressure on the steam bubbles effects the water levels in the boiler.

XXIII. Knowledge of the City of Seattle "Steam Engineer and Boiler Fireman License Law." Section 6.204.090 and chapter 6.230 of the Seattle Municipal Code.

1. Definitions.
2. License Required and Expiration.
3. Exemptions from License Requirements.
4. Grades of Licenses.
5. Maximum Capacity Allowable for Grades of Licenses.
6. Issuance of Licenses. Method and Requirements.
7. Special Licenses.
8. Department of Licenses and Consumer Affairs.
9. Licenses to be posted or carried.
10. Notice of place of employment.
11. Reporting of Defective Boilers.
12. Duties of Steam Engineers and Boiler Fireman.
13. Observation and Inspection of Boilers.

License Requirements for States and Cities   Chapter 6

14. Posting of Regulations.

EC; 5/89

opeiu8 afl/cio

1ST and 2ND - STUDY GUIDE FOR ADVANCED STEAM ENGINEER

NOTE: THE APPLICANT SHOULD BE INFORMED OF THE INFORMATION RECOMMENDED FOR THE BEGINNING STEAM FIREMAN AND THE INTERMEDIATE STEAM ENGINEER AS WELL AS THE FOLLOWING:

I. Sketch a water tube boiler to indicate the following:

1. Support.

2. Tube Banks and Water Walls.

3. Tube Headers.

4. Water Column.

5. Gage Glass.

6. Vent.

7. Boiler Steam Outlet Valve and Header Valve with Drain Connection.

8. Safety Valve(s).

9. Soot Blowers.

10. Baffles, and Flue Gas Flow.

11. Furnace Temperature at Burners, Furnace, Passes and Discharge.

Chapter 6   License Requirements for States and Cities

12. Air Preheaters.

13. Economizer.

14. Superheaters

15. Feedwater Connections to include Control Valve, Check Valve, and Supply or Isolation Valve.

16. Surface Blow Line.

17. Bottom Blow.

18. Breeching.

19. Location of Burners.

II. Step by step procedure to start a condensing steam turbine generator from completely down and cold to synchronizing the generator with an active bus and loading the generator..

III. Discuss the difference between loading an alternating current (AC) turbine-generator paralleled on the line and one not paralleled but carrying the complete connected load. Include the use and effect of the governor control and the voltage regulator in both cases.

IV. Discuss the difference between an impulse and reaction turbine, an extraction turbine, back pressure turbine, reheat turbine, compound turbine, and a re-entry turbine. Also general construction, dummy pistons, labyrinth rings, bearings, Kingsbury thrust bearings, lubrication, steam admission control, overspeed trip, and speed/load control. And, in addition, consider safety shut down controls involving electrical faults, low oil pressure, or overspeed trip-out.

V. Discuss feedwater treatment and means or tests to determine proper levels of treatment for desired results.

VI. Describe requirements to be met and how to conduct a hydrostatic test. Also discuss how to conduct an accumulation test and give the reason for requiring such a test.

VII. Discuss the relationship of current, voltage, resistance (IR=E), power (EI) or I^2R), alternating current (AC), phases, direct current (DC), magnetic circuit and relationship to AC and DC current, transformers, DC magnetic coils, generator excitation, how AC and DC current is generated, AC single phase and three phase motors, how rotation direction is controlled in single phase and three phase motors-AC, rotation direction obtained in DC motors, the use of collector rings, slip rings, commutators, induction, power factor, circuit breakers, time over-current relays, fuses, circuit breakers, differential relays, low voltage relays, process of synchronism, meggers, etc.

VIII. Discuss the operation and troubleshooting a continuously operating 500 SCFM or larger air compressor to include: loading control, temperature control, moisture removal, safety devices, air valves and determining faulty suction or exhaust valves. Include intake filters, aftercooler, intercooler, and receiver.

IX. Discuss steam tables and the use of mollier charts to graphically illustrate the characteristics of steam and water. Describe what is involved at critical pressure and temperature latent heat, sensible heat, superheat, steam density water density enthalpy of liquid, enthalpy of evaporation, enthalpy of superheated steam and changes brought by changes of pressure.

X. Discuss barometric condensers and surface condensers, the advantage of using a condenser on a steam turbine, and how the air is removed from a condenser.

XI. Discuss the combustion control process giving the requirements for efficient combustion and how it is adapted to natural gas or fuel oil of all grades. Include an explanation of metered controls to obtain proper fuel-air ratio throughout boiler load range.

XII. Define the following terms related to steam engineering.

1. British Thermal Unit (BTU)
2. Foot Pound.
3. Latent Heat.
4. Specific Gravity.
5. Mechanical Equivalent.
6. Inertia.
7. Convection.
8. Matter.
9. Momentum.
10. Siphon.
11. Clearance Pockets.
12. Intercooler.
13. Aftercooler.
14. Horsepower of Engine.
15. Axial Flow.
16. Dynamic.
17. Static.

18. Alkaline.

19. Atomic Weight.

20. Electron.

21. Ampere.

22. Volt.

23. Resistance.

24. Power.

25. Barometer.

26. Hydrometer.

27. Boyles Law.

28. Capillary.

29. Caustic Soda.

30. Oscillation.

31. Center of Gravity.

32. Centrifugal Force.

33. Density.

34. Back Pressure.

35. Enthalpy.

XIII. Discuss feedwater heater maintenance, operation and uses for deaerating and closed types.

XIV. Discuss ingredients and qualities of lubricating oils to include how they are filtered, applied and stored.

Chapter 6	License Requirements for States and Cities

XV. Discuss the various types of desuperheaters, location, and need.

XVI. Knowledge of the City of Seattle "Steam Engineer and Boiler Fireman License Law" Section 6.204.090 and Chapter 6.230 of the Seattle Municipal Code.

1. Definitions
2. License Requirements and Expiration.
3. Exemptions from License Requirements.
4. Grades of Licenses.
5. Maximum Capacity Allowable for Grades and Licenses.
6. Issuance of Licenses. Method and Requirements.
7. Special License.
8. Department of Licenses and Consumer Affairs.
9. Licenses to be Posted or Carried.
10. Notice of Place of Employment.
11. Reporting of Defective Boilers.
12. Duties of Steam Engineers and Boiler Fireman.
13. Observation and Inspection of Boilers.
14. Posting of Regulations.

XVII. Sketch a pressure reducing station showing all valves, gages, and safety devices. Discuss the relationship between the entering steam and the leaving steam.

ec;5/89

opeiu8 afl/cio

Spokane, Washington

SUMMARY OF QUALIFICATIONS:

1. Exam Location, Address and Phone Number

City of Spokane
City Hall
808 W. Spokane Falls Blvd.
Spokane, Washington 99201-3325
509/625-6102

2. Exam Days and Times

2nd Thursday of every month

3. Examination Cost

$24.00

4. License Cost and Renewal

First Class Engineer...............$48.00

Second Class Engineer..........$36.00

Third Class Engineer.............$30.00

Fireman..................................$24.00

5. Type of Test

All Written

6. Grades of Licenses

First Class Engineer

Second Class Engineer

Third Class Engineer

Fireman

7. Experience and Education Requirements

Chapter 6 License Requirements for States and Cities

First Class Engineer - 2 years practical experience (18 months in high pressure plant or 1 year of graduate school)

Second Class Engineer - 1 year practical experience in High Pressure plant

Third Class Engineer - 1 year experience as Fireman (must renew every year)

Fireman - 18 years old (no experience required)

8. U.S. Citizenship Required?

No

9. Local Residency Required?

No

10. Recognition of Licenses from other locales

Up to the Boiler Examiner

CODES for Spokane, Washington

10.29.020 Boilers - Operator's License.

No person may operate or have charge of a steam boiler of any size or a hot water heating boiler of five hundred thousand BTU or larger without a license, issued by property development services, of the appropriate class as follows:

A. A Fireman's license authorizes the holder to:

1. Have charge of and operate any specifically described and located steam boiler used for power not exceeding one million BTU, or

2. Have charge of and operate any specifically described and located fifteen-pound steam boiler or hot water boiler used for heating purposes only not exceeding four million BTU, or

License Requirements for States and Cities   Chapter 6

3. Act as a Fireman in any steam plant where a licensed Engineer is in charge.

4. With special permission of the boiler inspector, a Fireman may operate up to three low-pressure plants where the steam pressure does not exceed fifteen pounds per square inch used for steam heating purposes only.

B. A Third Class Engineer's license authorizes the holder to:

1. Have charge of and operate a steam plant not exceeding four million BTU, or

2. Have charge of and operate any size Low Pressure plant for heating purposes only, or

3. Have charge of an opposite shift to a Second-Class Engineer, or

4. Act as assistant to a First Class or Second Class Engineer.

C. A Second Class Engineer's license authorizes the holder to:

1. Have charge of and operate a steam plant not exceeding eight million BTU, or

2. Have charge of an opposite shift to a First Class Engineer, or

3. Act as an assistant to a First Class Engineer.

D. A First Class Engineer's license authorizes the holder to have charge of any steam plant.

E. Boiler Operators' licenses are class IIB licenses under Chapter 4.04. They are annual licenses which expire on December 31.

10.29.021 Boilers - Reporting hazards.

Every Engineer and Fireman is required to report to the owner of the premises and to the city's boiler inspector any damage to or malfunction of a boiler or related equipment which endangers safety.

10.29.022 Boilers - Operators on duty.

No licensed Boiler Operator may leave the boiler room area for longer than fifteen minutes when a boiler under his or her charge is being operated at a pressure greater than fifteen pounds per square inch (gauge) unless relieved by a properly licensed person. The operator may leave a boiler used for heating purposes only set to operate within the range of from one to fifteen pounds per square inch.

10.29.023 Boilers - Responsibilities of operators

In addition to the ongoing responsibility of every Boiler Operator to operate any boilers and steam plants in his or her charge in strict accordance with the boiler code (Chapter 11.03) and accepted safety standards, every operator is responsible for:

A. Maintaining sufficient water in the boiler;

B. Preventing the carrying of more pressure than authorized;

C. Preventing the sticking of the safety valve fast to the seat;

D. Remaining free from the influence of intoxicants or other drugs while on duty;

E. Displaying his or her license conspicuously;

F. Notifying the boiler inspector in writing whenever leaving or changing employment.

License Requirements for States and Cities　　　Chapter 6

Tacoma, Washington
SUMMARY OF QUALIFICATIONS:
1. Exam Location, Address and Phone Number
City of Tacoma
City Hall
Second Floor
747 Market Street, Room 408
Tacoma, Washington 98402-3769
206/591-5029

2. Exam Days and Times
Every Monday except Holidays - Held at 8:30 a.m.

3. Examination Cost
$55.00

4. License Cost and Renewal
$25.00 for initial license and renewal

5. Type of Test
Written (if you pass-take oral)

Class 1 & 2

6. Grades of Licenses
Class 1 Chief Operating Engineer

Class 2 Operating Engineer

Class 3 Boiler Fireman

Class 4 Boiler Fireman

Class 5 Boiler Fireman

7. Experience and Education Requirements
See Codes for Tacoma below

Chapter 6 License Requirements for States and Cities

8. U.S. Citizenship Required?

Not Specified

9. Local Residency Required?

No

10. Recognition of Licenses from other locales

Reciprocity from Seattle, WA.

CODES for Tacoma, Washington

Chapter 6.20

STEAM BOILERS-ENGINEERS AND FIREMEN LICENSING

6.20.010 License required.

It shall be unlawful to have charge of or operate, or permit anyone to have charge of or operate, any boiler without first obtaining a license issued as required by this chapter.

6.20.030 Classes of licenses.

There shall be five (5) grades of licenses to cover the operation and maintenance of boiler plants, such Grades of licenses to be designated and limited as follows:

A. CLASS 1 CHIEF OPERATING ENGINEER LICENSE

Shall entitle the holder thereof to take complete charge of the operation and maintenance of any boiler plant.

B. CLASS II OPERATING ENGINEER LICENSE

Shall entitle the holder thereof to operate or to have charge of the operation of any boiler plant while on duty, under the direct supervision of a Chief Operating Engineer. In plants where there is not a Licensed Chief Operating Engineer, the licensee shall be limited to operation of a boiler plant not exceeding an aggregate of 300 million BTU/hr input.

C. CLASS III BOILER FIREMAN LICENSE

Shall entitle the holder thereof to operate any boiler plant under the direct supervision of a licensed Chief Operating Engineer or Operating Engineer. When not working under the direct supervision of an Operating Engineer, the licensee shall be limited to operation of a boiler plant not exceeding an aggregate of 20 million BTU/hr input.

D. CLASS IV BOILER FIREMAN LICENSE

Shall entitle the holder thereof to operate any boiler plant under the direct supervision of a licensed Chief Operating Engineer or Operating Engineer. The licensee shall be limited to operation of a boiler plant not exceeding an aggregate of 5 million BTU/hr input.

E. CLASS V BOILER FIREMAN LICENSE

Shall entitle the holder thereof to operate any boiler not exceeding 1,000,000 BTU/hr input.

6.20.050 Special location license.

A. When approving of any license issued under this chapter, the Director of Tax and License may impose stated conditions or limitations to such license restricting the Licensee to the operation and maintenance of a certain class of boilers at a stated location. Such restrictions shall be based upon the applicant's qualifications and be reasonably related to the protection of the public in the safe operation and maintenance of boilers. The restriction may be removed by written and oral examination and must be removed prior to new employment of operating a boiler, or operating a boiler at a different location.

B. Any licensed Operating Engineer or Boiler Fireman whose license limits are changed by this amendatory chapter so that the capacity of the plant where he/she is employed exceeds his/her license limits shall, upon application, be certified by the Department of Tax and License for a provisional license of the same grade, with limits extended to apply only to the plant where he/she is so employed for a period not to exceed one (1) year.

6.20.060 Issuance of licenses.

Persons desiring a license described in Section 6.20.030 of this chapter shall make written application to the Department of Tax and License on the forms provided by that Department. Application shall include the applicant's full name and address, and if the applicant is an employee, the name and address of his/her employer. No person shall be eligible for examination for any class of license unless he/she possesses the following qualifications.

A. CLASS I CHIEF OPERATING ENGINEER LICENSE:

(1) The applicant shall furnish verification of an aggregate of five years experience operating high or Low Pressure boiler plants as a Class II Operating Engineer or equivalent experience, or (2) The applicant shall be a graduate from a recognized curriculum in technology or a recognized training program, and shall furnish verification of an aggregate of three years' experience in the care and operation of high or low pressure boilers as a Class II Operating Engineer.

B. CLASS II OPERATING ENGINEER LICENSE:

(1) The applicant shall furnish verification of an aggregate of three years' experience operating high or Low Pressure boilers, or (2) The applicant shall be a graduate from a recognized

curriculum in technology or a recognized training program, and shall furnish verification of an aggregate of two years' experience in operation of high or low pressure boilers.

C. CLASS III BOILER FIREMAN LICENSE:

(1) The applicant shall furnish verification of an aggregate of one year's experience as a boiler Fireman, or (2) The applicant shall be a graduate from a recognized curriculum in technology or a recognized training program and furnish verification of three months' experience as a boiler Fireman.

D. CLASS IV BOILER FIREMAN LICENSE and CLASS V BOILER FIREMAN LICENSE:

The applicant shall furnish to the City Examiner satisfactory evidence of qualifications for examination.

6.20.070 Application and examination for license.

Application for license shall be on a printed form furnished by the City of Tacoma. Any person making application for any grade of license as an operator of a boiler plant shall be required, before such license is issued, to submit to examination. The examination shall be both written and oral and shall consist of questions covering care, maintenance, operation, and construction of boilers, appurtenances, and auxiliaries applicable to the class of license for which such examination is held. The applicant shall pass the written portion with a grade of 70% and shall also satisfy the City Examiner by oral examination that he/she is competent to work as a Class I, II, III, IV, or V licensed operator or Fireman.

If, upon examination, any applicant fails to obtain a license, he/she shall not be eligible for re-examination until a period of 30 days has elapsed. A fee of Thirty Dollars ($30.00) shall be charged for each examination for all classes of licenses.

Chapter 6   License Requirements for States and Cities

If the applicant fails to pass the examination and believes an injustice has been done, he/she may appeal the decision by serving a written notice of his/her intention to appeal such decision to the Chief of the Building Divisions, Department of Public Works, within ten (10) days. The Appeals Board shall review the examination papers of said applicant, and if necessary submit the applicant to an additional oral examination and decide whether or not he/she is entitled to a license. The decision of the majority of the members of the Appeals Board shall control and said decision shall be final.

6.20.075 Reciprocity.

In lieu of a qualifying examination, the City Examiner (designated by the Chief of the Buildings Division to test candidates for Boiler Operator Licenses), may accept as evidence of meeting the ordinance requirements of Section 6.20.070, a valid and current license issued by the City of Seattle.

6.20.080 Appeals board.

The Chief of the Buildings Division, Department of Public Works, shall appoint three (3) qualified licensed Chief Operating Engineers and one (1) alternate of not less than five (5) years' experience in the operation of steam boiler plants, who have operated steam plants for the last three (3) years, and who live or work in the City of Tacoma, who shall constitute an Appeals Board. The Appeals Board shall carry out all the functions and duties enumerated herein, as well as generally advise the City on all matters relative to this chapter. The Chief Boiler Inspector for the City of Tacoma shall be an ex officio member of the Appeals Board and shall act as secretary.

6.20.090 Exemptions.

The following installations shall not require a licensed operator.

License Requirements for States and Cities   Chapter 6

A. Any boiler or steam engine subject to federal regulations.

B. Heating boilers in group R1 occupancies of less than six units, group R3 occupancies and group M occupancies as defined in the Building Code as now or hereafter amended.

C. Low-pressure hot water, low-pressure steam and hot water supply boiler plants having inputs of less than 2,500,000 BTU per hour.

D. Any boiler having an input of less than 100,000 BTU per hour.

E. Water Heaters.

6.20.100 Operating rules and regulations.

A. Every licensed Operating Engineer or Boiler Fireman, before operating any boiler, shall first examine the boiler inspection certificate issued for such boiler or boilers to see that the boiler inspection certificate is in force, and if the certificate has expired, he/she shall notify his/her employer. If the boiler inspection certificate has been expired for more than ninety (90) days, he/she shall notify his/her employer and then the City Boiler Inspector of the date of expiration. He/she shall note the pressure allowed by the certificate and shall test the operation of the boiler and its control and safety devices for proper operation.

B. Whenever the licensed Operating Engineer or Boiler Fireman believes any part of a boiler to be defective or in potentially unsafe condition, he/she shall report the fact to his/her employer. If immediate corrective action is not taken, he/she shall report such defective or potentially unsafe condition to the City Boiler Inspector.

Chapter 6 — License Requirements for States and Cities

C. The City Boiler Inspector shall thereupon investigate the same, and report any lack of proper care on the part of any licensed person to the employer and the Department of Tax and License. The Department of Tax and License shall record the facts on the records of the licensee.

D. The licensed Operating Engineer or Boiler Fireman in charge of any boiler shall report to his/her employer and to the City Boiler Inspector any damage or injury to any such boiler under his/her charge or care which affects the safe operation of such boiler. Failure to make such reports to his/her employer and the City Boiler Inspector shall be sufficient cause for the suspension or revocation of the license of the person in charge.

E. It shall be the duty of each licensed Operating Engineer and boiler Fireman to report serious negligence in the care of boilers to his/her employer and the City Boiler Inspector.

F. Operators' licenses shall be posted in a conspicuous place under glass on the wall of the Boiler Room of the plant where they are employed, and such shall be effective only for the operation of the plant where they are displayed.

G. Notice of Placement of Employment: Every licensed Operating Engineer or Boiler Fireman shall inform the City Boiler Inspector of any change of place of employment. Notice shall be given within twenty-four (24) hours after leaving and/or accepting a position. Such notice shall be in writing, addressed to the Chief Boiler Inspector, Buildings Division, 747 Market Street, Tacoma, Washington, 98402, giving the licensee's name, number and grade of license and the name and address of the plant of the last employment and of new employment.

H. A licensed operator shall assist the boiler inspector in the examination of any boiler under the operator's charge and

License Requirements for States and Cities   Chapter 6

point out all defects, damage, or repair known to him/her in said boiler or pressure vessel, and the license of any operator deliberately neglecting or refraining to do so shall be revoked by City.

6.20.110 Duties of licensed Engineers and Boiler Firemen.

A. A licensed Operating Engineer or Boiler Fireman shall perform the following duties in connection with his/her operation and maintenance of boilers:

1. Test the operation of the boiler and its control and safety devices on a routine basis in accordance with nationally recognized standards and/or boiler and control manufacturer's written recommendations;

2. Maintain and operate the equipment in a safe manner and according to nationally recognized standards such as those recommended by the American Society of Mechanical Engineers for boilers;

3. Prepare and maintain a boiler log book and record such pertinent boiler data as may be required by the boiler inspector and the senior license holder in charge of the boiler operation, such as repair, adjustment, installation of controls and safety devices.

B. Boiler Log Book, Boiler Data and Operating Limits. The licensed operator shall:

1. Prepare a boiler log book with his/her name on the front cover. The boiler log book shall be kept on the premises and be available for inspection by the Boiler Inspector.

2. Determine the proper light off, operating, shut down procedure and clearly set forth such procedures in the inside front cover of the boiler log book. Determine proper firing rate and the set point or operating limits of all safety devices required

Chapter 6 License Requirements for States and Cities

on automatic boilers by the code and clearly mark such set point or limits on the back cover of the boiler log book.

3. Enter pertinent boiler data required by the boiler owner in the boiler log book and list such entries on the inside back cover.

4. This list shall include such items as any unusual conditions observed, including safety shutdowns, repairs required, adjustments required and/or made. All entries shall be made in the boiler log book with permanent ink and shall include the signature of the person making such readings, observations, or adjustments. It shall be lawful to cross out words or sentences which should be changed or corrected, but erasures shall be prohibited.

5. Mark certain gauges on the faces thereof (not on the glass cover) with a conspicuous red area to indicate a potentially unsafe condition of the particular medium whose temperature, pressure, or vacuum is being measured, except that if due to the manner in which the gauge has been manufactured it is not practicable to remove the glass cover, the red area may be marked on the glass cover. These gauges shall be, as applicable, the boiler pressure gauge, oil temperature gauge, fuel oil suction line gauge, high and low gas pressure gauge, stack temperature gauge, and the windbox pressure gauge. In addition, the water gauge glass, if applicable, shall be red lined to indicate high and low safe water levels. Such gauges be observed by the Boiler Operator as often as safety requires but not less frequently than twice each day and such readings recorded in the boiler log book in accordance with subsection B3 of this section. His/her operating instructions shall specify that gauge observations in unsafe areas shall require immediate shutdown of the boiler and recording same in the boiler log book.

6. Attend any start up of an automatic boiler out of service after corrective work has been performed on the boiler, its firing equipment, or its control and safety devices, and remain in constant attendance until:

- (i) the boiler has reached its present operating range of pressure, and
- (ii) the primary controls and safety devices have been proved, and
- (iii) the boiler is acceptable for continued operation.

7. Be in attendance during light off of original boiler equipment being installed by and under the control of the boiler manufacturer or his/her representative, by a boiler installation contractor making such installation under a manufacturer's written instructions and recommendations, or by a boiler or burner installer making such installations under manufacturer's written instructions and recommendations. He/she shall be in attendance during light off following adjustment or authorized boiler or burner manufacturer alterations made by the above representative, contractor or installer within the guarantee or warranty time period during which time the representative contractor or installer is obligated to render such service provided, however, that such representative or installer shall furnish the Boiler Operator with recommended set points or operating limits of all control devices and recommended firing rates, as well as other pertinent data, in writing, and shall record all subsequent changes, adjustments, alterations, or recommendations in the boiler log book together with his/her signature.

8. Not allow adjustments by others without the authority of the licensed operator and shall be limited to:

(a) Restoring control devices to original factory operating conditions at the set point or within the operating limits determined by the licensed operator as set forth in the boiler log book; or

(b) Repair and/or adjustment of the burner system for viscosity changes or to correct fuel air ratios to restore proper operation at the firing rate indicated in the boiler log book by the licensed Boiler Operator; or

(c) Repair or adjustment of any other system not directly related to the primary safety controls or to the pressure vessel to restore such system to proper operating conditions. Entries of such repairs or adjustments shall be made in the boiler log book and shall include the name of the repair firm making such repairs or adjustments.

C. Attendance Requirement.

1. Nonautomatic boilers. No licensed Engineer or Boiler Fireman in charge of a boiler, boiler plant, or steam engine, for the operation of which this chapter requires a license of Class I,II,III,IV, or V shall leave the immediate vicinity thereof when such boiler or boiler plant is being operated. No Operating Engineer or Boiler Fireman licensed under this chapter, in charge of any boiler or steam plant shall leave the premises of his/her employment when such boiler or steam engine is being operated without first shutting off all sources of heat in the boiler, or being relieved by a person duly licensed under this chapter; provided that such attendance requirements shall not apply to the operation of small power boilers and power steam boilers having less than 1,000,000 BTU per hour input where such boilers are equipped with approved automatic burners and automatic burner safety controls in accordance with applicable provisions of the Mechanical Code as now or hereafter amended,

relating to oil and gas burners. For such boilers so equipped, the attendance requirements shall be the same as that set forth for power boilers in subsection C2 hereof.

2. Automatic boilers. The following provisions relating to the frequency of observation and/or inspection of boilers shall apply to the operation of automatic boilers:

(a) Low Pressure hot water heating boilers, low pressure steam heating boilers, hot water supply boilers with a capacity of 2,500,000 to 5,000,000 BTU per hour input: twice daily check by a licensed operator.*

(b) Low Pressure hot water heating boilers, low pressure steam heating boilers, hot water supply boilers with an input capacity over 5,000,000 BTU per hour but less than or equal to 20,000,000 BTU per hour: Checked by a licensed operator at a minimum of six (6) hour intervals.

(c) Low Pressure hot water boilers, low pressure steam heating boilers, hot water supply boilers with an input capacity of over 20,000,000 BTU per hour: Checked by a licensed operator at a minimum of two (2) hour intervals.

(d) Power hot water boilers, power steam boilers with a capacity over 100,000 BTU per hour input; checked by a licensed operator* at two (2) hour intervals.

(e) Small Power boilers: At least twice daily checked by a licensed operator*.

The above are the minimum attendance requirements and shall not preclude in any way additional checks being made to ensure safe operation of the boiler.

*Phrases as used in this section shall have the following meaning:

(1) "Checked by licensed operator" shall mean supervision of boiler with responsibility for proper operation and maintenance pursuant to the requirements of and the inspection of all controls and safety devices as follows: Examine each boiler and boiler log book in accordance with the frequency of examinations required above. His/her examination shall include the testing of all control devices required for automatic boiler by the Building Code and the testing of monitoring systems when used. He/she shall, in addition, inspect and test all other controls on the boiler and shall flush the low water fuel cut-offs, if applicable, to assure that all control devices are in safe and proper operation. He/she shall permit continued automatic boiler operation only if his/her examination, inspection, and testing indicate that the boiler is in a safe operating condition. No modification or revision to the boiler or its control devices shall be made except under his supervision.

(2) Approved monitoring system: Shall mean a monitoring system manufactured, installed, and maintained in a manner approved by the Chief of Buildings Division.

(3) "Twice daily check" means two (2) inspections of a boiler that are required to be recorded in the boiler log book by this chapter. The first check of the day shall be made not less than eight (8) hours after the last recorded check of the previous day; the second check of the day shall be made at least six (6) hours after the first recorded check of the day. This definition shall not preclude, in any way, additional checks being made to ensure safe operation of the boiler.

6.20.115 Discrimination.

No person, firm, or corporation shall discharge or in any manner discriminate against any employee because such employee has filed any complaint or instituted or caused to be instituted any proceeding, including inspections under or related to this chapter.

6.20.120 Suspension or revocation of license.

The city shall have the power to suspend or revoke licenses for cause. All charges against persons licensed under this chapter shall be filed in writing to the Appeals Board. After a hearing, the Appeals Board may suspend or revoke the license of such person. No license shall be revoked for any cause without first giving the accused party and opportunity to be heard in his/her own defense.

6.20.130 Employment of operators.

No person, firm, or corporation owning or using boilers in the City of Tacoma shall operate or employ any person as an Operating Engineer or Boiler Fireman to take charge of or operate any boiler who has not first obtained a license as herein provided.

6.20.140 License fees.

The license fee for Operating Engineers and Boiler Firemen subject to this chapter shall be Twenty-five Dollars ($25.00) per annum, payable in advance. The employer's name and address shall be shown on the application for renewal of the license for all classes of licenses.

Terms of all licenses issued pursuant to Chapter 6.20 are set forth in Chapter 6.02 of the Official Code of the City of Tacoma.

Licensing renewal provisions shall prevail as provided for in Section 6.02.050 of the Official Code of the City of Tacoma, with the exception of subsection (d) which shall, in respect to this Boiler Operator Code, read as follows: Any person who shall not secure a renewal within a period of forty-five (45) days following the expiration of the license shall forfeit any and all rights to the renewal thereof, but may apply for a new license and be re-examined.

6.20.150 Penalties.

Any person, firm, or corporation violating any of the provisions of this Code shall be deemed guilty of a misdemeanor, and each such person, firm, or corporation shall be guilty of a separate offense for each and every day or portion thereof during which any violation of any of the provisions of this Code is committed, continued, or permitted, and upon conviction of any such violation such person shall be punished by a fine of not more than five hundred dollars ($500.00) or by imprisonment for not more than one hundred eighty (180) days, or by both such fine and imprisonment.

6.20.160 Enforcement - Filing of charges.

The Chief of the Buildings Division shall assist the Director of Tax and License in the enforcement of this chapter, and in such connection they are hereby authorized to jointly promulgate rules and regulations as may be deemed necessary to provide the means for ensuring safe and proper installation, repair, use, and operation of boilers and steam engines.

West Virginia

No State or City License

License Requirements for States and Cities   Chapter 6

Wisconsin

Kenosha, Wisconsin
SUMMARY OF QUALIFICATIONS:

1. Exam Location, Address and Phone Number
City of Kenosha
City Hall
Room 111
625-52nd Street
Kenosha, Wisconsin 53140
414/656-8194

2. Exam Days and Times

Quarterly- January, April, July and October

Usually on 2nd Tuesday of month but not always the case.

3. Examination Cost

$10.00

4. License Cost and Renewal

Initial license.............$20.00

Renewal of license....$10.00

5. Type of Test

All Written for Third Class

Written and Oral for First and Second Class

6. Grades of Licenses

First Class Operator

Second Class Operator

Third Class Operator

7. Experience and Education Requirements

Chapter 6 — License Requirements for States and Cities

See Codes for Kenosha below

8. U.S. Citizenship Required?

Not Specified

9. Local Residency Required?

No

10. Recognition of Licenses from other locales

Up to Examining Board

CODES for Kenosha, Wisconsin

Chapter VIII

Stationary Engineer and Boiler Operator Requirements

8.02 OPERATIONAL REQUIREMENTS

No person shall operate, have control of, manage or permit to be operated any stationary boiler, or any portion of a steam plant or appliances connected therewith in the City, when working under pressure, except as follows:

1. Every manually controlled boiler must be operated by an Engineer or an assistant Engineer licensed under this chapter who is in full time attendance in the boiler room and who shall be responsible for implementation of proper maintenance and operating procedures for such boiler, including compliance with the Wisconsin Boiler and Unfired Pressure Vessel Code, as it may be amended from time to time.

2. No person owning or controlling any steam boiler, steam engine or steam turbine may authorize or permit any person who does not have a proper and valid license to operate, have control of, manage, or take charge of such boiler, steam engine or steam turbine or any part thereof.

3. Every automatically controlled boiler must be operated with respect to maintenance and operating procedures, under the direction and control of an Engineer or assistant Engineer licensed in accordance with this Chapter, who shall be responsible for implementation of proper maintenance and operating procedures for such boiler, including compliance with the Wisconsin Boiler and Unfired Pressure Vessel Code, as it may be amended from time to time; provided that any automatically controlled boiler on which the safety devices required by this ordinance have been properly installed.

4. Maximum Steam Pressure.

No boiler shall be operated at a pressure in excess of either the allowable working pressure stated on its current inspection certificate issued by a certified Inspector from the Department of Industry, Labor and Human Relations or the allowable working pressure determined by an insurance company inspector certified under the Wisconsin Boiler and Unfired Pressure Vessel Code, whichever is lower.

Safety Regulations.

When two or more boilers are connected, all valves shall be equipped with lock and chain, and when workmen are engaged in work in one of such boilers while the other boiler contains live steam, the valves must be securely locked so that they cannot be opened.

8.06 APPLICATION FOR LICENSE

All persons desiring to perform the duties of an Engineer in charge of stationary boilers or engines within the City of Kenosha shall make application to the Board of Examiners. The Board of Examiners shall examine the qualifications of the applicant, grant licenses or revoke the same. Every application for license shall be made on a form furnished by the City of Kenosha.

8.07 QUALIFICATIONS FOR APPLICANTS; FEES

The applicant must meet the requirements of the license for which the applicant is applying.

If the applicant meets the requirements prescribed in this Chapter, said applicant may be examined as to his fitness for the class of license requested.

The fee for such examination shall be as follows: Ten ($10.00) Dollars for Third Class; Ten ($10.00) Dollars for Second Class; and, Ten ($10.00) Dollars for First Class.

These fees shall be paid to the Department of Housing and Neighborhood Development.

8.08 GRANTING OF LICENSE; EXPIRATION

If the applicant, after examination by the Board of Examiners has passed the required examination by a grade of 80% he or she shall be granted the license for which they applied. All licenses shall expire on April 30 following granting of the original license or renewal thereof. Such license shall be annually renewed without further examination provided the said licensee presents his license to the City Clerk for renewal prior to the expiration date.

If the license is expired, the former licensee must appear before the Board of Examiners to be reexamined or in lieu thereof receive a certificate signed by two examiners to be presented to the City Clerk for the issuance of a license. The fee for a renewal certificate for an expired license shall be Ten ($10.00) dollars payable to the Department of Housing and Neighborhood Development upon Board approval; said renewal certificate fee may be waived by a majority vote of the

Board. All certificates for licenses must be presented to the City Clerk within ten (10) working days after their issuance. Any certificate not presented within such time shall become void.

The examination shall be conducted in the months of January, April, July and October on a date to be determined by the Board.

8.09 ISSUANCE OF LICENSE

All licenses issued under this chapter shall be issued by the City Clerk for a fee of Twenty ($20.00) Dollars for new licenses or Ten ($10.00) Dollars for renewal.

8.10 TEMPORARY LICENSE

Temporary licenses may be granted by the Secretary of the Board at his or her discretion if the secretary is satisfied that the applicant has the requisite training and experience for the class for which the applicant has applied for testing at the next test date. Temporary licenses shall expire on the date of the examination which immediately succeeds the application.

The temporary license, once granted, shall be issued by the City Clerk upon presentation of a letter from the Secretary of the Board.

8.11 CLASSES OF LICENSES

Licenses for the purpose specified in Section 8.02 of this Chapter shall be granted as follows:

1. Third Class License.

(a) Experience required. Applicant must have a minimum of six (6) months experience operating boilers and auxiliary equipment under supervision. Proof of this experience must be made available to the Board of Examiners in writing. This ex-

Chapter 6   License Requirements for States and Cities

perience requirement may be waived at the discretion of the majority of the Board.

(b) Examination required: Oral and written.

(c) Limits of Third Class Licenses: Due to limited experience for Third Class licenses, operators shall be limited to operation of:

Maximum Boiler Horse Power (unsupervised) 600 BHP.

2. Second Class License.

(a) Experience required: Two years as a Third Class Operator; however, this may be waived at the discretion of at least two Board members. Minimum time as Third Class: One year, although this may be waived if proof of experience is available.

(b) Examination required: Oral, written and proof of practical experience.

(c) Limits of Second Class License.

(1) Maximum Boiler Horse Power (unsupervised) 1000 BHP.

3. First Class License.

(a) Experience required: Two years as a Second Class operator; however, this may be waived if proof of years of experience is available.

(b) Examination required: Oral, written and proof of practical experience.

(c) Limits of First Class License:

(1) Boiler Horse Power Unlimited.

Assistant Stationary Engineer

License Requirements for States and Cities — Chapter 6

The holder of a Second class license may act as Assistant Stationary Engineer to a Stationary Engineer in charge of a boiler under a First Class license; and the holder of a Third Class license may act as Assistant Stationary Engineer to an Engineer in charge of a plant under a Second Class license.

8.13 REVOCATION OR SUSPENSION OF LICENSE

The Board of Examiners shall have power to suspend or revoke the license of an Engineer for failure to implement proper maintenance and operating procedures in accordance with Section 8.02 of this Chapter, for carrying a higher pressure of steam than that fixed under Section 8.02 hereof, for an unnecessary absence from his post of duty, or for any violation of the provisions of this chapter, or for any other neglect or incapacity such as intoxication or the use of drugs. No license shall be suspended or revoked without first giving the person holding such license an opportunity to be heard. When the license of an Engineer is revoked, no license shall be issued to the same person within thirty (30) days after revocation; nor within ninety (90) days after the second revocation; and for any subsequent revocation, the license shall be permanently revoked.

8.14 DISPLAY OF LICENSE

The license shall be displayed in a conspicuous place under glass in the boiler or power plant.

8.15 REPORTS BY LICENSED ENGINEERS

It shall be the duty of every licensed Engineer to report to the Chief of Inspection any defect in any steam boiler, engine or appurtenance under their care. Every licensed Engineer shall file written notice to the City Clerk of any change in employment which is within the terms of this Chapter and of any change of his home address. Such notice shall be given within ten (10) days after such commencement, termination or change.

Chapter 6　　　License Requirements for States and Cities

8.17 PENALTIES

Any person violating any of the provisions of this chapter shall upon conviction thereof be subject to a fine of not more than Five Hundred ($500.00) Dollars, and in default of payment of said fine and costs of prosecution, shall be committed to the County Jail for not more than thirty (30) days. Each day of violation shall be a separate offense.

Milwaukee, Wisconsin

SUMMARY OF QUALIFICATIONS:

1. Exam Location, Address and Phone Number

City of Milwaukee
10th Floor Municipal Building
841 N. Broadway
Room 1016
Milwaukee, Wisconsin 53202
414/278-2553

2. Exam Days and Times

Once a month on Friday for 1st, 2nd & 3rd Class Stationary Engineers at 12:30 p.m.

Three times a month - High Pressure and Low Pressure at 2 p.m.

3. Examination Cost

H.P., L.P., & 3rd Class.........$30.00

2nd & 1st Class....................$35.00

4. License Cost and Renewal

Initial license.............$25.00

Renewal of license....$25.00

5. Type of Test

License Requirements for States and Cities   Chapter 6

L.P. and H.P..........50 multiple choice questions

Engineers ....All Essay

6. Grades of Licenses

Low Pressure Boiler Operator

High Pressure Boiler Operator

Low Pressure Boiler Operator II

First Class Stationary Engineer

Second Class Stationary Engineer

Third Class Stationary Engineer

7. Experience and Education Requirements

Low Pressure Boiler Operator - 1 year experience operating steam boiler

High Pressure Boiler Operator -1 year experience operating steam boiler or completion of one year State of Wisconsin approved apprenticeship in power engineering.

Low Pressure Boiler Operator II - 6 months in plants containing steam boilers or, in lieu thereof, certification that the applicant has successfully completed a semester's course in stationary boiler operation practices at a school approved by the examining committee for Stationary Engineers and Boiler Operators. A Low Pressure Boiler Operator II shall be restricted to supervising Low Pressure plants located in apartment buildings.

First Class Stationary Engineer - 4 years experience operating high pressure steam boilers.

Second Class Stationary Engineer - 3 years experience operating high pressure steam boilers or completion of 3 years of

Chapter 6   License Requirements for States and Cities

State of Wisconsin approved apprenticeship in power Engineering.

Third Class Stationary Engineer - 2 years experience operating high pressure steam boilers.

8. U.S. Citizenship Required?

Not Specified

9. Local Residency Required?

No

10. Recognition of Licenses from other locales

They recognize the license for experience but you have to take the exam.

CODES for Milwaukee, Wisconsin

223-3. Stationary Engineer and Boiler Operator Licenses.

1. SCOPE.

Except as provided in sub. 2, this section shall apply to all steam boiler plants and steam engines and turbines:

(a) No person shall operate, manage, or take charge of any steam boiler, steam engine, or steam turbine regulated by this section without first procuring a Stationary Engineer or Fireman license.

(b) No person owning or controlling any steam boiler, steam engine or steam turbine may authorize or permit any person who does not have a proper and valid license to operate, have control of, manage, or take charge of such boiler, steam engine or steam turbine or any part thereof.

(c) Boilers which use liquids other than water shall be subject to the same license requirements as steam boilers as deter-

License Requirements for States and Cities   Chapter 6

mined by the boiler horsepower rating and safety valve setting of the boilers, except that a person having only a Low Pressure Fireman license shall not be permitted to operate such boilers.

2. EXCEPTIONS.

The license regulations of this section shall not apply to:

(a) Boiler plants consisting of one or more low pressure steam boilers with a total capacity of less than 15 boiler horsepower or 500,000 B.T.U. output as rated by boiler manufacturer rating plate or Wisconsin State Code.

(b) Steam boiler plants consisting only of one or more miniature steam boilers which are used as separate units.

(c) Steam engines or turbines which are supplied with steam from a miniature steam boiler.

(d) Locomotives used in interstate commerce.

(e) Low Pressure steam boiler heating plants in one and 2-family dwellings.

(f) Boilers and turbines operated at pressures exceeding 1200 psi by specifically trained personnel acting under the supervision of a licensed Boiler Operator and who is located on the plant site when approved by the department. The training program shall be approved by the examining committee for Stationary Engineers and Boiler Operators.

3. LICENSE CLASSIFICATION.

Six classes of Stationary Engineer or Boiler Operator licenses are established:

(a) First Class Stationary Engineer license.

(b) Second Class Stationary Engineer license.

(c) Third Class Stationary Engineer license.

(d) High Pressure Boiler Operator license.

(e) Low Pressure Boiler Operator license

(f) Low Pressure, Boiler Operator license II, apartment buildings only.

4. EXPERIENCE REQUIREMENTS:

Applicants for Stationary Engineer or Boiler Operator licenses shall have had the following experience as a Stationary Engineer, Engine Room Oiler or Boiler Operator in the operation of steam boilers or steam engines and turbines. Graduate Mechanical Engineers will be allowed one year's experience when applying for First, Second and Third Class Engineers licenses.

(a) First Class Stationary Engineer:

Four years in plants containing High Pressure steam boilers.

(b) Second Class Stationary Engineer:

Three years in plants containing High Pressure steam boilers or completion of 3 years of State of Wisconsin approved apprenticeship in power engineering.

(c) Third Class Stationary Engineer:

Two years in plants containing High Pressure steam boilers.

(d) High Pressure Boiler Operator:

One year in plants containing steam boilers or completion of one year State of Wisconsin approved apprenticeship in power engineering.

(e) Low Pressure Boiler Operator:

One year in plants containing steam boilers.

License Requirements for States and Cities Chapter 6

(f) Low Pressure Boiler Operator license II:

Six months in plants containing steam boilers or, in lieu thereof, certification that the applicant has successfully completed a semester's course in stationary boiler operation practices at a school approved by the examining committee for Stationary Engineers and Boiler Operators. A Low Pressure Boiler Operator II shall be restricted to supervising low pressure plants located in apartments buildings.

5. AGE REQUIREMENTS, EXPERIENCE VOUCHERS, FEES:

(a) Applicants for licenses shall be at least 18 years of age.

(b) Applications for licenses shall be made on forms furnished for such purposes by the commissioner.

(c) Applications shall be notarized and shall bear the signature of 2 adults vouching for the character and the experience of the applicant. Vouching signatures on applications for High Pressure Boiler Operator and Stationary Engineer licenses should contain the Engineer in charge of the high pressure plant where the applicant worked, whenever possible.

(d) For fees see s. 200-33.

7. INSPECTION OF EXAMINATION PAPERS.

(a) All persons who have taken the examination shall have the right to inspect their examination papers in the presence of a member of the staff of the commissioner once within a 2 week period after taking the exam.

(b) Examination papers can only be inspected by the applicant and members of the staff of the commissioner.

8. GRANTING OF LICENSES.

Chapter 6   License Requirements for States and Cities

(a) The commissioner shall upon notification by the examining committee of the persons who have been certified as First or Second Class Engineers issue licenses to the successful applicants within 7 days after certification. The license shall be sealed with the imprint of the department seal.

(b) The commissioner shall conduct written examinations, or oral when necessary, of applicants for Third Class Stationary Engineer licenses and High and Low Pressure Boiler Operator licenses.

(c) Applicants who successfully pass the examination and have the required experience shall be issued a license by the commissioner and sealed with the imprint of the department seal.

(d) Any applicant who has failed to pass an examination shall not be permitted to take another examination for a period of 30 days after the first failure, 60 days after the second failure, and one year after the third failure.

9. GENERAL OPERATION REGULATIONS.

(a) Notification of Absence by Employer.

In any case where a Stationary Engineer or Boiler Operator in charge of a plant is temporarily unable to report to duty because of sickness, personal injury or for any other reason or cause, the employer of the Stationary Engineer or Boiler Operator shall immediately obtain the services of a substitute Stationary Engineer or Boiler Operator who may have other employees to take charge of the plant. The employer shall notify the commissioner within 24 hours that a substitute Stationary Engineer or Boiler Operator has taken charge of the plant.

(b) Temporary Operation.

License Requirements for States and Cities — Chapter 6

If the employer is unable immediately to obtain the services of a Stationary Engineer or Boiler Operator having the class of license required for the plant, he or she may apply to the commissioner for a permit to allow one of the following persons to take charge of the plant for the following specified period of time:

(b-1) A Stationary Engineer or Boiler Operator having a lower class of license than that required for the plant for a period not to exceed 30 days.

(b-2) Any person who does not have the required boiler operation experience but can pass a written examination given by the commissioner on the operation of boilers and steam engines and turbines, for a period not to exceed 15 days.

(b-3) After the period of time specified in subd. 1 or 2 has lapsed, the employer shall provide the services of a Stationary Engineer or Boiler Operator having the proper class of license to take charge of the plant.

(c) Notice of Accidents by Operator.

Every licensed Stationary Engineer or Boiler Operator who is in charge of a boiler or steam engine or turbine shall notify the commissioner in writing of any accident in which such equipment is involved within 24 hours after the accident occurs, and report any serious defects or hazards in such equipment which might be contributory to a serious accident.

(d) Notice of Change in Employment.

Every person licensed under this section shall notify the commissioner when their place of employment as a Stationary Engineer or Boiler Operator changes within 3 days from the date of change.

(e) License to be Displayed.

Chapter 6 — License Requirements for States and Cities

Every person licensed under this section shall display their license certificate under glass or other transparent material in a conspicuous place in the boiler, engine or turbine room.

(f) Unlawful to Exceed Boiler Pressures.

It shall be unlawful to carry a higher steam pressure in any boiler than as determined by the commissioner in accordance with chs. ILHR 41 and 42, Wis. Adm. Code, as amended, and the condition of the boiler as found upon inspection.

(g) Daily Log.

Every Stationary Engineer or Boiler Operator licensed under this section shall keep a daily log in the boiler room to include the time and the day of checks being made on the boiler, engine or turbine.

License Requirements for States and Cities Chapter 6

Racine, Wisconsin

SUMMARY OF QUALIFICATIONS:

1. Exam Location, Address and Phone Number

City Hall, City Clerk's Office
Room 4, Basement
730 Washington Avenue
Racine, Wisconsin 53403-1184
414/636-9171

Test for the following cities:

Walworth, Burlington

2. Exam Days and Times

2nd and 4th Wednesday every month at 6:30 p.m.

3. Examination Cost

$10.00

4. License Cost and Renewal

Initial license.............$10.00

Renewal of license....$10.00

5. Type of Test

All Written

6. Grades of Licenses

Temporary Operator

Fourth Class Engineer

Third Class Engineer

Second Class Engineer

First Class Engineer

Chapter 6   License Requirements for States and Cities

Chief Engineer

7. Experience and Education Requirements

See Codes for Racine below

8. U.S. Citizenship Required?

Not Specified

9. Local Residency Required?

No

10. Recognition of Licenses from other locales

Not Specified

CODES for Racine, Wisconsin

Ordinance No. 31-92

To repeal and recreate Chapter 22.18 and amend Section 22.34.020 of the Municipal Code of the City of Racine relating to business - Stationary Engineers; Forfeitures.

The Common Council of the City of Racine do ordain as follows:

Part 1: Chapter 22.18 of the Municipal Code is hereby repealed and recreated to read as follows:

Chapter 22.18 Stationary Engineers

22.18.020 PROHIBITION.

No person shall operate, control, manage or commit to be operated any stationary boiler, engine or any portion of a steam plant or appliances connected therewith when working under pressure except those boilers exempt under ILHR 41.18, and except as follows:

(a) An engineer or an assistant engineer licensed under this chapter who is in full time attendance in the boiler room may operate a manually controlled boiler.

(b) An engineer or an assistant engineer holding an appropriate license under this chapter may operate an automatically controlled boiler providing that such person is periodically in attendance in the boiler room on at least a daily basis, excepting that for high pressure plants an engineer or assistant engineer shall be in constant attendance if any person or persons, other than boiler personnel, are present in the building in which the boiler is located.

22.18.030 BOARD OF EXAMINERS.

The Board of Examiners shall consist of three members who are residents of the City. The members of the Board shall be appointed by the Mayor, subject to confirmation by the Common Council. Appointments shall be made the third Tuesday of April. Members shall serve for a term of three years or until a successor has been duly appointed. No person shall be eligible for appointment unless such person holds a City of Racine Chief Engineer license and a NIULPE Examiner's commission.

22.18.040 POWERS OF THE BOARD.

The Board of Examiners shall adopt rules and regulations governing its operations under this chapter. The Board shall meet periodically as necessary to examine and determine the qualifications of applicants for licenses of engineers and for persons having charge of stationary steam boilers or engines as provided herein. Board members shall receive no salary.

22.18.050 LICENSE APPLICATION.

Chapter 6   License Requirements for States and Cities

A verified application for a license to perform work as an engineer in charge of stationary boilers or engines in the City of Racine shall be made to the Board of Examiners. Application shall be made on a form provided by the City Clerk which shall, in addition to information required thereon, include two signed references. The Clerk shall refer applications to the Board which shall examine the applicant and determine the applicant's fitness for a particular class of license.

22.18.060 QUALIFICATIONS; FEES.

(a) Applications for a license may be made by any person holding a National Institute for the Uniform Licensing of Power Engineers, Inc. (NIULPE) License of the category for which application is made. Excepting a fourth class engineer's license, no license shall be granted to an applicant unless such person has held the next inferior license and has been actively employed in duties thereunder for a period of one year. A qualified applicant who passes the Board's examination shall be granted a license which shall be valid for one year from the date of examination, upon payment of a $10.00 license fee and a $10.00 examination fee. Upon issuance of a certificate of licensing by the Board of Examiners, the applicant shall present it to the City Clerk for issuance of the license. Any certificate not presented within ten working days shall be void. The license shall be prominently displayed in a conspicuous place in the boiler room or power plant at the licensee's place of employment.

(b) An applicant for a fourth class engineer's license must demonstrate good mechanical aptitude, present evidence of successful completion of a stationary engineer's course at an accredited technical school and shall present evidence of successful completion of the NIULPE fourth class engineer's exam.

(c) A temporary operator's license may be applied for by any person who meets all qualifications under (b) above, excepting that only the first 10 week phase of the accredited course has been successfully completed and the NIULPE examination has not been taken. The temporary operator's license shall be issued by the City Clerk upon direction of the Board, and shall be valid for a period of nine months from date of issuance. A fee of $10.00 shall be paid to the City Clerk upon issuance of a temporary operator's license.

22.18.070 GRANTING OF LICENSE; EXPIRATION.

An applicant shall be granted a license by the Board of Examiners if the applicant is found to be qualified and if the applicant has passed his or her examination by the passing score specified on the examination. The license shall be annually renewed without further examination upon payment to the City Clerk of a fee of $10.00, provided the licensee presents his license to the City Clerk for renewal not later than ten business days after the expiration date. Members of the Board of Examiners shall be issued a renewal license without payment of a fee.

22.18.080 ISSUANCE OF LICENSES.

Upon authorization of the Board of Examiners, the City Clerk shall issue licenses hereunder. Not later than March 1 of each year, the City Clerk shall furnish the Board with a list of persons to whom licenses were issued during the previous calendar year.

22.18.090 CLASSES OF LICENSES.

Licenses of the class specified herein shall authorize a licensee to take charge of, operate and manage the equipment indicated, excepting that a holder of any class of license may operate equipment one class in excess of such license if such licen-

Chapter 6 — License Requirements for States and Cities

see is an assistant to a licensee holding the next superior license.

(a) Chief Engineer's License:

Any steam motive power or refrigeration plant.

(b) First Class Engineer's License:

Low pressure steam or hydronic plant of any size; a high pressure plant not exceeding 1500 BHP; a motive power plant not exceeding 7500 HP, or a refrigeration plant not exceeding 5000 tons.

(c) Second Class Engineer's License:

A low pressure steam or hydronic plant of any size; a high pressure plant not exceeding 500 BHP; a motive power plant not exceeding 2500 HP, or a refrigeration plant not exceeding 1000 tons.

(d) Third Class Engineer's License:

A low pressure steam or hydronic plant of any size; a high pressure plant not exceeding 200 BHP; a motive power plant not exceeding 1000 HP, or a refrigeration plant not exceeding 500 tons.

(e) Fourth Class Engineer's License:

A low pressure steam or hydronic plant; a high pressure steam plant not exceeding 100 BHP; a motive power plant not exceeding 500 HP, or a refrigeration plant not exceeding 100 tons.

(f) Temporary Operator's License:

A low pressure steam or hydronic plant not exceeding 150 BHP; a high pressure steam plant not exceeding 25 BHP; a motive power plant not exceeding 500 HP, or a refrigeration plant not exceeding 100 tons, for a period not to exceed nine months from date of issuance of Temporary Operator's License.

22.18.100   TEMPORARY ABSENCE.

In all cases of temporary absence an engineer with a lower grade license may take charge of a plant requiring a license of a higher grade, provided that such licensee may not act in said capacity for a period of more than 30 days, unless otherwise authorized by the Board of Examiners.

22.18.110   REVOCATION OF LICENSE.

In order to provide public health, safety and welfare, the Board of Examiners shall revoke any license issued under this chapter in the event that the licensee fails to implement proper maintenance and operating procedures in accordance with ILHR 41 and 42; or for an unnecessary absence from the licensee's post of duty; or for any violation of the substantive provisions of this chapter, or for any other neglect or incapacity which impairs the licensee's abilities to perform the duties required by ILHR 41 and 42 and this chapter. Within five days after such revocation, the licensee may petition the Board of Examiners for a hearing on the matter which shall be held within 30 days following such petition. Testimony shall be given under oath and the Board's witnesses and the licensee's witnesses shall testify under oath subject to cross examination. The Board shall make findings and conclusions in writing and provide them to the licensee within five days following such hearing.

22.18.120   RECORDS.

The Board of Examiners shall keep a record of all business transacted by it and shall submit a report annually to the Mayor and Common Council.

22.18.130   PENALTY.

Chapter 6 License Requirements for States and Cities

Upon conviction any person found to be in violation of this chapter shall forfeit not less than $25.00 nor more than $750.00.

Part 2:

Section 22.34.020 of the Municipal Code is hereby amended by adding the following:

Section Number: 22.18.020

Violation: Stationary Engineers: Prohibition

Forfeiture: $250.00

Part 3:

This ordinance shall take effect upon passage and the day after publication.

Passed by the Common Council: 10-20-92

Approved: N. Owen Davis, Mayor - 10-21-92

Attest: Karen Norton, City Clerk

Wyoming

No State or City License

Chapter 7
Regulations for Canadian Provinces and Territories
Canada
Province of Alberta
SUMMARY OF QUALIFICATIONS:

1. Exam Location, Address and Phone Number

Alberta Labour
Boiler and Pressure Vessel Safety
6th Floor, 10808-99 Avenue
Edmonton, Alberta, Canada T5K0G5
403/427-6855

Testing Sites: Edmonton, Calgary, Red Deer, Lethbridge, Grande Prairie, Medicine Hat, Fort McMurray and St. Paul

2. Exam Days and Times

Varies - Exam dates and times subject to change.

3. Examination Cost

$20.00

4. Certificate of Competency Cost and Renewal

From Schedule G-Fee Regulation

First Class Engineer......................$186.00

Second Class Engineer..................$138.00

Third Class Engineer.....................$ 96.00

Fourth Class Engineer...................$ 53.00

Fireman...$ 27.00

Chapter 7 Regulations for Canadian Provinces and Territories

Building Operator A........................$ 53.00

Building Operator B........................$ 27.00

Special Boiler Operator..................$ 27.00

5. Type of Test

All Written

6. Grades of Certificates

First Class Engineer

Second Class Engineer

Third Class Engineer

Fourth Class Engineer

Fireman

Building Operator "A"

Building Operator "B"

Special Boiler Operator

7. Experience and Education Requirements

See Regulations 10 through 17 for Alberta

8. Canadian Citizenship Required?

See Below-Regulation 21 and 22

Generally, must work or live in Province of Alberta.

Up to Chief Inspector.

9. Local Residency Required?

Generally, Yes - Up to Chief Inspector.

10. Recognition of Licenses from other locales

Accept interprovincial certificates for equal class of certificate with a pass on exam except Ontario and Quebec will accept certificate of 1 class lower with pass on exam.

RULES AND REGULATIONS for Province of Alberta

Province of Alberta

Boilers and Pressure Vessels Act

Engineers' Regulation

Being Alberta Regulation 319/75 with amendments up to and including Alberta Regulation 355/84

1. In these regulations:

(a) "Act" means the Boiler and Pressure Vessels Act;

(b) "Assistant Engineer" means a person who holds a certificate of competency permitting him to perform the functions of an assistant engineer and who takes charge of a section of a power plant under the supervision of a shift engineer;

(c) "Assistant Shift Engineer" means a person who holds a certificate of competency permitting him to perform the functions of an assistant shift engineer and who assists the shift engineer in supervising all aspects of the operation in a power plant;

(d) "Building Operator" means a person who holds a certificate of competency permitting him to perform the functions of a building operator;

Chapter 7 Regulations for Canadian Provinces and Territories

(e) "Chief Steam Engineer" means a person who holds a certificate of competency permitting him to perform the functions of a Chief Steam Engineer;

(e.1) "Fee Regulation" means the Fee Regulation (Alta. Reg. 353/84);

(f) "Fireman" means a person who holds a Fireman's Certificate of competency who takes charge of a power plant not exceeding 500 kW or who takes charge of a shift in the operation of a power plant not exceeding 1000kW;

(g) "Heating Surface" means any part of the surface of a boiler that is in contact with liquid under pressure on one side and the products of combustion on the other side;

(h) Repealed by AR 76/78;

(i) "Shift Engineer" means a person who holds a certificate of competency permitting him to perform the functions of a shift engineer and who has charge of a shift in a power plant under the supervision of a Chief Steam Engineer;

(i.1) "Special Boiler Operator" means a person who holds a certificate of competency permitting that person to perform the functions of a special boiler operator;

(j) "Special Oil Well Operator" means a person who holds a certificate of competency permitting him to perform the functions of a special oil well operator.

2(1) Pursuant to section 27, subsection (4) of the Act the following power plants are exempted from the provisions of section 27, subsections (2) and (3) of the Act:

(a) a power plant consisting of one or more coil type drumless boilers having an aggregate capacity not exceeding 5000

Regulations for Canadian Provinces and Territories   Chapter 7

kW and used for the sole purpose of underground thermal flooding in oilfields;

(b) a power plant operating above 103 kPa but not more than 140kPA steam pressure;

(c) a power plant not exceeding an aggregate capacity of 250 kW;

(d) a power plant not exceeding an aggregate capacity of 0.085 m.

(2) Pursuant to section 27, subsection (5) of the Act, the owner or person in charge of a power plant specified in this subsection shall hold a certificate of competency of the type specified as follows:

(a) with respect to a power plant referred to in subsection (1), clause (b), it shall be under the general supervision of the holder of not less than a Fourth Class Engineer's Certificate of Competency;

(b) with respect to a power plant referred to in subsection (1), clause (a), it shall be under the general supervision of the holder of a certificate of competency the classification of which qualifies him to have general supervision of the power plant;

(c) with respect to a power plant of over 20 kW but not exceeding an aggregate capacity of 250 kW, it shall be under the general supervision of the holder of not less than a Special Boiler Operator's Certificate of Competency;

(d) repealed by AR 76/78.

(2.1) Notwithstanding subsections (1) and (2), the provisions of section 27, subsection (2), clause (b) of the Act apply to:

Chapter 7   Regulations for Canadian Provinces and Territories

(a) a traction boiler

(b) a boiler operating in a parade or used in a display or for the purpose of entertainment, and

(c) a locomotive boiler to which the Act applies that is operating on a railroad.

(2.2) No owner or person in charge of a thermal liquid heater system shall operate it or cause it to be operated unless it is under the general supervision:

(a) where the heating surface of the system does not exceed 100 m^2 of the holder of a Fourth Class Engineer's Certificate of Competency, and

(b) where the heating surface of the system exceeds 100 m^2, the holder of a Third Class Engineer's Certificate of Competency.

(3) Pursuant to section 28, subsection (1) of the Act, no owner or person in charge of a heating plant exceeding 750 kW, having a capacity exceeding 0.085 m^3 and used primarily for the purpose of heating one or more buildings, shall operate it, or permit or cause it to be operated unless it is under the general supervision of the holder of a valid certificate of competency, issued pursuant to these regulations, the classification of which qualifies the holder to act as building operator of the heating plant,

(3.1) For the purposes of subsection (3), "general supervision" means supervision in accordance with the recommendations set out in the A.S.M.E. Boiler and Pressure Vessel Code, Section VI Recommended Rules for Care and Operation of Heating Boilers - 1980 as adopted under section 5(1)(d)(IX) of Alberta Regulation 227/75 as amended.

(4) Pursuant to section 28, subsection (1) of the Act, a heating plant not exceeding 750kW or having a capacity not exceeding 0.085 m^3, is not required to be operated under the supervision of the holder of a certificate of competency.

Certificates of Competency

3(1) The certificates of competency specified in these regulations shall be issued by the chief inspector upon a person satisfying the requirements therefor and paying the appropriate fee prescribed in the Schedule.

(2) The owner of person in charge of a power plant or heating plant shall post in a conspicuous place on the premises the certificates of competency required under the Act or the regulations.

(3) Where the certificate of competency of a person is required to be posted in more than one heating plant under subsection (2), that person may apply to the chief inspector for a duplicate certificate of competency.

(4) An application under subsection (3) must be accompanied by the fee for the issuance of a duplicate certificate of competency as specified in the Schedule.

(5) The chief inspector may issue a duplicate certificate of competency for the purposes of subsection (2) and shall indicate on the duplicate certificate of competency

(a) that it is a duplicate, and

(b) that it may be posted only in the building identified on its face.

Chapter 7 Regulations for Canadian Provinces and Territories

(6) A duplicate certificate issued under subsection (4) shall be posted only in the building identified on the duplicate certificate.

(7) Where a duplicate certificate of competency issued under subsection (4) is no longer required to be posted under the Act or the regulations for any reason the person named in the duplicate certificate shall return it to the chief inspector forthwith.

(8) The owner or person in charge of a power plant or heating plant shall provide the chief inspector with any information the chief inspector may require with respect to the qualifications of the personnel involved in the operation of the power plant or heating plant.

4(1) The following grades of a Building Operator's Certificate of Competency are hereby established:

(a) Building Operator A, and

(b) Building Operator B.

(2) The holder of a Fourth Class Engineer's Certificate of Competency issued prior to January 1, 1977 may be issued a Building Operator A Certificate of Competency, without examination, on payment of a fee in accordance with Schedule G of the Fee Regulation.

(3) The holder of a Fireman's Certificate of Competency issued prior to January 1, 1977 may be issued a Building Operator B Certificate of Competency, without examination, on payment of a fee in accordance with Schedule G of the Fee Regulation.

4.1 The Special Boiler Operator's Certificate of Competency is hereby established.

5(1) A First Class Engineer's Certificate of Competency qualifies the holder to:

(a) take charge of the general care and operation of any power plant as chief steam engineer, and to supervise the engineers in that plant, or

(b) take charge of a shift in any power plant as shift engineer.

(2) A Second Class Engineer's Certificate of Competency qualifies the holder to:

(a) take charge of the general care and operation of a power plant not exceeding 10,000 kW as chief steam engineer and to supervise the engineers in that plant, or

(b) take charge of the general care and operation of a power plant consisting of one or more coil type drumless boilers having an aggregate capacity not exceeding 15,000 kW when used for the sole purpose of underground thermal flooding in oil fields, as chief steam engineer and to supervise the engineers in that plant, or

(c) take charge of a shift in any power plant as shift engineer.

(3) A Third Class Engineer's Certificate of Competency qualifies the holder to:

(a) take charge of the general care and operation of a power plant not exceeding 5000 kW as chief steam engineer and to supervise the engineers in that plant, or

(b) take charge of the general care and operation of a power plant consisting of one or more coil type drumless boilers having an aggregate capacity not exceeding 10,000 kW when used for the sole purpose of underground thermal flooding in oil fields, as chief steam engineer and to supervise the engineers in that plant, or

Chapter 7   Regulations for Canadian Provinces and Territories

(c) take charge of a shift in a power plant not exceeding 10,000 kW as shift engineer, or

(d) take charge of a shift in a power plant consisting of one or more coil type drumless boilers having an aggregate capacity not exceeding 15,000 kW when used for the sole purpose of underground thermal flooding in oil fields, as shift engineer, or

(e) take charge of a section of any power plant as assistant engineer, under the supervision of the shift engineer in that plant, or

(f) take charge of the general care and operation of any power plant operating at a pressure of not more than 140 kPa.

(4) A Fourth Class Engineer's Certificate of Competency qualifies the holder to:

(a) take charge of the general care and operation of a power plant not exceeding 1000 kW as chief steam engineer and to supervise the engineers in that plant, or

(b) take charge of the general care and operation of a power plant consisting of one or more coil type drumless boilers having an aggregate capacity not exceeding 5000 kW when used for the sole purpose of underground thermal flooding in oil fields, as chief steam engineer, or

(c) take charge of a shift in a power plant not exceeding 5000 kW as shift engineer, or

(d) take charge of a shift in a power plant consisting of one or more coil type drumless boilers having an aggregate capacity not exceeding 10,000 kW when used for the sole purpose of underground thermal flooding in oil fields, as shift engineer, or

(e) take charge of a section of a power plant not exceeding 10,000 kW, as assistant engineer, under the supervision of the shift engineer in that plant, or

(f) take charge of the general care and operation of a power plant not exceeding 5000 kW and operating at a pressure not more than 140 kPa.

(5) A Fireman's Certificate of Competency qualifies the holder to:

(a) take charge of the general care and operation of a power plant not exceeding 500 kW, as fireman in charge and to supervise the firemen on shift in that plant, or

(b) take charge of a shift in a power plant not exceeding 1000 kW.

(6) A Special Oilwell Operator's Certificate of Competency qualifies the holder to:

take charge of a power plant operating on an oil drilling site having an aggregate capacity not exceeding 1000 kW.

(7) A Building Operator A Certificate of Competency qualifies the holder to:

exercise general supervision of any heating plant and take responsibility for its general care and operation.

(8) A Building Operator B Certificate of Competency qualifies the holder to:

exercise general supervision of a heating plant not exceeding 3000 kW and take responsibility for its general care and operation.

(9) A Certificate of Competency of a grade higher than fourth class qualifies the holder to:

Chapter 7 Regulations for Canadian Provinces and Territories

exercise general supervision of any heating plant and take responsibility for its general care and operation.

(10) A Special Boiler Operator's Certificate of Competency qualifies the holder to:

exercise general supervision of the specific power plant, not exceeding 250 kW in capacity, named on the Certificate of Competency.

6(1) Any certificate of competency may be issued to a person on a temporary basis, if:

(a) his employer applies for the temporary certificate of competency on the person's behalf, and

(b) his employer certifies that:

- (i) he cannot obtain the services of the holder of a certificate of competency as required by section 27 or 28 of the Act, or

- (ii) he require the temporary certificate of competency for:

    (A) holiday, emergency or sick relief purposes, or

    (B) to enable a person to operate under it for the purposes of training.

(2) Where the chief inspector is not satisfied that the person referred to in subsection (1) should be issued a temporary certificate of competency, he may require that person to take an examination as he considers necessary.

(3) A temporary certificate of competency issued by the chief inspector shall not be more than one grade higher than the certificate of competency held, to permit:

(a) a shift engineer of a power plant to act as chief steam engineer of the same plant, or

(b) an assistant engineer or assistant shift engineer to act as shift engineer in the same plant, or

(c) subject to clauses (a) and (b), any person to act in any other position as specified by the chief inspector.

(4) An employer shall apply for a temporary certificate of competency on a form prescribed by the chief inspector which shall contain a declaration that the person in whose favour the application is being made is, to the best of his knowledge, capable of acting in the capacity for which the temporary certificate of competency is being requested.

(5) Where the chief steam engineer is sick or expects to be absent from the power plant for which he is responsible for a period exceeding ninety-six hours, the employer or chief steam engineer shall apply to the chief inspector for a temporary certificate of competency to be issued in accordance with subsection (3), clause (a).

(6) The duration of any temporary certificate of competency is at the discretion of the chief inspector, but in no case shall a temporary certificate of competency be issued for a period longer than six months.

(7) Where a temporary certificate of competency is issued, the chief inspector may impose such conditions on the holder of the certificate of competency as he considers necessary which may include a condition as to the person by whom the holder of the temporary certificate of competency is to be employed while he holds the temporary certificate of competency.

7(1), (2) and (3) Repealed by AR 76/78.

Chapter 7   Regulations for Canadian Provinces and Territories

(4) Where a temporary certificate of competency is issued pursuant to these regulations, a fee is payable in respect thereof in accordance with Schedule G of the Fee Regulation.

8 and 9 Repealed by AR 76/78.

Qualifications and Examinations

10(1) To qualify to take a First Class Engineer's Certificate of Competency examination a candidate shall:

(a) hold a Second Class Engineer's Certificate of Competency or equivalent, and

(b) furnish evidence satisfactory to the chief inspector of employment for a period of

- (i) thirty months as chief steam engineer in a power plant having a capacity exceeding 5000 kW, or
- (ii) thirty months as chief steam engineer in a power plant consisting of coil type drumless boilers used for the sole purpose of underground thermal flooding having a capacity exceeding 10,000 kW, or
- (iii) thirty months as shift engineer in a power plant having a capacity exceeding 10,000 kW, or
- (iv) thirty months as shift engineer in a power plant consisting of coil type drumless boilers used for the sole purpose of underground thermal flooding having a capacity exceeding 15,000 kW, or
- (v) forty-five months as assistant shift engineer in a power plant having a capacity exceeding 10,000 kW, or

Regulations for Canadian Provinces and Territories   Chapter 7

- (vi) one-half that specified in subclauses (i), (ii), (iii), (iv) or (v) and in addition has been employed for a period of fifteen months in a pressure plant in an operating capacity approved by the chief inspector, or

- (vii) fifteen months as specified in subclauses (i), (ii), (iii), (iv) or (v) and in addition is the holder of a degree in mechanical engineering or equivalent from a university satisfactory to the chief inspector, or

- (viii) one-half that specified in subclauses (i), (ii), (iii), (iv) or (v) and in addition has been employed for a period of thirty-six months in a supervisory capacity satisfactory to the chief inspector on the design, construction, installation, repair, maintenance or operation of equipment to which the Act applies, or

- (ix) thirty months as inspector or boilers and pressure vessels under this Act.(2) Twelve months credit in lieu of power plant operating experience as specified in subsection (1), clause (b), subclauses (i), (ii), (iii), (iv) or (v) may be granted by the chief inspector on successful completion of a course in power engineering satisfactory to the chief inspector, leading towards a First Class Engineer's Certificate of Competency examination.

(3) Twelve months credit in lieu of pressure plant experience as specified in subsection (1), clause (b), subclause (vi) may be granted by the chief inspector on successful completion of a course in power engineering satisfactory to the chief inspector leading towards a First Class Engineer's Certificate of Competency examination.

Chapter 7 Regulations for Canadian Provinces and Territories

(4) The minimum educational requirements to qualify for a First Class Engineer's Certificate of Competency examination are at least 50% standing in Physics 30 or 32, Mathematics 20 or 22 and English 20 or 23, or equivalent, or a pass in Part "A" of a First Class Course in Power Engineering satisfactory to the chief inspector.

(5) The examination shall be divided into two parts, letter A and B and a candidate may:

(a) write any one or all papers for Part A at any scheduled examination after obtaining a Second Class Engineer's Certificate of Competency, or

(b) if he has obtained the experience specified in subsection (1), write the papers for both parts at the same sitting or write any or all the papers for Part B.

(6) The examination shall consist of questions relating to the subjects contained in the current reference syllabus as established by the chief inspector for the First Class Engineer's Certificate of Competency examination.

(7) to obtain a pass:

(a) when a candidate writes a single paper without having previously written any of the papers, the candidate must obtain 70% of the marks for the paper, or

(b) when a candidate writes more than one paper at the same sitting or when a candidate has previously received a pass in any paper or papers, the candidate must obtain an average of 70% of the total marks for the papers and not less than 60% of the marks for each paper.

11(1) To qualify to take a Second Class Engineer's Certificate of Competency examination, a candidate shall:

Regulations for Canadian Provinces and Territories   Chapter 7

(a) hold a Third Class Engineer's Certificate of Competency or equivalent, and

(b) furnish evidence satisfactory to the chief inspector of employment for a period of:

- (i) twenty-four months as chief steam engineer of a power plant having a capacity exceeding 1000 kW, or
- (ii) twenty-four months as chief steam engineer in a power plant consisting of coil type drumless boilers used for the sole purpose of underground thermal flooding having a capacity exceeding 5000 kW, or
- (iii) twenty-four months as a shift engineer in a power plant having a capacity exceeding 5000 kW, or
- (iv) twenty-four months as shift engineer in a power plant consisting of coil type drumless boilers used for the sole purpose of underground thermal flooding having a capacity exceeding 10,000 kW, or
- (v) thirty-six months as shift engineer in a power plant having a capacity exceeding 1000 kW, or
- (vi) twenty-four months as assistant engineer in a power plant having a capacity exceeding 10000 kW, or
- (vii) one-half of that specified in subclauses (i), (ii), (iii), (iv), (v) or (vi) and in addition has been employed for a period of twelve months in a pressure plant in an operating capacity satisfactory to the chief inspector, or
- (viii) twelve months as specified in subclauses (i), (ii), (iii), (iv), (v) or (vi) and in addition is the holder of a

Chapter 7 Regulations for Canadian Provinces and Territories

degree in mechanical engineering or equivalent from a university satisfactory to the chief inspector, or

- (ix) one-half that specified in subclauses (i), (ii), (iii), (iv), (v) or (vi) and in addition has been employed for a period of twenty-four months in supervisory capacity satisfactory to the chief inspector on the design, construction, installation, repair, maintenance or operation of equipment to which the Act applies.

(2) Nine months credit in lieu of power plant operating experience as specified in subsection (1), clause (b), subclauses (i), (ii), (iii), (iv), (v) or (vi) may be granted on successful completion of a course in power engineering satisfactory to the chief inspector leading towards a Second Class Engineer's Certificate of Competency examination.

(3) A candidate who is the holder of a diploma issued by an educational institution, after completing a two year day course in power plant engineering satisfactory to the chief inspector and who is the holder of an Alberta Third Class Engineer's Certificate of Competency, is qualified to take a Second Class Engineer's Certificate of Competency examination after obtaining one-half the qualifying experience as specified in subsection (1), clause (b), subclauses (i), (ii), (iii), (iv), (v) or (vi).

(4) Nine months credit in lieu of pressure plant experience as specified in sub-section (1), clause (b), subclause (vii) may be granted on successful completion of a course in power engineering satisfactory to the chief inspector, leading towards a Second Class Engineer's Certificate of Competency examination.

(5) The minimum educational requirements to take a Second Class Engineer's Certificate of Competency examination are at

Regulations for Canadian Provinces and Territories   Chapter 7

least a 50% standing in Science or Physics 20 or 22, Mathematics 20 or 22 and English 20 or 23, or equivalent, or a pass in Part "A" of a Second Class Course in power engineering, satisfactory to the chief inspector.

(6) The examination shall be divided into two parts, letter A and B, and a candidate may write:

(a) any one or all papers for Part A at any schedule examination after obtaining a Third Class Engineer's Certificate of Competency, or

(b) the papers for both parts at the same sitting or any or all papers for Part B if he has obtained the experience specified in subsection (1).

(7) The examination shall consist of questions relating to the subjects contained in the reference syllabus as established by the chief inspector for the Second Class Engineer's Certificate of Competency examination.

(8) To obtain a pass:

(a) when a candidate writes a single paper, without having previously written any of the papers, the candidate must obtain 70% of the marks for the paper, or

(b) when a candidate writes more than one paper at the same sitting, or when a candidate has previously received a pass in any paper or papers, the candidate must obtain an average of 70% of the total marks for the papers, and not less than 60% of the marks for each paper.

12(1) To qualify to take a Third Class Engineer's Certificate of Competency examination a candidate shall furnish evidence satisfactory to the chief inspector of employment for a period of:

Chapter 7 Regulations for Canadian Provinces and Territories

(a) twelve months as chief steam engineer in a power plant having a capacity exceeding 500 kW while holding a Fourth Class Engineer's Certificate of Competency, or equivalent, or

(b) twelve months as chief steam engineer in a power plant consisting of coil type drumless boilers used for the sole purpose of underground thermal flooding having a capacity exceeding 1000 kW while holding a Fourth Class Engineer's Certificate of Competency or equivalent, or

(c) twelve months as shift engineer in a power plant having a capacity exceeding 1000 kW while holding a Fourth Class Engineer's Certificate of Competency or equivalent, or

(d) twelve months as shift engineer in a power plant consisting of coil type drumless boilers used for the sole purpose of underground thermal flooding having a capacity exceeding 5000 kW while holding a Fourth Class Engineer's Certificate of Competency or equivalent, or

(e) twelve months as assistant engineer in a power plant having a capacity exceeding 5000 kW, while holding a Fourth Class Engineer's Certificate of Competency, or equivalent, or

(f) twenty-four months as building operator in a heating plant having a capacity exceeding 3000 kW while holding a Building Operator A Certificate of Competency and in addition has successfully completed a course in power engineering satisfactory to the chief inspector, leading to a Third Class Engineer's Certificate of Competency examination, or

(g) one-third that specified in clauses (a), (b), (c), (d) or (e) and in addition has been employed for a period of eight months in a pressure plant in an operating capacity satisfactory to the chief inspector, or

(h) one-half that specified in clauses (a), (b), (c), (d) or (e) and in addition is the holder of a degree in mechanical engineering or equivalent from a university satisfactory to the chief inspector, or

(h.i) one-third of the period specified in clauses (a), (b), (c), (d) or (e) and in addition has been employed for a period of two months in a pressure plant in an operating capacity satisfactory to the chief inspector while the holder of a degree in mechanical engineering or equivalent from a university satisfactory to the chief inspector, or

(i) not less than one-half that specified in clauses (a), (b), (c), (d) or (e) and in addition has been employed for a period of twelve months in a capacity satisfactory to the chief inspector, on the design, construction, installation, repair, maintenance or operation of equipment to which the Act applies.

(2) Six months credit in lieu of power plant operating experience as specified in subsection (1), clauses (a), (b), (c), (d) or (e) may be granted by the chief inspector on successful completion of a course in power engineering satisfactory to the chief inspector, leading towards a Third Class Engineer's Certificate of Competency examination.

(3) Six months credit in lieu of pressure plant experience as specified in subsection (1), clause (g) may be granted by the chief inspector on successful completion of a course in power engineering satisfactory to the chief inspector, leading towards a Third Class Engineer's Certificate of Competency examination.

(4) The minimum educational requirements to qualify to take a Third Class Engineer's Certificate of Competency examination are at least a 50% standing in Science or Physics 10 or 12,

Chapter 7   Regulations for Canadian Provinces and Territories

Mathematics 10 or 12 and English 10 or 13, or equivalent, or a pass in Part "A" of a Third Class Course in power engineering satisfactory to the chief inspector.

(5) The examination shall be divided into two parts, lettered A and B, and a candidate may write

(a) any one or all papers for Part A at any scheduled examination after obtaining a Fourth Class Engineer's Certificate of Competency, or

(b) the papers for both parts at the same sitting or any or all papers for Part B if he has obtained the experience specified in subsection (1).

(6) The examination shall consist of questions relating to the subjects contained in the current reference syllabus as established by the chief inspector for the Third Class Engineer's Certificate of Competency examination.

(7) To obtain a pass:

(a) when a candidate writes a single paper, without having previously written any of the papers, the candidate must obtain 60% of the marks for the paper, or

(b) when a candidate writes more than one paper at the same sitting, or when a candidate has previously received a pass in any paper or papers, the candidate must obtain an average of 60% of the total marks for the papers and not less than 50% of the marks for each paper.

(8) Candidates who do not hold a Fourth Class Engineer's Certificate of Competency and who qualify to take a Third Class Engineer's Certificate of Competency examination under section 20, subsection (2), may be granted:

(a) A Fourth Class Engineer's Certificate of Competency if they receive 50% of the total marks allotted for the examination, or

(b) a Fireman's Certificate of Competency if they receive 35% of the total marks allotted for the examination.

13(1) To qualify to take a Fourth Class Engineer's Certificate of Competency examination a candidate shall furnish evidence satisfactory to the chief inspector:

(a) of employment for a period of twelve months assisting in the operation of a power plant having a capacity exceeding 250 kW, or

(b) of being a holder of a degree in mechanical engineering or equivalent from a university satisfactory to the chief inspector, or

(c) of employment for a period of one-half that specified in clause (a) and in addition has been employed for a period of twelve months in a capacity satisfactory to the chief inspector on the design, construction, installation, repair, maintenance or operation of equipment to which the Act applies, or

(d) of employment for a period of twelve months in a pressure plant in an operating capacity satisfactory to the chief inspector, or

(e) of having successfully completed a vocational course in power engineering satisfactory to the chief inspector, or

(f) of employment for a period of twelve months in a heating plant exceeding 750 kW while the holder of a Building Operator's Certificate of Competency and in addition has successfully completed a course in power engineering satisfactory to the chief inspector, leading to a Fourth Class Engineer's Certificate of Competency examination.

(2) Six months credit in lieu of power or pressure plant operating experience as specified in subsection (1), clause (a) or (d) may be granted by the chief inspector on successful completion of a course in power engineering satisfactory to the chief inspector, leading towards a Fourth Class Engineer's Certificate of Competency examination.

(3) A candidate who has successfully completed the first full term of a two year day course in power plant engineering from an educational institution satisfactory to the chief inspector is qualified to take a Fourth Class Engineer's Certificate of Competency examination.

(4) A candidate who is the holder of a diploma in gas technology after completing a two year day course from an educational institution satisfactory to the chief inspector is qualified to take a Fourth Class Engineer's Certificate of Competency examination.

(5) The examination shall be divided into two parts, letter A and B and a candidate may write:

(a) Part A at any scheduled examination after:

- (i) obtaining a Fireman's Certificate of Competency,
- (ii) 6 months' employment as specified in subsection (1)(a) or (d), or
- (iii) successfully completing a course in power engineering satisfactory to the chief inspector, leading towards a Fourth Class Engineer's Certificate of Competency examination, or

(b) both parts at the same sitting or Part B if he has obtained the experience specified in subsection (1).

(6) The examination shall consist of questions relating to the subjects contained in the current reference syllabus as established by the chief inspector for the Fourth Class Engineer's Certificate of Competency examination.

(7) To obtain a pass:

(a) when a candidate writes a single paper, without having previously written any of the papers, the candidate must obtain 60% of the marks for the paper, or

(b) when a candidate writes more than one paper at the same sitting, or when a candidate has previously received a pass in any paper or papers, the candidate must obtain an average of 60% of the total marks for the papers and not less than 50% of the marks for each paper.

14(1) To qualify to take a Fireman's Certificate of Competency examination a candidate shall furnish evidence satisfactory to the chief inspector:

(a) of having acted as fireman operating any boiler for a period of at least six months, or

(b) of having successfully completed a vocational course in boiler operation satisfactory to the chief inspector.

(2) Three months credit in lieu of operating experience as specified in subsection (1) may be granted by the chief inspector upon successful completion of a course in boiler operation satisfactory to the chief inspector, leading towards a Fireman's Certificate of Competency examination.

Chapter 7   Regulations for Canadian Provinces and Territories

(3) To qualify for a Fireman's Certificate of Competency, a candidate must receive 50% of the total marks allotted for the examination.

(4) The examination shall consist of questions relating to the subjects contained in the current reference syllabus as established by the chief inspector for the Fireman's Certificate of Competency examination.

15(1) To qualify to take a Special Oil Well Operator's Certificate of Competency examination a candidate shall furnish evidence satisfactory to the chief inspector that he has:

(a) obtained six months experience in a power plant on an oil drilling site, or

(b) successfully completed a vocational course in boiler operation satisfactory to the chief inspector.

(2) To qualify for a Special Oil-Well Operator's Certificate of Competency a candidate must receive 50% of the total marks allotted for the examination.

(3) The examination shall consist of questions relating to the subjects contained in the current reference syllabus as established by the chief inspector for the Special Oil-Well Operator's Certificate of Competency examination.

16(1) To qualify to take a Building Operator A Certificate of Competency examination a candidate shall furnish evidence satisfactory to the chief inspector that the candidate:

(a) has been employed for a period of 12 months in the operation of a heating plant, or

(b) has successfully completed a vocational course in heating plant operation satisfactory to the chief inspector.

Regulations for Canadian Provinces and Territories   Chapter 7

(2) Six months credit in lieu of operating experience as specified in subsection (1) may be granted upon successful completion of a course leading towards a Building Operator A Certificate of Competency, satisfactory to the chief inspector.

(3) The examination shall be divided into 2 parts lettered A and B and a candidate may write:

(a) Part A at any schedule examination after:

- (i) obtaining a Building Operator B Certificate of Competency,
- (ii) 6 months' employment as specified in subsection (1)(a), or
- (iii) successfully completing a course satisfactory to the chief inspector leading towards a Building Operator A Certificate of Competency examination, or

(b) both parts at the same sitting or Part B if he has obtained the experience specified in subsection (1)

(4) The examination shall consist of questions relating to the subjects contained in the current reference syllabus as established by the chief inspector for the Building Operator A Certificate of Competency examination.

(5) To obtain a pass:

(a) when a candidate writes a single paper, without having previously written any of the papers, the candidate must obtain 50% of the marks for he paper, or

(b) when a candidate writes more than one paper at the same sitting, or when a candidate has previously received a pass in any paper or papers, the candidate must obtain an average of

Chapter 7 Regulations for Canadian Provinces and Territories

50% of the total marks for the papers and not less than 40% of the marks for each paper.

17(1) To qualify to take a Building Operator B Certificate of Competency examination, a candidate shall furnish proof of having been employed for a period of six months in the operation of any heating plant.

(2) Three months credit in lieu of operating experience as specified in subsection (1) may be granted upon successful completion of a course satisfactory to the chief inspector, leading towards a Building Operator B Certificate of Competency.

(3) To qualify for a Building Operator B Certificate of Competency, a candidate must receive 50% of the total marks allotted for the examination.

(4) The examination shall consist of questions relating to the subjects contained in the current reference syllabus as established by the chief inspector for the Building Operator B Certificate of Competency examination.

17.1(1) To qualify to take a Special Boiler Operator's Certificate of Competency examination, a candidate shall furnish proof of being currently employed in the operation of a power plant not exceeding 250 kW.

(2) Repealed AR 355/84.

(3) To qualify for a Special Boiler Operator's Certificate of Competency, a candidate must receive a 50% of the total marks allotted for the examination.

(4) The examination shall consist of questions relating to the subjects contained in the current reference syllabus as established by the chief inspector for the Special Boiler Operator's Certificate of Competency examination.

Regulations for Canadian Provinces and Territories   Chapter 7

18(1) The chief inspector may provide a person credit for taking courses in power engineering or heating plant operation that he considers satisfactory in lieu of practical experience, but that credit shall only be permitted once for each class of engineer's or building operator's examination.

(2) A candidate shall not receive credit for a course in power engineering satisfactory to the chief inspector in lieu of the specified minimum power plant operating experience as required for each class of engineer's examination.

(3) A candidate who has received a Fourth Class Engineer's Certificate of Competency by examination and who has successfully completed the first full term of a two year day course in power engineering satisfactory to the chief inspector is qualified:

(a) to undertake a Third Class Engineer's Certificate of Competency examination during the second full term of the course,

(b) to receive a Third Class Engineer's Certificate of Competency after passing that examination and obtaining 3 months of the experience specified in section 12(1)(a),(b),(c),(d), or (e) subsequent to successfully completing the second year of the course, and

(c) to write Part A, Second Class Engineer's Certificate of Competency examination after passing the examination referred to in clause (a).

19 The assessment of equivalent education which may be accepted in lieu of educational minimum requirements under these regulations shall be determined by the chief inspector.

20(1) When a candidate has experience made up in part as chief steam engineer, shift engineer, assistant shift engineer, as-

Chapter 7 Regulations for Canadian Provinces and Territories

sistant engineer, building operator, or other experience, the chief inspector may evaluate that experience.

(2) When a candidate can furnish proof of experience acceptable to the chief inspector with equipment to which the Act applies, other than that specifically mentioned in sections 10, 11, and 12, the chief inspector may allow such credit in lieu of the specified experience therein.

(3) When a power plant is in operation for only part of a year and the engineer is retained for the non-operational period and is employed on plant maintenance, the chief inspector may grant a credit of two-thirds of the maintenance time towards experience required for a higher level of examination.

(4) Credit for practical experience previously used in qualifying for an engineer's examination shall not be used again in qualifying for a higher level of examination.

(5) When a plant consists of a combination of a power plant and a heating plant, the boiler rating of each plant shall be considered separately when assessing the experience required to qualify to take an engineer's of building operator's examination.

21(1) Where the Chief Inspector considers that a person holds a certificate of competency from a jurisdiction outside Alberta equivalent to a certificate of competency specified in these regulations, the chief inspector may, upon application to him, issue an equivalent certificate of competency.

(2) Any certificate of competency issued pursuant to this section may be issued subject to such conditions as the chief inspector considers necessary.

Regulations for Canadian Provinces and Territories Chapter 7

(3) The chief inspector shall not issue a certificate of competency pursuant to this section until he is satisfied as to the applicant's identity, experience and qualifications and for that purpose may require such evidence as he considers necessary.

(4) A candidate from a jurisdiction outside Alberta who has passed any paper of an engineer's examination in that jurisdiction may be given credit by the chief inspector for having passed that paper.

22. At the discretion of the chief inspector, a certificate of competency of a class determined appropriate by the chief inspector may be issued by the chief inspector to the holder of a valid certificate of competency as an engineer from the government of Canada or any province or territory of Canada or any competent authority in any other jurisdiction who makes application therefor accompanied by evidence satisfactory to the chief inspector of the applicant's qualifications and identity.

Application and Conduct of Examinations

23(1) A candidate for examination shall apply on the form prescribed by the chief inspector at least twenty-one days before the date of examination.

(2) Applications for examination shall be submitted for approval to:

(a) the chief inspector in the case of candidates for any examination under these regulations, or

(b) the district inspector in the case of candidates for any examination other than examinations for First, Second, or Third Class Engineers' Certificates of Competency.

Chapter 7   Regulations for Canadian Provinces and Territories

(3) Originals of references or photo copies thereof vouching for the candidate's experience, ability and conduct shall accompany the application.

(4) Original documents shall be returned to the candidate after verification.

(5) The qualifications of a candidate relating to plant operation, engineering experience, ability and general conduct may be proved by references signed by the owner or chief steam engineer of the plant where the candidate was employed, but if such references are not available, a written statement may be accepted if it is made by a person who has personal knowledge of the facts to be established.

(6) Where a candidate for a Special Oil Well Operator, Fireman or Fourth Class Engineer's Certificate of Competency examination is unable to produce the statement referred to in subsection (5), a statutory declaration may be accepted if it is made by the candidate declaring that he has obtained the required operating experience to qualify him for the examination.

(7) Educational qualifications shall be vouched for by documents issued by the institution from which the candidate received training.

24(1) A candidate for examination shall appear at such place and time as the chief inspector may direct.

(2) Examination are under the direction of the chief inspector.

(3) Prior to the examination commencing, the existing certificate of competency held by a candidate must be presented to the person conducting the examination.

(4) Where a candidate for examination fails to appear at the time and place directed by the chief inspector and has not

Regulations for Canadian Provinces and Territories   Chapter 7

given notice of non-appearance or has not within 7 days after the examination date given a satisfactory reason for failure to appear to the chief inspector, the chief inspector may disqualify him from writing any examination for a period not exceeding 3 months from the date of the examination.

25(1) A candidate who is unable to write and who is qualified to take an examination may employ a person to write the examination.

(a) if the person selected is approved by the inspector designated to conduct the examination, and

(b) if the person selected signs a statement on the form prescribed by the chief inspector stating that he is not an engineer, nor has any knowledge of the construction or operation of boilers, pressure vessels, engines, or other equipment to which the Act applies and gives it to the person conducting the examination when the candidate takes the examination.

(2) The chief inspector may authorize an inspector to conduct an oral examination of any candidate who is unable to undertake a written examination, when a suitable amanuensis look up is not readily available.

(3) Notwithstanding subsections (1) and (2), in every case candidates for examination respecting First, Second or Third Class Engineers' Certificates of Competency must complete a written examination and the use of an amanuensis is not permitted.

26(1) The inspector conducting any examination under these regulations may declare a candidate to have failed an examination if:

Chapter 7   Regulations for Canadian Provinces and Territories

(a) formulae or other information not approved or authorized by the chief inspector have been added to or inserted into any published text of a book, table, regulation or code that is taken into the examination room;

(b) a candidate looks at or refers to any material not approved or authorized by the chief inspector during the examination;

(c) a candidate removes or attempts to remove any questions or part thereof from an examination room;

(d) a candidate copies from another candidate;

(e) a candidate communicates with another candidate in any manner during the examination.

(2) Any candidate who contravenes any provision of subsection (1) during an examination may be disqualified by the chief inspector from writing any further examination for a period not exceeding 12 months from the date of examination.

27   Every candidate for examination shall provide pens, ink, pencils, drawing instruments and such other equipment as may be required and permitted for use during the examination.

28   Every paper in an examination specified in these regulations shall be marked by an inspector.

29(1)   A candidate failing to pass a paper in any part of an examination for any class of certificate of competency specified in these regulations on three or more consecutive attempts shall not be examined again for a period of six months from the date of the latest examination attempted.

(2) Repealed by AR 402/82.

Regulations for Canadian Provinces and Territories   Chapter 7

Miscellaneous

30(1) Where calculations are made with respect to the application of this Act or these regulations, boiler rating shall be determined on the basis that:

(a) one square metre of heating surface equals 10 kilowatts,

(b) where electric power is used as the heat source, the boiler rating shall be the maximum kilowatt capacity of the heating element, or

(c) where neither of the above determinations are applicable, an hourly boiler output of 36 megajoules is equivalent to 10 kilowatts.

(2) The heating surface of a boiler shall be determined by computing the area of the surface involved in square metres and where a computation is to be made of a curved surface the surface having the greater radius shall be taken.

31(1) Where any certificate of competency is lost or destroyed, a duplicate certificate of competency may be issued upon evidence being furnished to the satisfaction of the chief inspector, that the original certificate of competency has been lost or destroyed.

(2) An application under subsection (1) must be accompanied by the fee for the issuance of a duplicate certificate of competency as specified in the Schedule.

32 Any persons acquiring any certificate of competency specified in these regulations other than the person whose name appears thereon shall send the certificate of competency to the chief inspector.

33 The chief steam engineer or building operator shall ensure that a log book is maintained to record any matters relating to the operation of the power plant, or heating plant, including a record of the testing and servicing of safety valves and other safety devices and controls.

ACTS for Province of Alberta

Revised Statutes of Alberta 1980, Chapter B-8 with amendments in force as of June 6, 1983

Consolidated September 13, 1984

Part 5
CERTIFICATES OF COMPETENCY
Application for and issue

34(1) A person may, in accordance with the regulations, apply for any certificate of competency specified in this section or any other certificate of competency prescribed in the regulations.

(2) There shall be the following certificates of competency;

(a) First Class Engineer's Certificate of Competency;

(b) Second Class Engineer's Certificate of Competency;

(c) Third Class Engineer's Certificate of Competency;

(d) Fourth Class Engineer's Certificate of Competency;

(e) Fireman's Certificate of Competency;

(f) Special Oil Well Operator's Certificate of Competency;

(g) Pressure Welder's Certificate of Competency;

(h) Building Operator's Certificate of Competency;

(i) Any other certificates of competency and any grade or class thereof that are prescribed in the regulations.

(3) When a person meets the qualifications required and passes any examinations required to be passed, he shall be granted the appropriate certificate of competency.

Authority of engineer's certificate

35. The holder of a certificate of competency the classification of which authorizes him to act as an engineer may sketch, construct, install, operate, repair and give advice on all things pertaining to a power plant in which he is employed but is not entitled to perform welding unless he holds a certificate of competency permitting him to do so.

Welding of boilers, etc.

36(1) No person shall:

(a) weld or offer to weld a boiler, pressure vessel or pressure piping system or a fitting unless that person is the holder of a certificate of competency and a valid performance qualification card issued pursuant to the regulations authorizing him to do that type of welding, or

(b) require, cause or permit the welding of a boiler, pressure vessel or pressure piping system or a fitting unless the person required, caused or permitted to do the welding is the holder of a certificate of competency and a valid performance qualification card issued pursuant to the regulations authorizing him to do that type of welding.

(2) No person shall alter or repair a boiler or pressure vessel by welding unless he is authorized to do so by an inspector.

Complaints, Investigations and Appeals

Chapter 7   Regulations for Canadian Provinces and Territories

Complaint against certificate holder

37(1) When an inspector or any other person is of the opinion that the holder of a certificate of competency:

(a) has acted in an incompetent manner,

(b) has acted in a grossly negligent or dangerous manner,

(c) is incapable of performing the duties that his certificate of competency authorizes or permits him to perform, or

(d) has acted in an improper manner,

He may make a complaint to the chief inspector.

(2) On receipt of a complaint the chief inspector shall make a preliminary investigation into the complaint.

Committee of inquiry

38(1) On the conclusion of a preliminary investigation the chief inspector shall:

(a) direct that no further action be taken, or

(b) make a report to the Minister recommending that a committee of inquiry be established if he considers that there is sufficient evidence to justify an inquiry into the conduct, capability or fitness of the person against whom the complaint was made.

(2) When the chief inspector:

(a) directs that no further action be taken under subsection (1)(a), he shall notify the person making the complaint in writing and give him the reason for his direction, or

(b) recommends to the Minister that a committee of inquiry be established under subsection (1)(b), he shall notify the person making the complaint in writing but not of the reasons for the recommendations.

(3) On receipt of a report pursuant to subsection (1)(b), the Minister shall establish a committee of inquiry consisting of one or more persons and, if more than one person is appointed, appoint one of them as chairman.

Procedure on inquiry

39(1) At least 14 days before a committee of inquiry commences hearings for the purpose of taking evidence or otherwise ascertaining facts, a notice in writing shall be served on the person whose conduct, capability or fitness is the subject of the inquiry.

(a) containing reasonable particulars of the matter to be investigated, and

(b) specifying the time and place of the hearing.

(2) The person who is the subject of the investigation is entitled to be represented by counsel.

(3) Testimony may be adduced before a committee of inquiry in any manner the committee considers proper and the committee is not bound by the rules of law concerning evidence applicable to judicial proceedings, but a certificate of competency shall not be cancelled or suspended on affidavit evidence alone.

(4) Any member of the committee of inquiry is empowered to administer an oath to any witness who is to give evidence before it.

Witnesses and documents

Chapter 7   Regulations for Canadian Provinces and Territories

40(1) The holder of a certificate of competency whose conduct, capability or fitness is being investigated is a compellable witness in any hearing of the committee of inquiry.

(2) A witness may be examined on oath on all matters relevant to the proceedings and shall not be excused from answering any question on the ground that the answer:

(a) might tend to incriminate him,

(b) might subject him to punishment under this Act or the regulations, or

(c) might tend to establish his liability,

- (i) to a civil proceeding at the instance of the Crown or of any person, or
- (ii) to prosecution under any statute,

but the answer so given, if it tends to incriminate him or to establish his liability to a civil proceeding, shall not be used or received against him in any civil proceedings, in any proceedings under section 51 or under any other Act of the Legislature.

(3) A notice to enforce:

(a) the attendance of a witness, and

(b) the production of books, papers and other documents,

before a committee of inquiry may be issued by the chairman of the committee of inquiry stating the time and place at which the witness is to attend or to produce the documents, if any, he is required to produce.

(4) A witness, other than the person whose conduct, capability or fitness is being investigated, who is served with a notice to attend or to produce documents, is entitled to demand and to

be paid the usual fees payable to witnesses in an action in the Court of Queen's Bench.

(5) A witness:

(a) who fails:

- (i) to attend before a committee of inquiry, or
- (ii) to produce any books, records, papers or other documents or things in obedience to a notice issued under this section,

(b) who fails in any other way to comply with a notice issued under this section, or

(c) who refuses to be sworn, or to answer any questions allowed by the committee of inquiry,

is liable to attachment on application to a judge of the Court of Queen's Bench.

(6) For the purpose of obtaining the testimony of a witness who is out of Alberta, a judge of the Court of Queen's Bench, on an application ex parte by the chairman of the committee of inquiry may direct the issue of a commission for the obtaining of the evidence of the witness and the commission shall be issued and evidence taken pursuant to the Alberta Rules of Court, which apply with all necessary modifications.

Failure to attend

41(1) If the person whose conduct, capability or fitness is the subject of an investigation does not attend the inquiry, the committee of inquiry may, on proof of personal service of the notice to attend, proceed with the subject matter of the inquiry in his absence and make its decision under section 42 without further notice to him.

Chapter 7　Regulations for Canadian Provinces and Territories

(2) The non-attendance or refusal to testify by the person who is the subject of the inquiry may be held to be an action in an improper manner within section 42(2)(d).

Disciplinary action

42(1) After completing its inquiry the committee of inquiry shall prepare a written report of the complaint investigated.

(2) The committee of inquiry may dismiss the complaint or give a warning or, if the committee of inquiry is of the opinion that the holder of a certificate of competency:

(a) has acted in an incompetent manner,

(b) has acted in a grossly negligent or dangerous manner,

(c) is incapable of performing the duties that his certificate of competency authorizes or permits him to perform, or

(d) has acted in a manner which is improper,

the committee of inquiry may, by order in writing,

(e) suspend the certificate of competency of the person investigated with or without conditions, or

(f) cancel the certificate of competency.

(3) On the committee of inquiry making its decision the chairman shall cause a notice in writing to be sent to:

(a) the person whose conduct was the subject of the investigation or his counsel, and

(b) the chief inspector,

notifying them of the decision and any order made in connection with it.

Appeal

43(1) A person who is the subject of an order of a committee of inquiry may, within 30 days of the date of the order, appeal to the Court of Queen's Bench.

(2) The appeal shall be commenced by:

(a) filing a notice of appeal with the office of the clerk of the Court at Edmonton or Calgary, and

(b) serving a copy of the notice of appeal on the chief inspector and any other persons the Court may direct,

within 30 days from the date on which the order was made by the committee of inquiry.

(3) The person appealing may, after commencing the appeal and on notice to the chief inspector, apply to the committee of inquiry or to a judge of the Court for an order suspending any order of the committee of inquiry pending disposition of the appeal.

Material in an appeal

44(1) The appeal shall be founded on a copy of the report of the committee of inquiry, a copy of the record, a copy of the findings and order, if any, of the committee of inquiry and a copy of the evidence, if any, received by the committee of inquiry, all of which shall be certified by the chairman of the committee.

(2) The chief inspector shall on request furnish to the appellant or to his counsel or agent the number of copies of the documents mentioned in subsection (1) requested, but not exceeding 9 in any case.

Procedure in appeal

45. Notwithstanding section 44, the Court may:

Chapter 7   Regulations for Canadian Provinces and Territories

(a) receive further evidence by oral evidence or by affidavit, or

(b) direct a trial to determine all or any of the matters in issue.

Court's powers

46(1) The Court, on hearing the appeal, may:

(a) make any other findings that in its opinion ought to have been made,

(b) quash, confirm or vary any order or finding of the committee of inquiry, or

(c) refer the matter back to the committee of inquiry for further consideration by it.

(2) The Court may make any award as to costs that it considers just in the circumstances.

Fee Regulations

Being Alberta Regulation 353/84 with amendments up to and including Alberta Regulation 200/92.

Schedule G

<u>**Engineer's Fees**</u>

1. The following fees are payable for the issuance of a certificate of competency, whether following an examination or not.

(a) First Class Engineer's Certificate of Competency..$186.00

(b) Second Class Engineer's Certificate of Competency..$138.00

(c) Third Class Engineer's Certificate of Competency..$ 96.00

(d) Fourth Class Engineer's Certificate of Competency..$ 53.00

(e) Fireman's Certificate of Competency.......$ 27.00

(f) Special Oil Well Operator's Certificate of Competency..$ 27.00

(g) Building Operator A Certificate of Competency..$ 53.00

(h) Building Operator B Certificate of Competency..$ 27.00

(i) Special Boiler Operator's Certificate of Competency..$ 27.00

2(1) Subject to subsection (2), the fees payable for the issuance of a temporary certificate of competency are the same as the fees payable under section 1, for the corresponding certificate of competency.

(2) Where the chief inspector is satisfied that a temporary certificate of competency is required for holiday, emergency or sick relief, the fee payable for the issuance of the temporary certificate of competency is 50% of the fee specified in subsection (1).

(3) The fee payable for a duplicate certificate of competency is $27.

3.1. The fee payable for writing an examination is $20.

4(1) If an inspector is requested to conduct an examination in a place where engineers are not normally examined, the in-

Chapter 7   Regulations for Canadian Provinces and Territories

spector may direct the person requesting the examination to pay a special examination fee under this section.

(2) The fee payable for a special examination is $310.00 for each regular working day.

(3) The person requesting the examination shall also pay the cost of the inspector's traveling and subsistence expenses incurred in conducting the special examination at the current rate provided in the Subsistence, Travel and Moving Expense Regulation under the Public Service Act.

Province of British Columbia

SUMMARY OF QUALIFICATIONS:

1. Exam Location, Address and Phone Number

Office Location

Boiler and Pressure Vessel Safety Branch
Suite 300
750 Pacific Blvd. South
Vancouver, British Columbia, Canada V6B 5E7

Testing Sites: Vancouver, Victoria, Terrace, Prince George, Nelson, Nanaimo, Kelowna and Fort St. John

2. Exam Days and Times

Varies - 3rd Class, 4th Class, and Boiler Operator are held every Wednesday of every month except when there are 1st and 2nd Class exams being held. 1st Class exams held 3 times a year. 2nd Class exams held 4 times a year.

3. Examination Cost

$36.00 document fee - pay once per class of certificate.

Papers are $24.00 each. The class you want to obtain depends on how many papers you write.

4. Certificate of Competency Cost and Renewal

From Schedule H - Examination and Licensing Fees - Power Engineers and Boiler Operators

1st Class Power Engineer.....$191.00

2nd Class Power Engineer.....$156.00

3rd Class Power Engineer.....$114.00

4th Class Power Engineer......$ 78.00

Boiler Operator......................$ 73.00

5. Type of Test

All Written and depending on examiner could be oral if person requires it.

6. Grades of Certificates

First Class Power Engineer

Second Class Power Engineer

Third Class Power Engineer

Fourth Class Power Engineer

Boiler Operator

7. Experience and Education Requirements

See Regulations 13 through 17.

8. Canadian Citizenship Required?

Yes

9. Local Residency Required?

Generally, must work or live in Province of British Columbia.

10. Recognition of licenses from other locales?

See Section 24 and 25 below.

Chapter 7   Regulations for Canadian Provinces and Territories

REGULATION for Province of British Columbia

Power Engineers and Boiler and Pressure Vessel Safety Act

REGULATION RESPECTING CERTIFICATES OF COMPETENCY, LICENSES AND REGISTRATIONS

PART 2

CLASSES AND CERTIFICATES OF COMPETENCY

Certificates of competency

2. (1) There shall be the following classes of final certificates of competency:

(a) First Class Power Engineer's certificate;

(b) Second Class Power Engineer's certificate;

(c) Third Class Power Engineer's certificate;

(d) Fourth Class Power Engineer's certificate;

(e) Boiler Operator's certificate;

(f) Limited certificate;

(g) Refrigeration Operator's certificate;

(h) Pressure Welder's certificate.

(2) There shall be the following classes of interim certificates of competency for each of the classes set out in subsection (1):

(a) Temporary "A" certificate;

(b) Temporary "B" certificate.

Duration of certificates

3. (1) A final certificate of competency shall be in force for the life of the person to whom it was issued unless it is under suspension, returned under section 26(2) or was revoked under the Act or the former Act.

(2) Subject to subsections (3) and (4), an interim certificate of competency shall be in force for the period set out on the certificate unless it is under suspension or was revoked under the Act or the former Act.

(3) No temporary "A" certificate of competency shall be in force for longer than 6 months from the date of issue.

(4) No temporary "B" certificate of competency shall be in force for longer than 30 days from the date of issue.

Renewal of interim certificates of competency

4. (1) On application of a holder of a temporary "A" certificate of competency, the director may renew the temporary "A" certificate once, for a period of not longer than 6 months.

(2) On application of a holder of a temporary "B" certificate of competency, the director may renew the temporary "B" certificate twice, for a period of not longer than 30 days each.

PART 3

POWERS CONFERRED BY CERTIFICATES OF COMPETENCY

Who may be in charge of a plant

5. Only a person who is entitled under section 6 or 7 to be a Chief Engineer, Assistant Chief Engineer, Shift Engineer, Assistant Shift Engineer or Assistant Engineer of or in charge of a

Chapter 7   Regulations for Canadian Provinces and Territories

plant shall be the Chief Engineer, Assistant Chief Engineer, Shift Engineer, Assistant Shift Engineer or Assistant Engineer of or in charge of the type of plant permitted under that section.

Scope of Power Engineer's certificates

6. (1) A First Class Power Engineer's certificate entitles the holder to be chief engineer of any plant.

(2) A Second Class Power Engineer's certificate entitles the holder to be:

(a) Chief Engineer of a power plant that has a heating surface of 1000 m^2 or less, and

(b) Shift Engineer of any power plant.

(3) A Third Class Power Engineer's certificate entitles the holder to be:

(a) Chief Engineer of a power plant that has a heating surface of 500 m^2 or less,

(b) Shift Engineer of a power plant that has a heating surface of 1000 m^2 or less,

(c) Chief Engineer of any low pressure steam plant, and

(d) in charge of any low pressure hot water plant or any low pressure organic fluid plant.

(4) A Fourth Class Power Engineer's certificate entitles the holder to be:

(a) Chief Engineer of a power plant that has a heating surface of 100 m^2 or less,

(b) Shift Engineer of a power plant that has a heating surface of 500 m^2 or less,

(c) Chief Engineer of a low pressure steam plant that has a heating surface of 300 m^2 or less,

(d) Shift Engineer of any low pressure steam plant, and

(e) in charge of a low pressure

- (i) hot water plant, or
- (ii) organic fluid plant that has a heating surface of 1000 m^2 or less.

(5) A person who holds a Power Engineer's certificate that is one class lower than the Power Engineer's certificate required under this section to be

(a) a Chief Engineer of a plant, may act as an Assistant Chief Engineer of the plant, or

(b) a Shift Engineer of a plant, may act as an Assistant Shift Engineer of the plant.

(6) A person who holds any class of Power Engineer's certificate may act as an Assistant Engineer.

(7) Notwithstanding subsections (2) to (4), the holder of a Second, Third or Fourth Class Power Engineer's certificate or a Boiler Operator's certificate may perform the duties for the size of plant authorized for the holder of a First, Second or Third Class Power Engineer's certificate respectively where the boilers in the plant are unfired boilers.

(8) The holder of any class of Power Engineer's certificate of competency may perform the duties authorized for the holder of a Power Engineer's certificate of a lower classification.

Scope of Boiler Operator's certificate

Chapter 7   Regulations for Canadian Provinces and Territories

7. A Boiler Operator's certificate or any class of Power Engineer's certificate entitles the holder to be:

(a) Chief Engineer in a:

- (i) power plant that has a heating surface of 50 m^2 or less, or

- (ii) low pressure steam plant that has a heating surface of 200 m^2 or less,

(b) Shift Engineer in a:

- (i) power plant that has a heating surface of 100 m^2 or less, or

- (ii) low pressure steam plant that has a heating surface of 300 m^2 or less, and

(c) In charge of a low pressure:

- (i) hot water plant, or

- (ii) organic fluid plant

that has a heating surface of 500 m^2 or less.

Scope of limited certificates

8. A limited certificate entitles the holder to operate the boilers named on the certificate:

(a) in a power plant that has a heating surface of 10 m^2 or less,

(b) in a low pressure steam plant that has a heating surface of 30 m^2 or less,

(c) in a low pressure

- (i) hot water plant, or

- (ii) organic fluid plant that has a heating surface of 300 m² or less, or

(d) notwithstanding this Part, in a power plant where he was qualified to operate the boiler but, as a result of an increase in the heating surface or change in the operating procedure, a higher class of Power Engineer's certificate would otherwise be required.

Scope of Refrigeration Operator's certificate

9. (1) A Refrigeration Operator's certificate or any class of Power Engineer's certificate entitles the holder to be in charge of a refrigeration plant.

(2) Only a person who holds a Refrigeration Operator's certificate or a Power Engineer's certificate may be in charge of a refrigeration plant.

Scope of Pressure Welder's certificate

10. (1) Subject to section 14 of the Act, a Pressure Welder's certificate entitles the holder to do the type of pressure welding that is set out in the certificate in order to construct, repair, alter or install equipment to which the Act applies.

(2) Only a person who holds a pressure welder's certificate may do pressure welding in relation to the construction, repair, alteration or installation of equipment to which the Act applies.

Interim certificates of competency

11. Notwithstanding this Part, an interim certificate of competency entitles the holder to perform the duties set out on the certificate.

PART 4

Chapter 7 Regulations for Canadian Provinces and Territories

REQUIREMENTS AND QUALIFICATIONS FOR CERTIFICATES OF COMPETENCY

General requirements

12. Under section 16 of the Act, every applicant for a certificate of competency shall:

(a) Repealed.(B.C. Reg. 142/90.)

(b) be able to write, read and speak English, and

(c) supply with his application a written statement from his employer or the Chief Engineer of the plant in which he is employed that:

- (i) is signed by the person giving the written statement,

- (ii) includes a description of the plant in which the applicant is employed, including the boiler heating surface and types of equipment in the plant, and

- (iii) sets out the knowledge, professional qualifications and experience of the applicant.

Qualifications for First Class Power Engineer's certificate

13. (1) Under section 16 of the Act, an applicant for a First Class Power Engineer's certificate shall

(a) hold a Second Class Power Engineer's certificate or an interprovincial Second Class Power Engineer's certificate of competency, and

(b) have been employed for a period of not less than

- (i) 36 months as Chief Engineer of a power plant that has a heating surface exceeding 500 m^2,

- (ii) 36 months as an Assistant Chief Engineer of a power plant that has a heating surface exceeding 1000 m^2,

- (iii) 48 months as a Shift Engineer of a power plant that has a heating surface exceeding 500 m^2,

- (iv) 48 months as an assistant shift engineer of a power plant that has a heating surface exceeding 1000 m^2, while holding a Second Class Power Engineer's certificate,

- (v) 24 months in the position and in the type of plant set out in subparagraph (i), (ii), (iii) or (iv), have successfully completed a first class power engineering course that has been approved by the director and have been employed for a period of not less than 36 months as a maintenance engineer or qualified mechanic, or

- (vi) 24 months in the position and in the type of plant set out in subparagraph (i), (ii), (iii) or (iv) and hold an engineering degree.

(2) Notwithstanding subsection (1) (b) (i) to (iv) but subject to section 22, where an applicant has successfully completed a first class engineering course that has been approved by the director, at a technical or vocational school or through a correspondence school, the required periods of employment referred to in subsection (1) (b) (i) to (iv) are reduced by one year.

Qualifications for Second Class Power Engineer's certificate

14. (1) Under section 16 of the Act, an applicant for a Second Class Power Engineer's certificate shall:

Chapter 7 Regulations for Canadian Provinces and Territories

(a) hold a Third Class Power Engineer's certificate or an interprovincial Third Class Power Engineer's certificate of competency, and

(b) have been employed for a period of not less than:

- (i) 36 months as a Chief or shift engineer of a power plant that has a heating surface exceeding 250 m^2,

- (ii) 48 months as an Assistant Shift Engineer of a power plant that has a heating surface exceeding 1000 m^2,

- (iii) one half of the period set out in subparagraph (i) or (ii) in the position and in the type of plant set out in that subparagraph, have successfully completed a Second Class Power Engineering course that has been approved by the director, and have been employed for a period of not less than 24 months as a qualified mechanic.

- (iv) one half of the period set out in subparagraph (i) or (ii) in the position and in the type of plant set out in that subparagraph and hold an engineering degree, or a diploma issued after completing a 2 year day course in power engineering that has been approved by the director, or

- (v) 48 months as a maintenance engineer of a pressure plant that has a heating surface exceeding 1000 m^2.

(2) Notwithstanding subsection (1) (b) (i) and (ii), but subject to section 22, where an applicant has successfully completed a Second Class Power Engineering course that has been approved by the director, at a technical or vocational school or through a correspondence school, the required periods of em-

ployment referred to in subsection (1) (b) (i) and (ii) are reduced by 9 months.

Qualifications for Third Class Power Engineer's certificate

15. (1) Under section 16 of the Act, an applicant for a Third Class Power Engineer's certificate shall:

(a) have been employed for a period of not less than:

- (i)   24 months as a power engineer of a power plant that has a heating surface exceeding 50 m^2,

- (ii) 48 months as a power engineer trainee of a power plant that has a heating surface exceeding 50 m^2 or as an assistant engineer of a plant that has a heating surface exceeding 1000 m^2,

- (iii) one half of the period set out in subparagraph (i) or (ii) in the position and in the type of plant set out in that subparagraph and have been employed for a period of not less than 12 months as a qualified mechanic.

- (iv) one half of the period set out in subparagraph (i) or (ii) in the position and in the type of plant set out in that subparagraph and hold an engineering degree,

- (v)   48 months as shift engineer of a low pressure steam plant that has a heating surface exceeding 300 m^2,

- (vi) 48 months in a plant, if the plant was approved by an inspecting Power Engineer, and hold a Fourth Class Power Engineer's certificate, or

- (vii) 36 months as a maintenance engineer of a high pressure plant that has a heating surface exceeding 50 m^2, and

(b) in the case of paragraph (a) (iii) to (vii), have successfully completed a Third Class Power Engineering course that has been approved by the director.

(2) Notwithstanding subsection (1) (a) (i) and (ii), but subject to section 22, where an applicant has successfully completed a Third Class Power Engineering course that has been approved by the director, at a technical or vocational school or through a correspondence school, the required periods of employment referred to in subsection (1) (a) (i) and (ii) are reduced by 6 months.

(3) Notwithstanding subsection (1), but subject to section 12, a Third Class Power Engineer's certificate shall be issued to a person who:

(a) meets the requirements of section 16 of the Act,

(b) holds a diploma issued after completing a 2 year day course in Power Engineering where:

- (i) the course has been approved, and
- (ii) final examinations for the course are supervised by the director, and

(c) has been employed for at least 3 months in a high pressure plant that has a heating surface of not less than 100 m^2.

Qualifications for Fourth Class Power Engineer's certificate

16. (1) Under section 16 of the Act, an applicant for a Fourth Class Power Engineer's certificate shall:

(a) have been employed for a period of not less than:

- (i) 12 months as a Power Engineer trainee in a power plant that has a heating surface exceeding 10 m^2,

Regulations for Canadian Provinces and Territories   Chapter 7

- (ii) 18 months as a Power Engineer trainee of:

    (A) a low pressure steam plant,

    (B) a low pressure hot water plant, or

    (C) a low pressure organic fluid plant that has a heating surface exceeding 200 m^2,

- (iii) 24 months:

    (A) as a qualified mechanic in, or

    (B) in the maintenance of a low pressure hot water plant that has a heating surface exceeding 500 m^2 and refrigeration equipment that

    (C) contains group 2 or 3 refrigerants and consumes more than 25 kWh, or

    (D) contains group 1 refrigerants and consumes more than 100 kWh,

- (iv) 24 months in industrial instrumentation in a power plant with a heating surface exceeding 1000 m^2,

- (v) 48 months as an apprentice in the trade of machinist, machinist fitter, electrical work, steam fitting and piping or industrial instrumentation,

- (vi) 24 months in the operation of a refrigeration plant that:

    (A) contains group 2 or 3 refrigerants and consumes more than 215 kWh, or

    (B) contains group 1 refrigerants and consumes more than 100 kWh,

Chapter 7   Regulations for Canadian Provinces and Territories

- (vii) 36 months in the operation of a diesel engine power plant, gas engine power plant or a similar type of power plant where the diesel engine, gas engine or similar type of power plant consumes 375 kWh or more,
- (viii) 12 months as an operator of a fuel burning process that is shown to the director to be a process the operation of which demonstrates the person's competence to hold a Fourth Class Power Engineer's certificate,
- (ix) 12 months in the design, construction, installation, repair, maintenance or operation of equipment to which the Act applies, or
- (x) one half of the period and in the type of position set out in subparagraph (i), (ii) or (iii) and hold an engineering degree, and

(b) in the case of subparagraphs (ii) to (ix), have successfully completed a fourth class power engineering course that has been approved by the director.

(2) Notwithstanding subsection (1) (a) (i), but subject to section 22, where an applicant has successfully completed a Fourth Class Power Engineering course that has been approved by the director, at a technical or vocational school or through a correspondence school, the required period of employed referred to in that subsection is reduced by 6 months.

(3) Notwithstanding subsection (1), but subject to section 12, a Fourth Class Power Engineer's certificate shall be issued to a person who:

(a) meets the requirements of section 16 of the Act, and

(b) has successfully completed the first year of a 2 year day course in power engineering, where

- (i) the course has been approved, and
- (ii) final examinations for the course are supervised by the director.

Qualifications for a Boiler Operator's certificate

17. (1) Under section 16 of the Act, an applicant for a Boiler Operator's certificate shall have been employed for a period of not less than:

(a) 8 months as a power engineer trainee in a high or low pressure steam plant with a heating surface of not less than 3 m^2, or

(b) one half of the period and in the type of position set out in section 16(1)(a) (iv), (v), (vi), (vii) or (viii), and have successfully completed a boiler operator's course that has been approved by the director.

(2) Notwithstanding subsection (1) (a), where an applicant has successfully completed a boiler operator's course that has been approved by the director, at a technical or vocational school or through a correspondence school, the required period of employment referred to in subsection (1) (a) is reduced by 4 months.

Limited certificate

18. Subject to section 12, a limited certificate shall be issued to an applicant who meets the requirements of section 16 of the Act and has demonstrated to an inspecting power engineer that he has a thorough knowledge of:

(a) the plant in which he is employed, and

(b) the requirements of the regulations.

Interim certificates

19. (1) Under section 16 of the Act, the owner or Chief Engineer of a plant may apply for a temporary "A" certificate on behalf of a person who has been employed in the plant for a period of at least 12 months where:

(a) the heating surface of the plant is increased, or

(b) the person employed in the plant needs more time to prepare for the examinations required under the Act because of a promotion.

(2) Under section 16 of the Act, the owner or Chief Engineer of a plant may apply for a temporary "B" certificate on behalf of a person employed in the plant where there is an emergency situation in the plant.

Qualifications for a Refrigeration Operator's certificate

20. Under section 16 of the Act, an applicant for a Refrigeration Operator's certificate shall:

(a) have been employed for a period of not less than one year assisting in the operation of a refrigeration plant that contains:

- (i) group 2 or 3 refrigerants and consumes more than 25kWh, or

- (ii) group 1 refrigerants and consumes more than 100 KWh, or

(b) have assisted in the operation of a refrigeration plant referred to in paragraph (a) for a period of not less than 6 months and have successfully completed a course that has been approved by the director.

Regulations for Canadian Provinces and Territories   Chapter 7

Qualifications for a pressure welder's certificate

21. Under section 16 of the Act, an applicant for a pressure welder's certificate shall have taken a performance qualification test that conforms to section 3.5 of CSA Standard B51-M, 1981 Edition.

Reduction of period of employment

22. For the purposes of sections 13 to 16, an applicant for a certificate of competency who:

(a) holds an engineering degree, or a diploma issued after completing a 2 year day course in Power Engineering, and

(b) has taken a course referred to in section 13(2), 14(2), 15(2), or 16(2), may reduce the period that he is required to be employed in a plant under those sections by the amount permitted either because he:

(c) holds an engineering degree, or a diploma issued after completing a 2 year day course in Power Engineering, or

(d) has taken a course referred to in paragraph (b), but he cannot reduce the required period of employment by both periods.

Examinations - power of examiner

23. (1) An examiner may cancel an examination for an applicant or fail that applicant where the applicant has contravened a directive of the director respecting examinations.

(2) Where an examiner is unable to evaluate an applicant's knowledge adequately through a written examination, the director may require that the applicant for a certificate of competency take an oral examination.

Chapter 7 Regulations for Canadian Provinces and Territories

Requirements for a person who holds a certificate from outside the Province

24. (1) Notwithstanding sections 13 to 16, a person who holds an interprovincial or a provincial certificate of competency may be issued a power engineer's certificate of competency

(a) of the same class for which he holds an interprovincial or a provincial certificate of competency where he

- (i) subject to subsection (2) (b), writes the examination for that class of power engineer's certificate of competency, and

- (ii) has, in the province in which he holds the interprovincial or provincial certificate of competency, the equivalent work experience and educational qualifications that are required to obtain a Power Engineer's certificate of competency of that class under this regulation, or

(b) one class lower than that for which he holds a provincial certificate where he has, in the province in which he holds a provincial certificate of competency, the equivalent work experience and educational qualifications that are required to obtain a Power Engineer's certificate of competency of that lower class under this regulation.

(2) The director may exempt from taking the written examination the holder of

(a) a provincial certificate of competency applying for a power engineer's certificate of competency one class lower than that for which he holds a provincial certificate of competency, or

(b) an interprovincial certificate of competency.

Requirements for a person who holds a marine certificate

25. (1) An applicant for a Power Engineer's certificate of competency who holds a Marine Engineer (steam) certificate of competency issued by the government of Canada may be issued a power engineer's certificate of competency one class lower than the grade of certificate which he holds, where he has passed a written examination on the Act and the regulations made under it.

(2) An applicant for a Power Engineer's certificate of competency issued by the government of Canada may be issued a Power Engineer's certificate of competency one class lower than the grade of certificate which he holds, where he has passed a written examination for that class of Power Engineer's certificate of competency.

Rules respecting issue of certificates of competency

26. (1) Where a certificate of competency has been lost or destroyed, the director shall issue a duplicate copy of the certificate of competency to the holder of it.

(2) Before a certificate of competency is issued to an applicant, the director shall ensure that all certificates of competency previously issued to the applicant are returned to the director.

(3) No person shall make a false representation for the purpose of procuring a certificate of competency.

(4) A holder of a certificate of competency:

(a) shall ensure that no person forges or fraudulently alters his certificate of competency, and

Chapter 7 Regulations for Canadian Provinces and Territories

(b) shall not use a certificate of competency after it has been altered.

Posting of certificates of competency

27. When the holder of a certificate of competency is employed in a plant, he shall post the certificate of competency at the place in the plant required by the inspecting power engineer.

Suspension and revocation of certificates of competency

28. (1) The director may revoke a certificate of competency where the holder of the certificate of competency does not meet the requirements for issuing the certificate of competency.

(2) The director may suspend or revoke a certificate of competency where it is proven that the holder :

(a) was under the influence of alcohol or a drug while working,

(b) was incompetent or negligent in carrying out his duties,

(c) Has made false statements to an inspecting Power Engineer, Chief or shift engineer or owner of a plant in respect of the carrying out of his duties under the Act,

(d) has willfully destroyed equipment under his supervision,

(e) has allowed another person to operate under his certificate,

(f) leaves his plant without ensuring that:

- (i) all equipment in the plant to which the Act applies is in a safe condition, or

- (ii) a Power Engineer who is permitted under the regulations to take his position in the plant, is at the plant to relieve him of his duties, or

(g) commits any act that is hazardous to the plant or persons in the plant or performs his duties in a way that is hazardous to the public.

Supervision in power plants

29. (1) Where 2 or more shift engineers are required, each shift engineer shall hold the certificate of competency required under section 6 or 7.

(2) Where a Power Engineer has the written permission of the director, he may be the Chief Engineer of 2 or 3 low pressure steam plants.

(3) Where 2 or more Power Engineers are employed in a plant, the owner or person who is in charge of the plant shall designate one of the power engineers as Chief Engineer and the other Power Engineers shall be under the direction of the Chief Engineer with respect to their duties as power engineers.

Duty to inspect and report

30. (1) A Power Engineer shall thoroughly inspect every boiler, pressure vessel, pressure piping system, fitting and all refrigeration equipment as soon as practicable after he comes on shift.

(2) Where a Power Engineer discovers that any part of the boiler, pressure vessel, pressure piping system, fitting or refrigeration equipment is unsafe, he shall immediately report this to the owner or the person in charge of the plant.

(3) The owner or person in charge of a plant shall notify an inspecting Power Engineer immediately on becoming aware of

anything that renders the equipment or may render the equipment in the plant unsafe.

Absence of Chief Engineer

31. A Chief Engineer shall ensure that a Power Engineer holding a certificate of not less than that required for a shift engineer act as Chief Engineer when he is away from the plant.

SCHEDULE H - EXAMINATION AND LICENSING FEES - POWER ENGINEERS - BOILER OPERATORS - REFRIGERATION OPERATORS

1. Evaluation of Credentials..................................$36

Power Engineers - All Classes

Boiler Operators

Refrigeration Operators

2. Administration and Invigilation of Examinations

Power Engineers - All Classes

Boiler Operators

Refrigeration Operators

For each paper..$24

3. Granting of Certificate of Competency (Interprovincial Program)

Power Engineers

1st Class..$191

2nd Class...$156

3rd Class..$114

4th Class..$ 78

Boiler Operator and Refrigeration Operator............$ 73

4. Issuing Duplicate Certificate of Competency

All Classes..$ 72

Issuing Duplicate Wallet Card Certificate of Competency

All Classes..$ 6

5. Temporary "A" Certificate of Competency

All Classes..$120

Temporary "B" Certificate of Competency

All Classes..$ 72

6. Limited Certificate of Competency

All Categories...$ 60

Chapter 7   Regulations for Canadian Provinces and Territories

Province of Manitoba

SUMMARY OF QUALIFICATIONS:

1. Exam Location, Address and Phone Number

Department of Labor
401 York Ave.
Room 500
Winnipeg, Manitoba, Canada R3C OP8

2. Exam Days and Times

Main cities to test at: Winnipeg, Thompson, The Pas, Flin Flon, Brandon

May test at up to 20 locations per month.

3. Examination Cost

$20.00

4. Certificate of Competency Cost and Renewal

Renewal only - $20.00

See Power Engineers Regulations - Fees - Part I and Part II below)

5. Type of Test

All Essay - 3 1/2 hour test.

6. Grades of Certificates

First Class Certificate

Second Class Certificate

Third Class Certificate

Fourth Class Certificate

Fifth Class Certificate

Special Boiler Operator Certificate

Regulations for Canadian Provinces and Territories — Chapter 7

7. Experience and Education Requirements

See Power Engineers Regulations-Qualifications of Applicants 8(2) through 8(7)

8. Canadian Citizenship Required?

Yes, generally - up to Chief Inspector

9. Local Residency Required?

Yes, generally must be a resident to write exam - up to Chief Inspector

10. Recognition of licenses from other locales?

See The Power Engineers Act - 7(2) & 7(3) - Exception to examination requirement.

ACTS AND REGULATIONS for Province of Manitoba

Chapter P95

The Power Engineers Act

Her Majesty, by and with the advice and consent of the Legislative Assembly of Manitoba, enacts as follows:

Prohibition against operating.

4. No Person shall operate, or permit or employ any other person to operate, any plant unless the person or other person, as the case may be, who is operating the plant:

(a) is the holder of a valid and subsisting certificate authorizing him to operate that plant or the class of plant in which that plant is classified; and

(b) operates the plant in compliance with the provisions of this Act and the regulations; and the plant complies with the provisions of this Act and the regulations.

Chapter 7   Regulations for Canadian Provinces and Territories

Additional requirements.

5. The Lieutenant Governor in Council may, by regulation, set out requirements, in addition to the requirements of the Act, respecting any plant or class of plant, or the operation of any plant or class of plant, or any person operating a plant.

Classification of plants.

6(1) The Lieutenant Governor in Council may, by regulation, classify plants into various classes, and specify any factor required to be taken into account in rating a plant for the purpose of classification.

Classes of certificates.

6(2) The Lieutenant Governor in Council may, by regulation,

(a) classify certificates into various classes; and

(b) specify the class of certificate that is required to operate any plant or class of plant classified under subsection (1).

Issue of certificates.

7(1) The minister may issue a certificate of any of the classes of certificates set out in the regulations to any person who:

(a) submits to the minister an application therefor on a form prescribed and supplied by the minister, together with such proofs in support as the regulations may require;

(b) remits to the minister such application fee as the regulations may provide;

(c) subject to subsections (2), (3) and (4), takes and passes an examination;

(d) where the person takes an examination under clause (c), remits to the minister such examination fee as the regulations may provide; and

(e) satisfies such requirements, in addition to the requirements of clauses (a), (b), (c) and (d), and possesses such qualifications, as the regulations may prescribe.

Exception to examination requirement.

7(2) Any person who complies with clauses (1) (a), (b) and (e) and is the holder of and submits to the minister:

(a) a valid and subsisting certificate of the same class as that of the certificate for which the person is applying; or

(b) a valid and subsisting foreign certificate issued in a reciprocating jurisdiction and being of the same class as or of a class equivalent to that of the certificate for which the person is applying; is not required to take an examination under clause (1) (c).

Discretion to waive examination.

7(3) The minister may, in his discretion, waive the examination required under clause (1) (c) in the case of any person who:

(a) complies with clauses (1) (a), (b) and (e); and

(b) is the holder of and submits to the minister a valid and subsisting foreign certificate issued in a foreign jurisdiction other than a reciprocating jurisdiction and being a certificate of the same class as or of a class equivalent to that of the certificate for which the person is applying.

Prerequisites for examination.

7(4) Where any person is required to take an examination under clause (1) (c), the examination shall not be administered

Chapter 7 Regulations for Canadian Provinces and Territories

unless and until the person complies with clauses (1) (a), (b), (d) and (e).

Examinations.

7(5) The minister shall cause any examination required under clause (1) (c) to be set, administered and marked. R.S.M. 1987 Corr.

Form of certificate.

7(6) The minister may prescribe the form of any certificate.

Expiry of certificates.

7(7) Every certificate expires on a date provided in the regulations.

Reciprocating jurisdictions.

7(8) For the purposes of subsections (2) and (3), the minister may designate any province, state, country or other jurisdiction outside of Manitoba as a reciprocating jurisdiction.

Refusal of certificate.

8. Where the minister refuses to issue a certificate to any person, the minister shall, forthwith upon making the refusal, cause to be served upon the person a notice of refusal stating the refusal and the reason for the refusal.

Suspension of certificate.

9(1) Where the minister believes on reasonable grounds that any person holding a certificate

(a) obtained the certificate through a false or misleading statement; or

(b) has failed to observe a provision of this Act or the regulations; or

(c) is an alcoholic or drug addict, or has operated a plant while under the influence of alcohol or any drug; or

(d) is in a condition that renders it or is likely to render it unsafe for the person to operate a plant; or

(e) has assumed charge of a plant, knowing that he is incapable of operating the plant safely; or

(f) has operated a plant, or has conducted himself while operating a plant, in a manner that creates or is likely to creates an unsafe condition in the plant; or

(g) has committed an indictable offense; or

(h) is guilty of conduct that in the opinion of the minister is contrary to the public interest or the interests of safety; the minister may forthwith suspend the certificate by serving the person with a notice of suspension.

Notice of suspension.

9(2) A notice of suspension served under subsection (1) shall state the suspension, the date when the suspension becomes effective and the reason for the suspension.

Surrender of certificate.

9(3) Any person whose certificate is suspended under subsection (1) shall, forthwith upon the suspension becoming effective, surrender or deliver the certificate to the minister to be retained by the minister pending the holding of a hearing and the making of an order under section 10.

Power Engineers Regulation

Classes of certificates

Chapter 7   Regulations for Canadian Provinces and Territories

4(1) Certificates shall be of eight classes, corresponding respectively to the eight classes of plants set out in subsection 3(1), and a certificate of any class authorizes the holder thereof to operate any plant or class of plant specified in section 5 for that class of certificate, and to perform therein any function and duty authorized or required by that section.

4(2) In addition to the classes of certificates for which provision is made in subsection (1), there shall be:

(a) a Special Qualification Class of certificate; and

(b) an Endorsed Qualification Class of certificate.

Certificate authority

5(1) A person holding a First Class Certificate:

may act as Chief Engineer or Shift Engineer in any plant or class of plant.

5(2) A person holding a Second Class Certificate:

may act as:

(a) Chief Engineer in a Second Class Plant; or

(b) Shift Engineer in any plant or class of plant.

5(3) A person holding a Third Class Certificate:

may act as:

(a) Chief Engineer in a Third Class Plant;

(b) as Shift Engineer in a Second Class Plant; or

(c) Assistant Shift Engineer in a First Class Plant.

5(4) A person holding a Fourth Class Certificate may:

(a) act as Chief Engineer in a Fourth Class Plant;

(b) act as Shift Engineer in a Third Class Plant;

(c) act as Assistant Shift Engineer in a Second Class Plant; or

(d) be in charge of a Refrigeration Class Plant.

5(5) A person holding a Fifth Class Certificate may:

(a) be in charge of the operation of a Fifth Class Plant; or

(b) act as Shift Engineer in a Fourth Class Plant.

5(6) A person holding a Special Boiler Operator Class Certificate may:

(a) be in charge of the operation of a Special Boiler Class Plant; or

(b) act as Shift Engineer in a Fourth Class Plant.

5(7) A person holding a Refrigeration Class Certificate may:

act as Engineer in a Refrigeration Class Plant.

5(8) A person holding a Steam Traction Engine Class Certificate may:

act as operator of a Steam Traction Engine Class Plant.

5(9) A person holding a Special Qualification Class Certificate may:

(a) operate any plant or class of plant specified in the certificate, except a plant or class of plant set out in subsections (1) to (8); and

(b) perform in any plant or class of plant any function that is specified in the certificate, except a function set out in subsections (1) to (8).

Chapter 7 Regulations for Canadian Provinces and Territories

5(10) A person holding an Endorsed Qualification Class Certificate may:

during the period specified in the certificate and only within the plant in which the person is employed, perform any function the certificate authorizes the person to perform, as if the person has fulfilled the experience requirements for the certificate.

Supervision requirements

6(1) Subject to subsection (2) and except as otherwise permitted by the Act or this regulation, a plant as classified in subsection 3(1) shall not operate unless a Power Engineer of the required class for the plant is present and on duty in the operating area.

6(2) While supervising a plant, a Power Engineer may not leave the operating area for more than twenty minutes at a time.

6(3) The operation of a First, Second, Third or Fourth Class Plant shall be under the supervision of a Chief Engineer holding a certificate of a class not lower than is required under this regulation for the operation of the plant.

6(4) The operation of a plant during each shift shall be under the constant supervision of a Shift Engineer, or the Chief Engineer of the plant who may act as Shift Engineer; and the Shift Engineer shall hold a certificate of a class not more than one class below that required to be held by the Chief Engineer for the plant.

6(5) Where an explosion or other accident occurs in a plant, the Chief Engineer of the plant shall immediately report or cause to be reported to the minister, the explosion or accident and any damage to the plant or injury to persons.

6(6) A Shift Engineer in charge of a plant that is controlled through a central control station shall test the automatic safety controls at least once each day the plant is in operation.

6(7) A Shift Engineer in charge of a plant shall maintain a written log showing, for each day of operation:

(a) each check of the plant carried out by him or her and the time and date thereof;

(b) The results of any tests of automatic safety controls;

(c) any abnormal condition in the plant, and the time and date when it is first observed; and

(d) any order given respecting the operation of the plant and the time and date thereof;

and the entries for each shift shall be signed by the Shift Engineer.

6(8) The owner of a plant shall ensure that current copies of The Steam and Pressure Plants Act and The Power Engineers Act, and regulations made under both Acts, are kept with the written log.

6(9) Where an apparently unsafe condition arises in a plant, the Power Engineer in charge of the plant at the time shall:

(a) where the unsafe condition is an emergency, take immediate steps to rectify it;

(b) make and maintain a written record of the occurrence of the unsafe condition and of any step taken to rectify it; and

(c) where the unsafe condition is not rectified within a reasonable time, notify the minister of the unsafe condition.

Exemption regarding constant supervision

Chapter 7   Regulations for Canadian Provinces and Territories

7(10) In the absence of the Chief Engineer or Shift Engineer from a plant, the minister may in writing authorize a Power Engineer holding a certificate of a class not less than one class below that required under this regulation for the Chief Engineer or Shift Engineer of the plant, to act as Chief Engineer or Shift Engineer of the plant for such limited period of time as the minister may specify in the authorization.

7(11) Where the classification of a plant is changed, a Chief Engineer or any Shift Engineer who has been employed in the plant as a Power Engineer for a period of not less than 12 months immediately preceding the change may apply to the minister for an Endorsed Qualification Class Certificate that is not more than one class higher than the certificate currently held by the Chief Engineer or Shift Engineer and, where such Endorsed Qualification Class Certificate is issued, it shall be valid only for the plant and for such period of time as may be specified in the certificate.

7(12) Where a Special Qualification Class Certificate is issued under subsection 5 (9), the certificate is valid for a period not exceeding 12 months in the plant specified in the certificate.

Qualification of applicants

8(1) The requirements and qualifications set out in this section are in addition to any others made under clause 7 (1) (e) of the Act.

An applicant for a First Class Certificate must hold a valid and subsisting Second Class Certificate and:

(a) since the issue of the Second Class Certificate, must have acted:

- (i) as Chief Engineer of a Second Class Plant for a period of not less than two and one-half years,
- (ii) as Shift Engineer in the operation of a First Class Plant for a period of not less than two and one-half years, or
- (iii) as Assistant Shift engineer of a First Class Plant for a period of not less than three and one-half years, assisting in supervising all aspects of the shift operation;

(b) must have acted in one of the capacities described in clause (a) for a period of not less than 15 months and must be a graduate engineer; or

(c) must have successfully completed and passed a First Class Course in power engineering approved by the minister, and acted in any of the capacities described in clause (a) for a period of not less than one-half of the time specified in the clause.

8(3) An applicant for a Second Class Certificate must hold a valid and subsisting Third Class Certificate and:

(a) since the issue of the Third Class Certificate, must have acted:

- (i) as Chief Engineer in the operation of a Third Class Plant for a period of not less than two years,
- (ii) as Shift Engineer in the operation of a Second Class Plant for a period of not less than two years,
- (iii) as Assistant Shift Engineer in the operation of a First Class Plant for a period of not less than two years, or

Chapter 7   Regulations for Canadian Provinces and Territories

- (iv) as Shift Engineer in the operation of a Third Class Plant for a period of not less than three years;

(b) must have acted in any of the capacities described in clause (a) for a period of not less than one-half of the time specified in clause (a) for that capacity and completed at least two years in a supervisory capacity on the repair, design, construction, installation, operation or maintenance of plant equipment;

(c) must have acted in any of the capacities described in clause (a) for a period of not less than one year and must be a graduate engineer; or

(d) must have successfully completed and passed a Second Class Course in Power Engineering approved by the minister, and acted in any of the capacities described in clause (a) for a period of not less than one-half of the time specified in the clause.

8(4) An applicant for a Third Class Certificate must hold a valid and subsisting Fourth Class Certificate and:

(a) since the issue of the applicant's Fourth Class Certificate, must have:

- (i) acted as Chief Engineer in a Fourth Class Plant for a period of not less than one year,
- (ii) acted as Shift Engineer in a Third Class Plant for a period of not less than one year,
- (iii) had charge of the operation of a low pressure heating plant developing more than 3000 kW (300 boiler horsepower) for a period of not less than two years, or

Regulations for Canadian Provinces and Territories   Chapter 7

- (iv) acted as Assistant Shift Engineer in a Second Class Plant for a period of not less than one year;

(b) must have completed at least one year on the design, construction, installation, repair, maintenance or operation of plant equipment, and had experience in assisting in any of the capacities described in clause (a) for a minimum period of not less than one-half of the time specified in the clause for that capacity;

(c) must have had experience in assisting in the operation of a First, Second or Third Class Plant for a period of not less than two and one-half years;

(d) must have completed and passed a Third Class Course in Power Engineering approved by the minister and acted in any of the capacities described in clause (a) for a period of not less than one-half of the time specified in the clause; or

(e) must be a graduate engineer and have acted in any of the capacities described in clause (a) or (b) for a period of not less than one-half of the time specified in the clause for that capacity.

8(5) An applicant for a Fourth Class Certificate must have:

(a) at least one year of experience in assisting in the operation of a high pressure plant developing not less than 250 kW (25 boiler horsepower);

(b) have completed and passed a Fourth Class Course in Power Engineering approved by the minister, and acted in the capacity described in clause (a) for a period of not less than one-half year;

(c) be a graduate engineer;

Chapter 7 Regulations for Canadian Provinces and Territories

(d) have not less than six months of the experience specified in clause (a) and, been employed for a period of not less than 12 months on the design, construction, installation, repair, maintenance or operation of plant equipment.

(e) have been in charge of a low pressure heating plant developing not less than 750 kW (75 boiler horsepower) for not less than 24 months while holding a Fifth Class Certificate or Special Boiler Operator Certificate; or

(f) have at least 24 months experience in a low pressure plant developing not less than 1500 kW (150 boiler horsepower) while holding a Fifth Class Certificate or Special Boiler Operator Certificate.

8(6) An applicant for a Fifth Class Certificate must have:

(a) not less than six months of experience in assisting in the operation of a steam or hot water boiler developing over 200 kW (20 boiler horsepower);

(b) completed and passed the Basic Building Operation Course approved by the minister and have not less than three months of experience in a heating plant developing over 200 kW (20 boiler horsepower); or

(c) completed and passed the Advance Building Operation Course or Building System Technician Course approved by the minister, and have not less than one month of experience thereafter in a heating plant developing over 200 kW (20 boiler horsepower).

8(7) An applicant for a Special Boiler Operator Class Certificate must have:

(a) not less than six months of experience assisting in the operation of a high pressure plant developing over 50 kW (five boiler horsepower); or

Regulations for Canadian Provinces and Territories   Chapter 7

(b) completed and passed the Special Boiler Operator Course approved by the minister, and have not less than three months of experience thereafter in the operation of a high pressure plant capable of developing over 50 kW (five boiler horsepower).

8(8) An applicant for a Refrigeration Class Certificate must have:

(a) not less than one year of experience in assisting in the operation of a refrigeration plant developing over 175 kW (17.5 boiler horsepower); or

(b) completed and passed a course approved by the minister and assisted in the operation of a refrigeration plant developing over 175 kW (17.5 boiler horsepower) for a period of not less than six months.

8(9) An applicant for a steam traction engine class certificate must have:

(a) experience in operating a steam traction engine; and

(b) be competent in the operation of a steam traction engine.

8 (10) Where an applicant has completed a course, other than a course approved by the minister under this section, and passed a final examination in that course, the practical experience requirements shall be reduced proportionately to the value of the course; but time credit allowable under this subsection shall not exceed one year for a First Class Certificate, nine months for a Second Class Certificate, and six months for a Third or Fourth Class Certificate.

8(11) For the purpose of this section, any experience that was obtained by an applicant more than 10 years before the date of the application shall be deemed not to be experience.

Chapter 7　　Regulations for Canadian Provinces and Territories

8(12) The qualifying period for an applicant to write a class of examination, except refrigeration, shall include experience in a steam plant to the satisfaction of the minister.

8(13) For the purpose of this regulation, 1800 hours of operation constitute one year of experience.

8(14) An applicant who fails an examination may apply for re-examination after a period of not less than 90 days elapses from the date of the first examination.

8(15) Where an applicant fails to pass an examination on three attempts, the minister may require the applicant to prove further knowledge or experience before the applicant is eligible for re-examination.

Proofs

9. The following proof shall be submitted in support of an application for a certificate:

(a) where the applicant holds a certificate or foreign certificate, a certified copy of the certificate or foreign certificate; and

(b) where the applicant claims to have experience required under the regulation, a letter of verification from each employer in whose employ the experience was obtained.

Certificate expiry and renewals

10(1) Every certificate issued under this regulation expires on December 31st in the year following its issue.

10(2) a person who holds a valid and subsisting certificate, but is not employed as a power engineer in Manitoba owing to retirement, illness or residence outside the province, may renew the certificate at three year intervals on payment of one-third the applicable renewal fees.

Regulations for Canadian Provinces and Territories Chapter 7

10(3) Where a person referred to in subsection (2) undertakes employment as a Power Engineer in Manitoba for any period of time, on a permanent, temporary, full-time or part-time basis, the person's certificate expires;

(a) on December 31st following the day the employment begins, where less than 12 months have elapsed between the date of issue or renewal of the certificate and the date the employment begins; and

(b) when the employment commences, where more than 12 months elapse between the date of issue or renewal of the certificate and the date the employment begins.

10(4) Subject to subsection (5), where a person who previously held a certificate fails to renew the certificate for three or fewer years and thereafter seeks to renew the certificate, the person shall pay, in addition to the fee payable for the renewal of the certificate, the applicable renewal fee for each of the years of non-renewal.

10(5) Subject to subsection (2), where a person fails to renew a certificate for a period of more than three consecutive years, the person shall, before a renewal certificate is issued to the person,

(a) pass a new examination; or

(b) satisfy the minister that he or she has been continuously employed as a Power Engineer at the level of his or her certificate during the period of non-renewal and pay, in addition to the renewal fee, an amount equal to double the applicable renewal fee for each year during which the person failed to renew the certificate.

Fees

11(1) The fees payable under the Act shall be as set out in the Schedule, and shall be paid by an applicant at the time an application is made.

11(2) Where an applicant for a certificate fails to appear for an examination at a time and place prescribed by the minister, the minister may order that any fee paid in respect of the examination be forfeited.

Repeal

12. The Power Engineers Regulation, Manitoba Regulation 107/87 R, is repealed.

SCHEDULE (Section 11)

FEES

PART I

Certificate

1. The fee an examination or re-examination for a Power Engineer's certificate of a class of certificate referred to in subsections 8 (2) to (9) is $20.00 per paper.

Renewal of Certificate

2. The fee for a renewal of a Power Engineer's certificate of a class referred to in subsections 8 (2) to (9) is $20.00 per year.

Certificate without examination

3. The fee for a certificate that is not a renewal certificate and that is issued without an examination is the same as the appropriate renewal fee for that class of certificate.

PART II

1 Appeal of examination result (per paper)......$40.00

2 Special Class Certificate...........................$50.00

Regulations for Canadian Provinces and Territories   Chapter 7

3   Duplicate Wallet Card....................................$10.00

4   Duplicate Certificate......................................$30.00

Province of New Brunswick
SUMMARY OF QUALIFICATIONS:

1. Exam Location, Address and Phone Number

Office Location

Advanced Education & Labour

P.O. Box 6000

Fredricton, New Brunswick, Canada E3B 5H6

506/453-2336

Exam Locations

Exams held at three local colleges in Fredericton 2nd Tuesday of every month at 8:30 a.m.

Exams can also be taken at St. Johns Community College and Bathhurst Community College - dates and times vary.

2. Exam Days and Times

In Fredericton - 2nd Tuesday of every month at 8:30 a.m.

3. Examination Cost

$17.00

4. Certificate of Competency Cost and Renewal

First issue.........................$45.00

Renewal of certificate.......$28.00

5. Type of Test

All Written

6. Grades of Certificates

First Class Stationary Engineer

Chapter 7   Regulations for Canadian Provinces and Territories

Second Class Stationary Engineer

Third Class Stationary Engineer

Fourth Class Stationary Engineer

7. Experience and Education Requirements

See Below

8. Canadian Citizenship Required?

No

9. Local Residency Required?

No

10. Recognition of licenses from other locales?

Yes, reciprocal with other provinces and territories in Canada for same class of certificate except Ontario and Quebec - reciprocal for one class lower of certificate.

ACTS AND REGULATIONS for Province of New Brunswick

New Brunswick Regulation 84-175

Under the Boiler and Pressure Vessel Act (O.C. 84-607)

CLASSES OF STATIONARY ENGINEER'S LICENSES

12. The classes of Stationary Engineer's licenses required to operate or have charge of a heating plant or power plant or a class of heating plants or power plants are as follows:

(a) First Class Stationary Engineer's License, authorizing the holder to be:

The Chief Stationary Engineer or Shift Engineer of any heating plant or any power plant;

(b) Second Class Stationary Engineer's License, authorizing the holder to be:

- (i) a Chief Stationary Engineer of a power plant not exceeding one thousand therm hour,
- (ii) a Chief Stationary Engineer of any heating plant, or
- (iii) a Shift Engineer of any heating plant or any power plant;

(c) Third Class Stationary Engineer's License, authorizing the holder to be:

- (i) a Chief Stationary Engineer of a power plant or a heating plant not exceeding four hundred therm-hour,
- (ii) a Chief Stationary Engineer of any low pressure heating plant,
- (iii) a Shift Engineer of a power plant not exceeding seven hundred therm hour,
- (iv) a Shift Engineer of any heating plant, or
- (v) an Assistant Shift Engineer of any power plant; and

(d) Fourth Class Stationary Engineer's License, authorizing the holder to be:

- (i) a Chief Stationary Engineer of a high pressure heating plant not exceeding two hundred therm hour,
- (ii) a Chief Stationary Engineer of a low pressure heating plant not exceeding four hundred therm hour,

- (iii) a Shift Engineer of a power plant not exceeding four hundred therm hour,
- (iv) an Assistant Shift Engineer of a heating plant not exceeding seven hundred therm hour,
- (v) a Shift Engineer of a high pressure heating plant or a power plant not exceeding four hundred therm hour,
- (vi) a Shift Engineer of a low pressure heating plant not exceeding four hundred therm hour, or
- (vii) an assistant shift engineer of any low pressure heating plant.

QUALIFICATIONS OF CANDIDATES

13. A candidate for a class of stationary engineer's license issued under the Act shall complete and file an application form with the Chief Inspector.

14. A candidate for a First Class Stationary Engineer's License shall:

(a) have held a Second Class Stationary Engineer's License for at least two years, and

(b) have a total of six years' practical operating experience in a high pressure heating plant or a power plant, two years of which experience were spent in a high pressure heating plant or a power plant having a therm hour rating greater than seven hundred.

15. A candidate for a Second Class Stationary Engineer's License shall:

(a) have held a Third Class Stationary Engineer's License for at least two years, and

(b) have a total of four years' practical operating experience in a high pressure heating plant or a power plant, one year of which experience was spent in a heating plant or a power plant having a therm hour rating greater than four hundred.

16. A candidate for a Third Class Stationary Engineer's License shall:

(a) have held a Fourth Class Stationary Engineer's License for at least one year, and

(b) have a total of two years' practical operating experience in a heating plant or a power plant, one year of which experience was spent in a heating plant or a power plant having a therm hour rating greater than two hundred.

17. A candidate for Fourth Class Stationary Engineer's License shall:

(a) have at least six months' practical operating experience in a heating plant or power plant under the direct supervision of a licensed stationary engineer, or

(b) have completed a course of instruction approved by the Stationary Engineers Board.

18(1) The following may be granted such time in lieu of practical operating experience as the Stationary Engineers Board deems fair and reasonable:

(a) a graduate engineer;

(b) a person having special engineering training in a recognized university or technical institution;

Chapter 7 Regulations for Canadian Provinces and Territories

(c) a person having completed a course in stationary engineering satisfactory to the Stationary Engineers Board; and

(d) a person having experience in the construction or repair of boilers.

18(2) A person who holds a certificate as a First Class or a Second Class Marine Engineer, combination or steam, may be a qualified candidate for any class of stationary engineers license which the Stationary Engineers Board deems fair and reasonable.

18(3) A person who has served or is serving in the Armed Forces may be a candidate for any class of stationary engineers license which in the opinion of the Stationary Engineers Board is fair and reasonable, having regard to his classification and practical operating experience in connection with a heating plant or power plant.

GENERAL REQUIREMENTS

19(1) Except as otherwise provided in the Act and this Regulation, a Stationary Engineer holding a Stationary Engineers License of the class prescribed for the heating plant or power plant shall be in attendance at all times while the boiler is in operation.

19(2) During the temporary period of absence of the Stationary Engineer holding a Stationary Engineers License of the class prescribed for any heating plant or power plant, the Chief Inspector may authorize, in writing, a Stationary Engineer holding a Stationary Engineers License of not more than one class lower to act in his stead for a period of ninety days or less.

19(3) Where one or more holders of any class of Stationary Engineers license are employed to operate a heating plant or power plant on each shift, the employer shall designate one of them as having charge of the heating plant or power plant.

19(4) No employer shall permit an employee who is employed by him as a Stationary Engineer to engage during working hours in any labour or pursuit not immediately connected with the operation of the heating plant or power plant that would interfere with the safe operation of the heating plant or power plant.

19(5) If a Stationary Engineer engages during working hours in a labour or pursuit not immediately connected with the operation of a heating plant or power plant, the chief inspector shall determine whether such labour or pursuit interferes with the safe operation of the heating plant or power plant and his decision is final.

20(1) Every Stationary Engineers License shall expire on the thirty-first day of December of the year for which it was issued.

20(2) The Stationary Engineers Board may renew a Stationary Engineers license from year to year without examination upon application of the holder thereof and upon payment of the prescribed fee.

DUTIES OF CHIEF STATIONARY ENGINEER

21. In addition to the powers and duties prescribed by the Act, a Chief Stationary Engineer shall:

Chapter 7   Regulations for Canadian Provinces and Territories

(a) take all measures necessary to maintain the heating plant or power plant in a safe operating condition and notify the owner of the measures taken,

(b) maintain discipline among the persons employed in the heating plant or power plant or under his control or supervision,

(c) direct and supervise shift engineers in their work and duties for the safe operation of the heating plant or power plant,

(d) ensure that an accurate record of matters that may affect the safety of the heating plant or power plant is made and maintained at all times as required, and

(e) ensure that all operational and maintenance work on the plant is performed in accordance with safe operating procedures and acceptable engineering practices.

DUTIES OF SHIFT ENGINEER

22. In addition to the powers and duties prescribed by the Act, a shift engineer shall:

(a) under the direction and supervision of the Chief Stationary Engineer be responsible for:

- (i) the safe operation of the heating plant or power plant, and
- (ii) the supervision of other employees on his shift who are under his control;

(b) maintain close watch on the condition and repair of all equipment in the heating plant or power plant and report to the Chief Stationary Engineer any condition that may impair the safety of the heating plant or power plant;

(c) take all measures that are necessary to prevent any immediate danger to the heating plant or power plant;

(d) ensure that an accurate record of matters that may affect the safety of the heating plant or power plant is made and maintained at all times during the shift period as required; and

(e) ensure that all maintenance and operational work performed on the heating plant or power plant is in accordance with safe operating procedures and acceptable engineering practices.

DUTIES OF ASSISTANT SHIFT ENGINEER

23. The Assistant Shift Engineer, under the direction and supervision of the Chief Stationary Engineer of the Shift Engineer, as the case may be, shall be responsible for:

(a) the safe operation of a particular section of the heating plant or power plant;

(b) ensuring that an accurate record of matters that may effect the safety of that section of the heating plant or power plant is made and maintained at all times during the shift period;

(c) performing maintenance and operational work on the heating plant or power plant as directed by the Chief Stationary Engineer or the Shift Engineer; and

(d) the work performance of apprentice shift engineers.

LOG BOOK

Chapter 7 Regulations for Canadian Provinces and Territories

24. The owner of a heating plant or power plant shall provide a log book, in a form approved by the Chief Inspector, for use in the heating plant or power plant.

25. The person in charge of a shift in a heating plant or power plant shall record in the log book in respect to his shift:

(a) the date, the number or designation of the shift and his name;

(b) any change from normal operating procedure and the time of such change;

(c) any special instructions that may have been given to achieve the change referred to in paragraph (b) and the name of the person who gave the instructions;

(d) any unusual or abnormal condition observed in the heating plant or power plant and the time thereof;

(e) repairs to any part of the heating plant or power plant and the time such repairs were commenced and, if completed on his shift, the time thereof;

(f) the time of commencing and terminating his shift;

(g) the testing and recording of all safety controls; and

(h) the testing of all safety valves.

26(1) No person shall deface, damage or destroy a log book.

26(2) No person shall remove the log book from the heating plant or power plant without the permission of the owner.

26(3) The owner shall ensure the log book is kept accessible in the heating plant or power plant for at least one year after the last entry therein and shall produce the log book upon the request of a boiler inspector.

Regulations for Canadian Provinces and Territories   Chapter 7

FEES

27. The fees for the purposes of this Regulation are as follows:

(a) for each examination paper..........................$17.00;

(b) for each re-examination paper......................$17.00;

(c) for the issuance of a Stationary Engineer's License under subsection 8(5) of the Act..$45.00;

(d) for the renewal of a Stationary Engineer's License..$28.00;

(e) subject to paragraph (f), for a duplicate Stationary Engineer's License..$28.00;

(f) for a duplicate Stationary Engineer's License if obtained at the time of issue of the Stationary Engineer's License...$17.00;

(g) for the inspection and approval of a guarded plant, per hour or any part of an hour..$45.00 (minimum charge shall be $45.00);

(h) for affixing a seal to a boiler in accordance with section 8, per hour or any part of an hour.............................$45.00, (minimum charge shall be $45.00);

(i) for the issuance of a certificate of competency under subsection 27(1) of the Act..$28.00;

(j) for the renewal of a certificate of competency..$28.00; and

(k) for a duplicate certificate of competency.........$28.00.

Chapter 7 Regulations for Canadian Provinces and Territories

Province of Newfoundland
SUMMARY OF QUALIFICATIONS:
1. Exam Location, Address and Phone Number
Dept. of Labor and Employment Relations
Occupational Health and Safety Branch
Public Safety Diviwon
P.O. Box 8700
St. Johns, Newfoundland, Canada A1B 4J6
709/729-5505

Testing Sites: St. Johns (at Confederation Building), Grand Falls, Corner Brook, Labrador City

2. Exam Days and Times

3 times per year in February. June and November at all locations

3. Examination Cost

Power Engineer First Class Part A..........$40.00

Power Engineer First Class Part B..........$40.00

Power Engineer Second Class Part A......$30.00

Power Engineer Second Class Part B......$30.00

Power Engineer Third Class Part A.........$20.00

Power Engineer Third Class Part B.,.......$20.00

Power Engineer Fourth Class..................$20.00

Fireman...$15.00

4. Certificate of Competency Cost and Renewal
See Below - Regulations-FEES-31.(2) & (3)

A certificate of competency is issued free when a candidate passes the appropriate examination.

The fee for renewal of a certificate for a Power Engineer or Operator is $10.00.

5. Type of Test

All long answer for every class except 4th class which has multiple choice and long answer

6. Classes of Certificates of Competency

Power Engineer First Class

Power Engineer Second Class

Power Engineer Third Class

Power Engineer Fourth Class

Fireman

7. Experience and Education Requirements

See Below - Regulations-Examinations 13. to 18.

8. Canadian Citizenship Required?

Generally, Yes.

9. Local Residency Required?

Generally, Yes.

10. Recognition of licenses from other locales?

See Below - Acts - 34. (1) and (2)

ACTS AND REGULATIONS for Province of Newfoundland

The Boiler and Pressure Vessel and Compressed Gas Regulations, 1983

Newfoundland Regulation 33/83

CLASSIFICATIONS

Chapter 7 Regulations for Canadian Provinces and Territories

12. Certificates of Competency in the form prescribed by the Director are classified as:

(a) Power Engineer First Class;

(b) Power Engineer Second Class;

(c) Power Engineer Third Class;

(d) Power Engineer Fourth Class;

(e) Fireman;

EXAMINATIONS

13. (1) A candidate for examination for a Certificate as Power Engineer First Class shall in addition to the experience required for qualification as a Power Engineer Second Class, furnish proof that:

(a) he has been employed for a period of thirty months as the Chief Power Engineer in a power plant having a capacity exceeding 12000 kilowatts;

(b) he has been employed for a period of thirty months as a Shift Engineer in a power plant having a capacity exceeding 24000 kilowatts;

(c) he has been employed for a period of forty-two months as an assistant shift engineer assisting in supervising all aspects of the shift operation in a plant having a capacity exceeding 24000 kilowatts;

(d) he is a professional engineer and has been employed for a period of fifteen months as specified in paragraphs (a), (b) or (c); or

(e) he has been employed for a period of one-half that specified in paragraphs (a), (b) or (c) and in addition has been employed for a period of thirty-six months as an inspector under the Act; and

(f) he holds a valid Power Engineer's Certificate of Competency Second Class.

(2) One year's credit in lieu of operating experience as specified in paragraphs (a), (b) and (c) of subsection (1) may be granted on successful completion of an approved course in power engineering leading towards a Power Engineer's Certificate of Competency First Class.

(3) The minimum educational requirements to qualify for a first class power engineer's examination shall be as determined by the institution providing the approved course.

14. (1) The examination shall be divided into two parts "A" and "B" and a candidate may:

(a) write the papers for both parts at the one time;

(b) write the papers for Part "A" at any scheduled examination after he has obtained a Power Engineer's Certificate of Competency Second Class; or

(c) write the papers for Part "B" at a time subsequent to writing the papers for Part "A" and completion of the required operating experience.

(2) The examination shall consist of questions relating to the subjects contained in the current reference syllabus as established by the Director for the first class engineer's examination, and

(a) Part "A" shall consist of four papers, and

(b) Part "B" shall consist of four papers.

(3) Candidates shall be allowed three and one-half hours to write each paper.

(4) Where a candidate writes Part "A" and Part "B" of the examination at the one time, he must obtain not less than sixty per centum of the marks for each paper and an average of seventy per centum of the total marks.

(5) Where a candidate writes Part "A" and Part "B" of the examination separately, he must obtain not less than sixty per centum of the marks of each paper and an average of seventy per centum of

(a) the marks for Part "A", and

(b) when Part "B" is written, seventy per centum of the total marks for Part "B".

(6) A candidate will not be required to be re-examined in any paper in which he has received a mark not less than seventy per centum of the total marks allotted for the paper.

15. (1) A candidate for examination of a certificate as a Power Engineer Second Class shall, in addition to the experience required for qualification as a Power Engineer Third Class, furnish proof that:

(a) he has been employed for a period of twenty-four months as the Chief Power Engineer of a power plant having a capacity exceeding 3600 kw;

(b) he has been employed for a period of twenty-four months as a shift engineer in a power plant having a capacity exceeding 10000 kw and not exceeding 20000 kw;

(c) he has been employed for a period of thirty-six months as a shift engineer in a power plant having a capacity exceeding 3600 kw and not exceeding 12000 kw;

Regulations for Canadian Provinces and Territories   Chapter 7

(d) he has been employed for a period of twenty-four months as an assistant shift engineer in a power plant having a capacity exceeding 24000 kw;

(e) he is a professional engineer and has been employed for a period of twelve months as specified in paragraphs (a), (b), (c) or (d); or

(f) he has been employed for a period of one-half that specified in paragraphs (a), (b), (c) or (d) and in addition has been employed for a period of twenty-four months in an approved supervisory capacity on the design, construction, installation, repair, maintenance or operation of equipment to which the Act applies; and

(g) he holds a valid Power Engineers Certificate of Competency Third Class.

(2) Nine months credit in lieu of operating experience as specified in paragraphs (a), (b), (c) or (d) of subsection (1) may be granted on successful completion of an approved course in Power Engineering leading towards a Power Engineer's Certificate of Competency Second Class.

(3) The minimum educational requirements to qualify for a Second Class engineer's examination shall be as specified by the institute or school supplying the approved course.

(4) The examination shall be divided into two parts, "A" and "B", and a candidate may:

(a) write the papers for both parts at the one time;

(b) write the papers for Part "A" at any scheduled examination after he has obtained a Power Engineer's Certificate of Competency Third Class; or

Chapter 7   Regulations for Canadian Provinces and Territories

(c) write the papers for Part "B" at a time subsequent to writing the papers for Part "A" and completion of the required operating experience.

(5) The examination shall consist of questions relating to the subjects contained in the current reference syllabus as established by the Director for the Second Class engineer's examination.

(a) Part "A" shall consist of three papers; and

(b) Part "B" shall consist of three papers.

(6) Candidates shall be allowed three and one-half hours to write each paper.

(7) Where a candidate writes Part "A" and Part "B" of the examination at the one time, he must obtain not less than sixty per centum of the marks for each paper and an average of seventy per centum of the total marks.

(8) Where a candidate writes Part "A" and Part "B" of the examination separately, he must obtain not less than sixty per centum of the marks of each paper and an average of seventy per centum of:

(a) the marks for Part "A" and

(b) when Part "B" is written, seventy per centum of the total marks for Part "B".

(9) A candidate will not be required to be re-examined in any paper in which he has received a mark not less than seventy per centum of the total marks allotted for the paper.

16.(1) A candidate for examination for a certificate as a Power Engineer third class shall, in addition to the experience required for qualification as a Power Engineer fourth class show proof that:

Regulations for Canadian Provinces and Territories   Chapter 7

(a) he has been employed for a period of twelve months as the Chief Power Engineer in a power plant having a capacity exceeding 1200 kilowatts while holding a Power Engineer's Certificate of Competency Fourth Class;

(b) he has been employed for a period of twelve months as a shift engineer in a power plant having a capacity exceeding 2400 kilowatts while holding a Power Engineer's Certificate of Competency Fourth Class;

(c) he has been employed for a period of twelve months as an assistant shift engineer in a power plant having a capacity exceeding 12000 kilowatts while holding a Power Engineer's Certificate of Competency Fourth Class;

(d) he has been employed for a period of twenty-four months as the Chief Power Engineer or a Shift Engineer in a heating plant having a capacity exceeding 9600 kilowatts while holding a Power Engineer's Certificate of Competency Fourth Class; or

(e) he is a professional engineer and has been employed for a minimum period of not less than one-half that specified in paragraphs (a), (b), (c) or (d).

(2) Six months credit in lieu of operating experience as specified in paragraphs (a), (b), (c) or (d) of subsection (1) may be granted on successful completion of an approved course in Power Engineering leading towards a Power Engineer's Certificate of Competency Third Class.

(3) The minimum educational requirements to qualify for a Third Class Power Engineer's examination shall be as specified by the institute or school providing the approved course of study.

Chapter 7 Regulations for Canadian Provinces and Territories

(4) To qualify for a Power Engineer's certificate of competency Third Class, a candidate must receive sixty per centum of the total marks allotted for examination.

(5) The examination shall consist of questions relating to the subjects contained in the current reference syllabus as established by the Director for the Third Class Power Engineer's Examination.

(6) A Third Class examination shall consist of four papers.

(7) A candidate will not be required to be re-examined in any paper in which he has received a mark of not less than sixty per centum of the total marks allotted for the paper.

17. (1) To qualify for a Fourth Class Power Engineer's examination, a candidate shall furnish proof that:

(a) he has been employed for a period of twelve months as a Fireman of a power plant having a capacity of not less than 600 kilowatts;

(b) he has been employed for a period of twenty-four months as a Chief Power Engineer in a heating plant of not less than 1800 kilowatts while holding a Fireman's Certificate of Competency or equivalent;

(c) he has been employed for a period of twenty-four months as a Fireman or shift engineer in a heating plant of not less than 3600 kilowatts while holding a Fireman's Certificate of Competency or equivalent;

(d) he has successfully completed an approval course in Power Engineering; and

(e) he has a minimum standard of education, as required by the institute or school providing the approved course, or its equivalent.

(2) Three months credit in lieu of operating experience as specified in paragraphs (a), (b) or (c) of subsection (1) may be granted on successful completion of an approved correspondence or part-time course in Power Engineering leading towards a Power Engineer's Certificate of Competency Fourth Class.

(3) To qualify for a Power Engineer's Certificate of Competency Fourth Class, a candidate must receive sixty per centum of the total marks allotted for the examination but if he receives fifty per centum of the marks allotted, he may be granted a Fireman's Certificate of Competency.

(4) The examination shall consist of questions relating to the subjects contained in the current reference syllabus as established by the Director for the Fourth Class Engineer's examination.

(5) A Fourth Class examination shall consist of two papers.

(6) A candidate will not be required to be re-examined in any paper in which he has received a mark of not less than sixty per centum of the total marks allocated for the paper.

18. To qualify for a Fireman's Examination, a candidate shall:

have one year or more operating experience on a heating boiler or more than 1500 kilowatts on a regular shift in the boiler room.[8]

23. An applicant for an examination or re-examination for a certificate of competency shall:

---

[8] 19-22 Not Applicable

Chapter 7   Regulations for Canadian Provinces and Territories

(a) complete a statement in the form prescribed by the Director stating accurately all operating experience and training he has acquired relative to the examination for which he is a candidate;

(b) certify that the statements in his application form are true;

(c) where required by the Director, produce references from his present and past employers, if any, as to the length and nature of his service with them;

(d) write the answers to the examination questions presented by the examiner and obtain at least sixty centum in the examination, unless otherwise stated; and

(e) pay the appropriate fee as prescribed in section 31.

24. Examinations for certificates of competency are subject to the following conditions:

(a) examination questions shall be furnished only at the time of the examination;

(b) no one shall be allowed in an examination room during an examination except persons whose duties necessitate their presence;

(c) answers to the examination questions shall be in writing;

(d) examination candidates shall provide themselves with pens, ink, pencils, drawing instruments and other permissible equipment which may be required during the examination;

(e) candidates may use in the examination any regulations which are in force, tables of logarithms, decimal equivalents, areas of circle, steam tables, calculators and slide rules, but marks shall not be given for answers in which candidates do not show the calculations by which the questions were answered;

(f) candidates shall use sketches whenever possible in order to supplement written answers;

(g) candidates may ask an examiner to explain any questions which they do not understand;

(h) candidates shall return all questions papers, together with their answers, working and scrap papers, to the examiner before leaving the examination room;

(i) during an examination, a candidate may not talk to other candidates, write on paper other than that supplied by the examiners, copy any of the examination questions, leave the examination room without the permission of the examiner or use any personal notes or books other than those approved by the examiner under paragraph (e);

(j) a certificate shall not be issued to any candidate who removes or copies with the intent to remove from the examination room any questions given in the examination and the Director may, upon recommendation of the Examiner, cancel any certificate issued prior to the discovery of such removal or intended removal;

(k) the candidate has successfully completed the technical course prescribed by the Director; and

(l) if a candidate fails his examination, sixty days shall elapse before he may apply for re-examination or such other time as may be set at the discretion of the Director.

25. (1) Where an applicant for a Certificate of Competency has failed on three occasions to pass an examination required by the Director, and a period of one year has elapsed since the applicants third attempt to pass any ex-

amination, the applicant may, with permission of the Board, re-write the examination.

(2) Candidates who apply for an examination or re-examination may obtain the prescribed application forms from the department.

(3) The holder of a Certificate of Competency shall renew that certificate before the thirty-first day of December in each year.

(4) A duplicate certificate of competency may be issued where a certificate has been mutilated or under such other circumstances as the Director may see fit.

(5) A Power Engineer who holds a Second Class "A" or First Class "A" Certificate of Competency issued under the Boiler and Pressure Vessel Regulations 1963, shall have their certificates validated for operation in a heating plant only.

CANCELLATION OR SUSPENSION OF CERTIFICATES OF COMPETENCY

26. The Director may cancel or suspend a certificate of competency if the Director is satisfied that the holder of the certificate:

(a) if incompetent or negligent in the discharge of his duties;

(b) has obtained his certificate of competency through misrepresentation or fraud;

(c) has maliciously destroyed his employer's property;

(d) allows or has allowed another person to operate under his certificate of competency;

(e) has attempted, by impersonating another person, to secure falsely a certificate of competency for a person other than himself;

(f) absents himself from the pressure plant in which he is employed before ascertaining that he is relieved of his duties by a proper person or that all machinery and boilers are properly shut down and left in a safe condition;

(g) performs or has performed other related duties within or outside of the pressure plant where, in the opinion of the Board, those unrelated duties are inconsistent with safe practice in the operation of the pressure plant;

(h) verifies the application form of a candidate for examination without personally knowing the truth of the written statements of operating experience therein contained; or

(i) refuses to give such information as the Board or an inspector deems necessary in carrying out the requirements of the Act and these Regulations.

TEMPORARY CERTIFICATES

27. (1) A temporary certificate of competency may be issued to a person on a temporary basis if:

(a) his employer applies for the temporary certificate of competency on the persons behalf; and

(b) his employer certifies that he cannot obtain the services of the holder of a certificate of competency as required by the Act.

(2) Where the Director is not satisfied that the person referred to in subsection (1) should be issued a temporary certificate of competency it may require that person to take such examination as it considers necessary.

Chapter 7 Regulations for Canadian Provinces and Territories

(3) A temporary certificate issued by the Director shall not be more than one grade higher than the certificate of competency held, to permit:

(a) a shift engineer of a power plant to act as Chief Power Engineer of the same plant; or

(b) an assistant shift engineer to act as a shift engineer in the same plant.

(4) An employer shall apply for a temporary certificate of competency on a form prescribed by the Director, which shall contain a declaration that the person in whose favour the application is being made is, to the best of his knowledge, capable of acting in the capacity for which the temporary certificate of competency is being requested.

(5) Where the chief Power Engineer is sick or expects to be absent from the registered plant for a period exceeding ninety-six hours, the employer or chief Power Engineer shall apply for a temporary certificate of competency to be issued in accordance with paragraph (a) of subsection (3).

(6) The duration of any temporary certificate of competency is at the discretion of the Director but in no case shall a temporary certificate of competency be issued for a period longer than six months.

(7) Where a temporary certificate of competency is issued, the Director may impose such conditions on the holder of the certificate of competency as it considers necessary, which may include a condition as to whom the holder of the temporary certificate of competency is to be employed by while he holds the temporary certificate.

(8) Where a temporary certificate of competency is issued pursuant to these regulations a fee is payable in accordance with subsection (6) of section 31.

RESTRICTED CERTIFICATE

28. (1) On application from an employer the Director may issue a Certificate of Authorization to an employee with long time service in a registered plant.

(2) The Certificate is restricted to that registered plant and is non-transferable.

(3) The level of such certification shall not exceed a Certificate of Competency Second Class.

GENERAL REQUIREMENTS

29. (1) A candidate shall at least fifteen days before the date fixed for examination, submit an application as prescribed by the Director, together with the appropriate fee as prescribed in section 31 and copies of any testimonials vouching for the candidate's experience.

(2) The qualification of a candidate relating to plant operation, engineering experience, ability and general conduct may be proved by testimonials signed by the owner or Chief Power Engineer of the plant in which he has been employed, but if such testimonials are not available, statutory declarations may be accepted if they are made by responsible persons who have personal knowledge of the facts which are to be established.

(3) Educational qualifications shall be vouched for by documents issued by the institution in which the candidate received his training.

(4) The Director shall decide whether testimonials or statutory declarations furnished with an application are acceptable.

Chapter 7   Regulations for Canadian Provinces and Territories

(5) The examination for all certificates of competency shall be held in such city or town as the Director may direct.

(6) A Power Engineer holding a certificate issued under these regulations shall have it framed and protected by glass and shall keep it posted in some conspicuous place in the engine room or boiler room or in a place within the registered plant directed by an inspector.

(7) Nothing in these regulations shall be construed to permit any person to use a name, title or description which will lead to the belief that the person is qualified or entitled to practice professional engineering as defined in section 3 of The Newfoundland Professional Engineering (Amendment) Act.

(8) The owner of a registered plant shall not require or permit a Power Engineer or Operator in charge of the plant to perform any duty not related to the operations of the plant, where an inspector is of the opinion that the performance of that duty may endanger the safety of the plant.

(9) A Power Engineer or Operator holding a current certificate of competency of the class required for the registered plant shall be in attendance at all times while the plant is in operation except as otherwise provided for in the Act and the regulations.

(10) No owner shall permit any person to operate or take charge of a registered plant who is not the holder of a valid and current Certificate of Competency of the class required for that plant by these regulations.

(11) No owner shall permit any person to operate or take charge of a registered plant who is not the holder of a valid and current Certificate of Competency of the class required for that plant by these regulations.

(12) The holder of a Power Engineers certificate of competency under these regulations may operate, repair and give advice on all things pertaining to any steam plant, pressure vessel or refrigeration plant or any machine, equipment or accessories, of which his certificate lawfully gives him charge pursuant to the Act.

(13) Where a Power Engineer or Operator is required to hold a certificate under the Act or the regulations, no other certificate or license shall be required of him in any city, town or village within the Province.

(14) A graduate of an approved course of study in Power Plant Engineering at the technologist level may be eligible for a Power Engineers examination Third Class.

(15) An applicant for a Power Engineer's Certificate of Competency Fourth Class who has successfully completed an approved Power Engineer's course may qualify for experience as required by Sections 13 to 22.

(16) A person who successfully completes an approved apprenticeship program may be eligible for a Power Engineer's examination for the Fourth Class level.

(17) A person permanently residing in the province who holds a First Class, Second Class, Third Class or Fourth Class Certificate of current issue by any other province of Canada, may obtain an equivalent certificate subject to conditions as the Director may prescribe.

(18) A person permanently residing in the province who holds a First Class, Second Class, Third Class or Fourth Class Steam Certificate issued under The Canada Shipping Act, may qualify for examination in the equivalent class subject to conditions as the Director may prescribe.

Chapter 7   Regulations for Canadian Provinces and Territories

(19) A person permanently residing in the province who has completed a Power Engineering course under the Canadian Armed Forces program may be qualified for an examination subject to the conditions as the Director may prescribe.

(20) A certificate of competency which has not been renewed for ten consecutive years shall be cancelled.

(21) The owner or chief Power Engineer of a registered plant shall not sign any references or application forms required by these regulations for a candidate for a pressure plant unless he personally knows the truth of the statements contained in the references or applications where he is required to know the truth of those statements.

(22) A person shall be deemed to have the experience prescribed in Sections 13 to 22 only if such experience has been acquired immediately prior to the date of his application or within such longer period as may be determined by the Director to be reasonable.

(23) A candidate for a Power Engineer's examination may, if he so declares, write each paper of Part A or Part B separately provided the papers are written in the sequence as laid down in the syllabus and he meets the requirements of Sections 13 to 17.

(24) When a registered plant is in operation for only part of a year, and the Power Engineer is retained for the non-operational period and is employed on plant maintenance, the Director may grant a credit of one half of the maintenance time toward experience required for a higher level of examination.

(25) Credit for practical experience previously used in qualifying for a Power Engineer's examination shall not be used again in qualifying for a higher level of examination.

POWER ENGINEERS DUTIES

30. (1) In addition to any powers and duties prescribed by the Act and the Analysis Series Power Engineer as published by Canada Manpower and Immigration, a Chief Power Engineer or Chief Operator shall:

(a) take all measures necessary to maintain the plant in safe operating condition and shall notify the owner of the measures taken;

(b) maintain discipline among the persons employed in the plant or under his control or supervision;

(c) direct and supervise shift engineers or shift operators, as the case may be, in their work and duties for the safe operation of the plant;

(d) ensure that an accurate record of matters that may affect safety of the plant is made and maintained at all times as required in section 45;

(e) ensure that all operational and maintenance work on the plant is performed in accordance with safe operating procedures and acceptable engineering practices.

(2) In addition to the powers and duties prescribed by the act and the Analysis Series Power Engineer as published by Canada Manpower and Immigration a shift engineer shall:

(a) under the direction and supervision of the Chief Power Engineer be responsible for the safe operation of the plant and supervision of other employees on his shift who are under his control;

(b) maintain close watch on the condition and repair of all equipment in the plant and report to the chief Power Engineer any condition that may impair safety of the plant;

Chapter 7   Regulations for Canadian Provinces and Territories

(c) take all measures that are necessary to prevent any immediate danger;

(d) ensure that an accurate record of the plant is made and maintained at all times during the shift period required;

(e) ensure that all maintenance and operational work performed on the plant is in accordance with safe operating procedures and acceptable engineering practices and in accordance with the Act and regulations.

(3) In addition to the powers and duties prescribed by the Act and the Analysis Series Power Engineer as published by Canada Manpower and Immigration, the assistant shift engineer shall:

(a) under the direction and supervision of the Chief Power Engineer or the shift engineer, be responsible for the safe operation of some particular section of the plant and ensure that an accurate record of matters that may affect the safety of that section of the plant is made and maintained at all times during the shift period; and

(b) perform maintenance and operational work on the plant as may be directed by the Chief Power Engineer or shift engineer.

FEES

31. (1) An applicant for examination as a Power Engineer, Fireman or Operator shall pay the fee as listed herein:

(a) Power Engineer First Class Part A...................$40.00

(b) Power Engineer First Class Part B...................$40.00

(c) Power Engineer Second Class Part A................$30.00

(d) Power Engineer Second Class Part B................$30.00

(e) Power Engineer Third Class Part A..................$20.00

(f) Power Engineer Third Class Part B..................$20.00

(g) Power Engineer Fourth Class..........................$20.00

(h) Fireman...$15.00

(2) A certificate of competency is issued free when a candidate passes the appropriate examination.

(3) The fee for renewal of a certificate for a Power Engineer or Operator is $10.00.

(4) Where the holder of a Certificate of Competency fails to renew the certificate before the first day of February next following the expiration of the certificate the fee is $20.00 and the fee shall then increase at the rate of $10.00 per year until such time as the certificate is renewed or cancelled.

(5) Notwithstanding subsection (4), Power Engineers who are sixty-five years of age or over and retired may receive the annual renewal certificate without fee and upon application.

(6) The fee for a temporary certificate valid for the period as stated in section 27 and section 55 shall be:

(a) First Class Power Engineer........................$80.00

(b) Second Class Power Engineer....................$60.00

(c) Third Class Power Engineer.......................$40.00

(d) Fourth Class Power Engineer.....................$20.00

(e) Fireman Certificate....................................$20.00

(7) The fee for a duplicate certificate under subsection (4) of section 25 is $10.00.

Chapter 7   Regulations for Canadian Provinces and Territories

(8) The fee for a certificate issued in lieu of another certificate is the same as that prescribed in subsection (1).

(9) Where a holder of a certificate of competency fails to pay his annual fee before April first in any year, he shall stand suspended until the fee in respect of that year is paid and any arrears of fees are paid.

(10) No person who is the holder of a certificate of competency shall act in that capacity in any year in respect of which his annual fee remains unpaid.

(11) The fee for sample examination papers when and if they become surplus shall be as follows:

   (a) First Class.................................$8.00

   (b) Second Class............................$6.00

   (c) Third Class................................$4.00

   (d) Fourth Class.............................$3.00

(12) Where at the request of the owner of a pressure plant, a candidate for a certificate of competency is examined at a place other than that designated by the Director, the owner shall, in addition to the fees prescribed by these regulations, pay the cost of transportation and subsistence charges of the examiner or member of the committee who conducts the examination, together with a charge of seventy-five dollars for each day thereof during which the examiner is engaged.

LOG BOOK

45. (1) The owner shall provide a log book for use in his plant and the Chief Power Engineer shall see that the person in charge of each shift records in the log book:

Regulations for Canadian Provinces and Territories   Chapter 7

(a) the date, the number and designation of the shift and his name;

(b) any change from normal operating procedure and the time of such change;

(c) any special instructions which may have been given to achieve the change referred to in paragraph (b) and the name of the person who gave the instruction;

(d) any unusual or abnormal condition observed in the plant and the time thereof;

(e) repairs to any part of the plant and the time such repairs were commenced and if completed on this shift, the time thereof; and

(f) the time of commencing and terminating his shift.

(2) An inspector has the right to examine any log book and give such instructions relating thereto as he thinks fit and he may seize a log book as evidence in the case of an investigation.

(3) The owner shall ensure that the log book is kept accessible in the plant for at least one year after the last entry therein, and shall produce the log book for examination upon request of the inspector.

SCOPE OF EMPLOYMENT

50. (1) The holder of a Power Engineer's Certificate of Competency First Class may be employed as:

a Chief Power Engineer or shift engineer of any registered pressure plant of unlimited kilowatt capacity.

(2) The holder of a Power Engineer's Certificate of Competency Second Class may be employed as:

Chapter 7   Regulations for Canadian Provinces and Territories

(a) a Chief Power Engineer of:

- (i) a registered power plant not exceeding 24000 kilowatts; or
- (ii) a registered combined pressure plant not exceeding 48000 kilowatts where boiler capacity accounts for not more than 24000 kilowatts; and

(b) a Shift Engineer of:

- (i) a registered power plant of unlimited kilowatt rating,
- (ii) a registered combined pressure plant of unlimited kilowatt rating.
- (iii) a registered heating plant.

(3) The holder of a Power Engineer's Certificate of Competency Third Class may be employed as:

(a) a Chief Power Engineer of:

- (i) a registered power plant not exceeding 12000 kilowatts,
- (ii) a registered combined pressure plant not exceeding 24000 kilowatts where the boiler capacity accounts for not more than 12000 kilowatts,
- (iii) a heating plant of unlimited kilowatt capacity,
- (iv) a registered refrigeration plant of unlimited kilowatts capacity, or
- (v) a registered compressed gas plant of unlimited kilowatt capacity, and

(b) a Shift Engineer of:

- (i) a registered power plant not exceeding 24000 kilowatts,
- (ii) a registered combined pressure plant not exceeding 48000 kilowatts where boiler capacity accounts for not more than 24000 kilowatts, or
- (iii) a registered heating plant of unlimited kilowatt capacity.

(c) an assistant shift engineer in any registered plant of unlimited kilowatt capacity.

(4) The holder of a Power Engineers Certificate of Competency Fourth Class may be employed as:

(a) a Chief Power Engineer of:

- (i) a registered power plant not exceeding 3600 kilowatts,
- (ii) a registered combined pressure plant not exceeding 7200 kilowatts where the boiler capacity accounts for not more than 3600 kilowatts; and refrigeration not more than 300 kilowatts,
- (iii) a registered refrigeration plant not exceeding 12000 kilowatts,
- (iv) a registered refrigeration plant not exceeding 300 kilowatts,or
- (v) a registered compressed gas plant of unlimited kilowatt capacity, and

(b) a Shift Engineer of:

- (i) a registered power plant not exceeding 12000 kilowatts,

Chapter 7   Regulations for Canadian Provinces and Territories

- (ii) a registered combined pressure plant not exceeding 24000 kilowatts where boiler capacity accounts for not more than 12000 kilowatts;
- (iii) a registered heating plant of 24000 kilowatts capacity,
- (iv) a registered refrigeration plant of unlimited kilowatt capacity, or
- (v) a registered compressed gas plant of unlimited kilowatt capacity, and

(c) an assistant shift engineer of:

- (i) a registered power plant not exceeding 24000 kilowatts, or
- (ii) a registered combined pressure plant not exceeding 36000 kilowatts, where boiler capacity accounts for not more than 24000 kilowatts.

(5) The holder of a Fireman's Certificate of Competency may be employed as:

(a) a chief fireman of a registered heating plant not exceeding 2400 kilowatts; and

(b) a shift fireman of:

- (i) a registered power plant not exceeding 3600 kilowatts, or
- (ii) a registered heating plant not exceeding 12000 kilowatts.

AN ACT RESPECTING BOILERS, PRESSURE VESSELS AND COMPRESSED GAS

RECIPROCAL CERTIFICATES

34. (1) Where an engineer who holds a valid certificate of professional qualification that has been issued by another province of Canada, the Government of Canada or a competent authority in another jurisdiction applies for a certificate of competency, the director may, upon payment of the prescribed fee and subject to the conditions that the director may prescribe, issue a certificate of competency.

(2) The director may arrange with the appropriate authority of another province for the issuing of interprovincial engineers' certificates upon the terms and conditions that may be agreed upon.

TEMPORARY CERTIFICATE

35. Where in the opinion of the director there are exceptional circumstances that warrant it, the director may examine an applicant for a certificate of competency and if satisfied that the applicant is competent may issue a temporary certificate of competency to the applicant.

DISPLAY CERTIFICATE

36. The holder of a certificate of competency shall post that certificate in a conspicuous place where he or she is engaged in his or her duties.

PAYMENT

37. Where this Act provides for the issue of a certificate upon payment of the prescribed fee, the certificate may be issued before the payment but is not valid until the payment is made.

Chapter 7   Regulations for Canadian Provinces and Territories

Northwest Territories

SUMMARY OF QUALIFICATIONS:

1. Exam Location, Address and Phone Number

Majority of the exams are administered in Yellowknife (head office), however, exams have been administered in the communities when requested.

Office Location

Department of Safety and Public Services

Safety Division/Boiler Section

P.O. Box 1320

Yellowknife, Northwest Territories, Canada, X1A 2L9

403/873-7475

2. Exam Days and Times

No fixed days

3. Examination Cost

From Section 5. (1) The following fees are payable for the examination required for the following certificates of qualification or for a certificate issued under section 45:

(a) Class 1 Operating Engineer's certificate of qualification, (i) where the examination is written at one sitting........$240.00

(ii) Where all papers of an examination are not written at one sitting for each paper..$ 30.00

(b) Class 2 Operating Engineer's certificate of qualification,

(i) where the examination is written at one sitting....$180.00

(ii) where all papers of an examination are not written at one sitting, for each paper...$ 30.00

(c) Class 3 Operating Engineer's certificate of qualification,

(i) where the examination is written at one sitting...$120.00

(ii) where all papers of an examination are not written at one sitting, for each paper...$ 30.00

(d) Class 4 Operating Engineer's certificate of qualification,

(i) where the examination is written at one sitting..$ 60.00

(ii) where all papers of an examination are not written at one sitting, for each paper..$ 30.00

(e) Class 5 Operating Engineer's certificate of qualification..$ 40.00

4. Certificate of Competency Cost and Renewal

No charge for initial certificate

Renewal of certificate...$75.00

5. Type of Test

All written, the number of questions vary between class of licenses.

6. Grades of Certificates

1st Class Operating Engineer

2nd Class Operating Engineer

3rd Class Operating Engineer

4th Class Operating Engineer

5th Class Operating Engineer

7. Experience and Education Requirements

See below

8. Canadian Citizenship Required?

No

9. Local Residency Required?

No

10. Recognition of licenses from other locales?

Chapter 7   Regulations for Canadian Provinces and Territories

See Below - Regulation Section 44. (1) through (4)
Reciprocal from other provinces and territories except Ontario and Quebec.

ACT AND REGULATIONS for Northwest Territories

BOILERS AND PRESSURE VESSELS ACT

CONSOLIDATION OF BOILERS AND PRESSURE VESSELS REGULATIONS

PLANTS

28.(1) Subject to subsection (2), first, second, third and fourth class plants require continuous supervision under subsection 38(2) of the Act.

(2) Where there are no persons in a building in which a first, second, third or fourth class plant is contained, the plant may be left without continuous supervision.

29.(1) In every plant, the owner shall designate one person who is the holder of the required certificate of qualification to be Chief Operating Engineer.

(2) The chief operating engineer is responsible for the proper care and safe operation of the boilers, pressure vessels, piping, engines and auxiliaries under his or her charge and shall report all accidents to the chief inspector and to the owner.

30. The Chief Operating Engineer shall ensure that a log book is maintained to record all matters relating to the operation of a plant, including a record of the testing and servicing of safety valves and other safety devices and controls.

PART II - QUALIFICATIONS, EXAMINATIONS AND CERTIFICATION OF OPERATING ENGINEERS

31.(1) The chief inspector shall issue a certificate of qualification to a person who satisfies the requirements for the certificate of qualification.

(2) The owner or person in charge of a plant shall provide the chief inspector with any information the chief inspector may require with respect to the qualifications of the personnel involved in the operation of the plant.

32. (1) A certificate of qualification, other than a temporary certificate, expires three years after the day it is issued, unless sooner cancelled.

(2) A certificate of qualification may be limited as to purpose or area and, where the certificate is temporary, as to time.

(3) A certificate of qualification may be renewed without examination, on application and on payment of the fee set out in the Schedule, within one year after the day the certificate expires.

(4) A holder of a certificate of qualification shall requalify for a certificate unless it is renewed in accordance with this section.

(5) The holder of a non-standardized certificate of a certain class of certificate of qualification that was issued before these regulations came into force is deemed, for the purposes of these regulations, to be the holder of a certificate of qualification of that class until the non-standardized certificate expires or is cancelled.

(6) The holder of a Building Operator A Certificate of Qualification issued before these regulations come into force may be issued a Class 5 operating engineer's certificate of qualification, without examination, on application for renewal and on

Chapter 7   Regulations for Canadian Provinces and Territories

payment of the fee set out in the Schedule within one year after the day the Building Operator A Certificate of Qualification expires.

33.(1) The owner or person in charge of a plant shall post the certificates of qualification required under the Act or these regulations in a conspicuous place in the boiler room.

(2) Where the certificate of qualification of a person is required to be posted in more than one boiler room under subsection (1), that person may apply to the chief inspector for a duplicate certificate of qualification.

(3) An application under subsection (2) must be accompanied by the fee for the issuance of a duplicate certificate of qualification set out in the Schedule.

(4) The chief inspector may issue a duplicate certificate of qualification for the purposes of subsection (1) and shall indicate on the duplicate certificate of qualification:

(a) that it is a duplicate; and

(b) that it may be posted only in the plant identified on its face.

(5) A duplicate certificate issued under subsection (4) may be posted only in the plant identified on the duplicate certificate.

(6) Where a duplicate certificate of qualification issued under subsection (4) is no longer required to be posted under the Act or these regulations, the person named in the duplicate certificate shall, without delay, return it to the chief inspector.

34. Where a certificate of qualification is lost or destroyed, the chief inspector may issue a duplicate certificate on the re-

quest of the holder of the certificate of qualification and the payment of the fee set out in the Schedule.

35.(1) A Class 1 Operating Engineer's certificate of qualification entitles the holder to:

(a) exercise general supervision of and be responsible for a first class plant as chief operating engineer and to supervise the shift engineers in that plant; and

(b) operate a first class plant as a shift engineer.

(2) A Class 2 operating engineer's certificate of qualification entitles the holder to:

(a) exercise general supervision of and be responsible for a second class plant as chief operating engineer and to supervise the shift engineers in that plant; and

(b) operate a first class plant as a shift engineer.

(3) A Class 3 operating engineer's certificate of qualification entitles the holder to:

(a) exercise general supervision of and be responsible for a third class plant as chief operating engineer and to supervise the shift engineers in that plant;

(b) operate a second class plant as shift engineer; and

(c) operate a first class plant as assistant shift engineer under the supervision of the shift engineer of that plant.

(4) A Class 4 operating engineer's certificate of qualification entitles the holder to:

(a) exercise general supervision of and be responsible for a fourth class plant as Chief Operating Engineer and to supervise the shift engineers in that plant;

(b) operate a third class plant as a shift engineer; and

Chapter 7   Regulations for Canadian Provinces and Territories

(c) operate a second class plant as assistant shift engineer under supervision of the shift engineer of that plant.

(5) A Class 5 operating engineer's certificate of qualification entitles the holder to:

(a) exercise general supervision of and be responsible for a fifth class plant and to supervise the shift engineers in that plant;

(b) operate a fourth class plant as a shift engineer; and

(c) operate a third class plant as assistant shift engineer under supervision of the shift engineer of that plant.

36.(1) Where the Chief Operating Engineer of a plant is sick or expects to be absent from the plant or which he or she is responsible,

(a) the employer or Chief Operating Engineer shall notify the chief inspector of the absence or expected absence and of the name of the person proposed to replace the chief operating engineer; and

(b) the employer may apply to the chief inspector for a temporary certificate of qualification for the person proposed as a replacement.

(2) Where the absence or expected absence referred to in subsection (1) is greater than 96 hours, the notification and application referred to in that subsection must be in writing.

(3) an application referred to in paragraph (1) (b) must contain a declaration by the employer, or an agent of the employer where the employer is a corporation, that the person proposed as a replacement is, to the best of the knowledge of the declarant, capable of acting in the capacity for which the certificate of qualification is requested.

(4) The chief inspector may issue a temporary certificate of qualification to a person proposed as a replacement where:

(a) the chief inspector is satisfied the person has sufficient experience to operate a plant of the type or class specified; and

(b) the class of plant is no more than one grade higher than the class of plant the person is entitled to operate under his or her certificate of qualification.

(5) Where a temporary certificate of qualification is issued, the chief inspector:

(a) shall specify the date of expiry of the certificate, and

(b) may impose such conditions on the holder of the certificate as the chief inspector considers necessary.

(6) A temporary certificate of qualification that expires within 96 hours after its issue need not be in writing.

37.(1) To qualify to take an examination for a Class 1 operating engineer's certificate of qualification, a candidate must:

(a) hold a Class 2 Operating Engineer's certificate of qualification; and

(b) furnish evidence satisfactory to the chief inspector of employment for a period of:

- (i) 30 months as Chief Operating Engineer in a second class plant,
- (ii) 30 months as shift engineer in a first class plant,
- (iii) 45 months as assistant shift engineer in a first class plant,

Chapter 7 Regulations for Canadian Provinces and Territories

- (iv) one-half the number of months in a position specified in subparagraph (i), (ii) or (iii) and, in addition, employment for a period of 15 months in a compressor plant in an operating capacity approved by the chief inspector,

- (v) 15 months in a position specified in subparagraph (i), (ii) or (iii) and, in addition, that the candidate holds a degree in Mechanical Engineering from a university approved by the chief inspector or has, in the opinion of the chief inspector, the equivalent of such degree, or

- (vi) one-half the number of months in a position specified in subparagraph (i), (ii) or (iii) and, in addition, employment for a period of 36 months in a supervisory capacity satisfactory to the chief inspector in relation to the design, construction, installation, repair, maintenance or operation of regulated equipment.

(2) The chief inspector may grant 12 months credit in lieu of the employment specified in subparagraph (1) (b) (i), (ii) or (iii) on the candidate's successful completion of a course in power engineering that is satisfactory to the chief inspector and that leads to a Class 1 operating engineer's certificate of qualification examination.

(3) The chief inspector may grant 12 months credit in lieu of the compressor plant experience specified in subparagraph (1) (b) (iv) on the candidate's successful completion of a course in power engineering that is satisfactory to the chief inspector and that leads to a Class 1 operating engineer's certificate of qualification examination.

(4) The examination must be divided into two parts, lettered A and B, and a candidate may:

Regulations for Canadian Provinces and Territories   Chapter 7

(a) write any one or all papers for Part A at a scheduled examination after obtaining a Class 2 operating engineer's certificate of qualification;

(b) write any one or all papers for Part B after successfully completing the papers for Part A, providing the qualifying experience has been obtained as specified in subsection (1); or

(c) write the papers for both parts at the same sitting, providing the qualifying experience has been obtained as specified in subsection (1).

(5) the examination must consist of questions relating to the subjects contained in the current reference syllabus as adopted by the chief inspector for the Class 1 operating engineer's certificate of qualification examination.

(6) To obtain a pass and a certificate of qualification a candidate must,

(a) when he or she writes a single paper at one sitting, obtain 70% of the marks for the paper; or

(b) when he or she writes more than one paper at the same sitting, obtain an average of 70% of the total marks for the papers written at the sitting and not less than 60% of the marks for each paper.

38.(1) To qualify to take an examination for a Class 2 Operating Engineer's certificate of qualification, a candidate must:

(a) hold a Class 3 operating engineer's certificate of qualification; and

(b) furnish evidence satisfactory to the chief inspector of employment for a period of:

Chapter 7   Regulations for Canadian Provinces and Territories

- (i) 24 months as a Chief Operating Engineer of a third class high pressure plant,

- (ii) 24 months as a shift engineer in a second class plant,

- (iii) 36 months as a shift engineer in a third class high pressure plant,

- (iv) 24 months as an assistant shift engineer in a first class plant,

- (v) one-half the number of months in a position specified in subparagraph (i), (ii), (iii) or (iv) and, in addition, employment for a period of 12 months in a compressor plant in an operating capacity satisfactory to the chief inspector,

- (vi) 12 months in a position specified in subparagraph (ii, (ii), (iii) or (iv) and, in addition, that the candidate holds a degree in Mechanical Engineering from a university approved by the chief inspector or has, in the opinion of the chief inspector, the equivalent of such a degree,or

- (vii) one-half the number of months in a position specified in subparagraph (i), (ii), (iii) or (iv) and, in addition, employment for a period of 24 months in a supervisory capacity satisfactory to the chief inspector in relation to the design, construction, installation, repair, maintenance or operation of regulated equipment.

(2) The chief inspector may grant nine months credit in lieu of the employment specified in subparagraph (1)(b)(i), (ii), (iii) or (iv) on the candidate's successful completion of a course in

power engineering that is satisfactory to the chief inspector and that leads to a Class 2 operating engineer's certificate of qualification examination.

(3) A candidate who is the holder of a diploma issued by an educational institution after the candidate completed a two year day course in power engineering satisfactory to the chief inspector and who is the holder of a Class 3 Operating Engineer's certificate of qualification, is qualified to take a Class 2 Operating Engineer's certificate of qualification examination after obtaining one-half the qualifying experience specified in subparagraph (1)(b)(i), (ii), (iii) or (iv).

(4) The chief inspector may grant nine months credit in lieu of the compressor plant experience specified in subparagraph (1)(b)(v) on the candidate's successful completion of a course in Power Engineering that is satisfactory to the chief inspector and that leads to a Class 2 Operating Engineer's certificate of qualification examination.

(5) The examination must be divided into two parts, lettered A and B, and a candidate may write:

(a) any one or all papers for Part A at a scheduled examination after obtaining a Class 3 Operating Engineer's certificate of qualification;

(b) any one or all papers for Part B after successfully completing the papers for Part A, providing the qualifying experience has been obtained as specified in subsection (1); or

(c) all the papers for both parts at the same sitting, providing the qualifying experience has been obtained as specified in subsection (1).

Chapter 7 Regulations for Canadian Provinces and Territories

(6) The examination must consist of questions relating to the subjects contained in the reference syllabus as adopted by the chief inspector for the Class 2 Operating Engineer's certificate of qualification examination.

(7) To obtain a pass and a certificate of qualification a candidate must,

(a) when he or she writes a single paper at one sitting, obtain 70% of the marks for the paper; or

(b) when he or she writes more than one paper at the same sitting, obtain an average of 70% of the total marks for the papers, and not less than 60% of the marks for each paper.

39.(1) To qualify to take an examination for a Class 3 Operating Engineer's certificate of qualification, a candidate must:

(a) hold a Class 4 Operating Engineer's certificate of qualification; and

(b) furnish evidence satisfactory to the chief inspector of employment for a period of:

- (i) 12 months as Chief Operating Engineer in a fourth class plant,
- (ii) 18 months as shift engineer in a third class plant,
- (iii) 24 months as assistant shift engineer in a second class plant,
- (iv) one-third the number of months in a position specified in subparagraph (i), (ii) or (iii) and, in addition, employment for a period of eight months in a compressor plant in an operating capacity satisfactory to the chief inspector,

- (v) one-half the number of months in a position specified in subparagraph (i), (ii) or (iii) and, in addition, that the candidate is the holder of a degree in Mechanical Engineering or the equivalent of such a degree from a university satisfactory to the chief inspector,

- (vi) one-third the number of months in a position specified in subparagraph (i), (ii) or (iii) and, in addition, employment for a period of two months in a compressor plant in an operating capacity satisfactory to the chief inspector during which periods the candidate held a degree in mechanical engineering from a university approved by the chief inspector or had, in the opinion of the chief inspector, the equivalent of such a degree, or

- (vii) one-half the number of months in a position specified in subparagraph (i), (ii) or (iii) and, in addition, employment for a period of 12 months in a capacity satisfactory to the chief inspector on the design, construction, installation, repair, maintenance or operation of regulated equipment.

(2) The chief inspector may grant six month credit in lieu of the employment specified in subparagraph (1)(b)(i), (ii) or (iii) on the candidate's successful completion of a course in Power Engineering that is satisfactory to the chief inspector and that leads to a Class 3 Operating Engineer's certificate of qualification examination.

(3) The chief inspector may grant six months credit in lieu of the compressor plant experience specified in subparagraph (1)(b) (v) on the candidate's successful completion of a course in Power Engineering that is satisfactory to the chief inspector

Chapter 7 Regulations for Canadian Provinces and Territories

and that leads to a Class 3 Operating Engineer's certificate of qualification examination.

(4) The examination must be divided into two parts, lettered A and B, and a candidate may write:

(a) any one or all papers for Part A at a scheduled examination after obtaining a Class 4 operating engineer's certificate of qualification;

(b) any one or all papers for Part B after successfully completing the papers for Part A, providing the qualifying experience has been obtained as specified in subsection (1); or

(c) all the papers for both parts at the same sitting, providing the qualifying experience has been obtained as specified in subsection (1).

(5) The examination must consist of questions relating to the subjects contained in the current reference syllabus as adopted by the chief inspector for the Class 3 Operating Engineer's certificate of qualification examination.

(6) To obtain a pass and a certificate of qualification the candidate must,

(a) when he or she writes a single paper at one sitting, obtain 60% of the marks for the paper; or

(b) when he or she writes more than one paper at the same sitting, obtain an average of 60% of the total marks for the papers and not less than 50% of the marks for each paper.

40.(1) To qualify to take an examination for a Class 4 Operating Engineer's certificate of qualification, a candidate must:

(a) hold a Class 5 operating engineer's certificate of qualification; and

(b) furnish evidence satisfactory to the chief inspector of employment for a period of:

- (i) 12 months as chief operating engineer in a fifth class plant,
- (ii) 18 months as a shift engineer in a fourth class plant,
- (iii) 24 months as an assistant shift engineer in a third class plant,
- (iv) one-half the number of months in a position specified in subparagraph (i), (ii) or (iii) and, in addition, employment for a period of six months in a compressor plant in an operating capacity satisfactory to the chief inspector,
- (v) one-half the number of months in a position specified in subparagraph (i), (ii) or (iii) and, in addition, that the candidate is the holder of a degree in Mechanical Engineering or the equivalent of such a degree from a university satisfactory to the chief inspector.
- (vi) one-third the number of months in a position specified in subparagraph (i), (ii) or (iii) and, in addition, employment for a period of two months in a compressor plant in an operating capacity satisfactory to the chief inspector during which periods the candidate held a degree in Mechanical Engineering from a university approved by the chief inspector or had, in the opinion of the chief inspector, the equivalent of such a degree, or

Chapter 7   Regulations for Canadian Provinces and Territories

- (vii) one-half the number of months in a position specified in subparagraph (i), (ii) or (iii) and, in addition, employment for a period of 12 months in a capacity satisfactory to the chief inspector on the design, construction, installation, repair, maintenance or operation of regulated equipment.

(2) The chief inspector may grant six months credit in lieu of the employment specified in subparagraph (1) (b) (i), (ii) or (iii) on the candidate's successful completion of a course in Power Engineering that is satisfactory to the chief inspector and that leads to a Class 4 Operating Engineer's certificate of qualification examination.

(3) The chief inspector may grant six months credit in lieu of the compressor plant experience specified in subparagraph (1) (b) (iv) on the candidate's successful completion of a course in power engineering that is satisfactory to the chief inspector and that leads to a Class 4 operating engineer's certificate of qualification examination.

(4) The examination must be divided into two parts, lettered A and B, and a candidate may write:

(a) Part A at a scheduled examination after obtaining a Class 5 operating engineer's certificate of qualification;

(b) Part B after successfully completing Part A, providing the qualifying experience has been obtained as specified in subsection (1); or

(c) both parts at the same sitting, providing the qualifying experience has been obtained as specified in subsection (1).

(5) The examination must consist of questions relating to the subjects contained in the current reference syllabus as adopted

by the chief inspector for the Class 4 Operating Engineer's certificate of qualification examination.

(6) To obtain a pass and a certificate of qualification a candidate must:

(a) when he or she writes a single paper at one sitting, obtain 60% of the marks for the paper;or

(b) when he or she writes more than one paper at the same sitting, obtain an average of 60% of the total marks for the papers and not less than 50% of the marks for each paper.

41.(1) To qualify to take an examination for a Class 5 Operating Engineer's certificate of qualification, a candidate must:

furnish evidence satisfactory to the chief inspector that the candidate has been employed for a period of 12 months in the operation and maintenance of a boiler plant to which the Act applies.

(2) The chief inspector may grant six months credit in lieu of the employment specified in subsection (1) on the candidate's successful completion of a course that is satisfactory to the chief inspector and that leads to a Class 5 Operating Engineer's certificate of qualification.

(3) The candidate shall write all papers for the examination at one sitting.

(4) The examination must consist of questions relating to the subjects contained in the current reference syllabus established by the chief inspector.

(5) To obtain a pass, the candidate must obtain not less than 60% of the marks for each paper.

Chapter 7   Regulations for Canadian Provinces and Territories

(6) A certificate of qualification issued under this section is a non-standardized certificate.

42.(1) Subject to subsection (2), the chief inspector may grant credit to a person who has passed courses in Power Engineering that the chief inspector considers satisfactory in lieu of practical experience.

(2) Where a person is granted credit for a course for one level of examination, the chief inspector shall not grant credit to the person for that course to qualify for a higher level of examination.

43. (1) Where a candidate has experience made up in part as chief operating engineer, shift engineer or assistant shift engineer or any other experience, the chief inspector may evaluate that experience.

(2) Where a candidate furnishes proof of experience with regulated equipment, which experience is acceptable to the chief inspector and other than that specified in subsections 37(1), 38(1) and 39(1), the chief inspector may grant credit in lieu of the specified experience.

(3) Where a compressor plant is in operation for only part of a year and the operating engineer is retained for the non-operational period and is employed on plant maintenance, the chief inspector may grant credit of two-thirds of the maintenance time towards experience required for an examination.

(4) Where a person is granted credit for practical experience for a certain operating engineer's examination, the chief inspector shall not grant credit to the person for that experience to qualify for a higher level of examination.

Regulations for Canadian Provinces and Territories   Chapter 7

44.(1) Where a person holds a certificate of qualification from a jurisdiction outside the Territories that is, in the opinion of the chief inspector, equivalent to a certificate of qualification issued under these regulations, the chief inspector may, on application and payment of the prescribed fee, issue a certificate of qualification under these regulations.

(2) A certificate of qualification issued under this section may be issued subject to such conditions as the chief inspector considers necessary.

(3) The chief inspector shall not issue a certificate of qualification under this section until the chief inspector is satisfied as to the applicant's identity, experience and qualifications and, for that purpose, may require such evidence as the chief inspector considers necessary.

(4) A candidate from a jurisdiction outside the Territories who has passed any paper of an Operating Engineer's examination in that jurisdiction may be given credit by the chief inspector for having passed that paper.

45.(1) At the discretion of the chief inspector, a non-standardized certificate of a class determined appropriate by the chief inspector may be issued by the chief inspector to a person who:

(a) holds a valid non-standardized certificate as an Operating Engineer that is issued by the Government of Canada, the government of a province or the Yukon Territory or a competent authority in any other jurisdiction; and

(b) makes application for a non-standardized certificate accompanied by evidence satisfactory to the chief inspector of the applicant's qualifications and identity.

Chapter 7   Regulations for Canadian Provinces and Territories

(2) An application made under paragraph (1) (b) must be accompanied by the fee set out in the Schedule.

46. (1) A candidate for examination shall apply to write an examination in the approved form at least 21 days before the date of examination.

(2) An application for examination must be submitted for approval to the chief inspector for the examination to be written together with:

(a) photocopies of the references or of written statements referred to in sub-section (4);

(b) the fee set out in the Schedule; and

(c) where educational qualifications are required, documents in support of the candidate's educational qualifications.

(3) The chief inspector shall return any original document to the candidate after verification.

(4) The qualifications of a candidate relating to plant operation, engineering experience, ability and general conduct may be proved by references signed by the owner or Chief Operating Engineer of the plant where the candidate was or is employed, but if such references are not available, the chief inspector may, in lieu of references, accept a written statement made by a person who has personal knowledge of the facts to be established.

(5) Educational qualifications must be confirmed by documents issued by the institution from which the candidate received training.

47.(1) A candidate for examination shall appear at such place and time as the chief inspector may direct.

(2) An examination must be given under the direction of the chief inspector.

48.(1) The inspector conducting an examination under these regulations may declare a candidate to have failed the examination if:

(a) formulas or other information have been added to or inserted without the approval or authorization of the chief inspector into a published text of a book, table, regulation, code or standard that is taken into the examination room.

(b) the candidate looks at or refers to material during the examination that is not allowed into the examination room by the chief inspector.

(c) the candidate removes or attempts to remove a questions or part of a questions from an examination room;

(d) the candidate copies from another candidate; or

(e) the candidate communicates with another candidate in any manner during the examination.

(2) A candidate who has been declared to have failed an examination under subsection (1) may be disqualified by the chief inspector from writing a further examination for a period not exceeding 12 months after the date of examination.

49. Each paper of an examination must be marked by an inspector.

50. A person who receives or acquires a certificate of qualification issued under these regulations, other than the person whose name appears on the certificate, shall send the certificate of qualification to the chief inspector.

Chapter 7   Regulations for Canadian Provinces and Territories

51. (1) Subject to the discretion of the chief inspector, a candidate who fails an examination may not attempt the examination again until he or she furnishes evidence satisfactory to the chief inspector of additional experience of not less than:

(a) 30 days, where the candidate has failed the examination once or twice; or

(b) three months, where the candidate has failed the examination more than twice.

(2) The chief inspector may dispose of a paper written by a candidate for an examination on the expiry of 30 days after the candidate has been notified that he or she passed the paper or failed the paper, as the case may be.

SCHEDULE

5. (1) The following fees are payable for the examination required for the following certificates of qualification or for a certificate issued under section 45:

(a) Class 1 Operating Engineer's certificate of qualification,

(i) where the examination is written at one sitting...........$240

(ii) where all papers of an examination are not written at one sitting, for each paper..$ 30

(b) Class 2 Operating Engineer's certificate of qualification,

(i) where the examination is written at one sitting..........$180

(ii) where all papers of an examination are not written at one sitting, for each paper..$ 30

(c) Class 3 Operating Engineer's certificate of qualification,

(i) where the examination is written at one sitting............$120

(ii) where all papers of an examination are not written at one sitting, for each paper...$ 30

(d) Class 4 Operating Engineer's certificate of qualification,

(i) where the examination is written at one sitting.............$ 60

(ii) where all papers of an examination are not written at one sitting, for each paper...$ 30

(e) Class 5 operating engineer's certificate of qualification..$ 40

(2) The fee payable for a temporary certificate of qualification is...$ 50

(3) The fee payable for renewal of a certificate of qualification under section 32 is...$ 75

(4) The following fee is payable for a certificate of qualification issued under subsection 44(1):

(a) Class 1..$120

(b) Class 2..$ 90

(c) Class 3..$ 60

(d) Class 4..$ 40

(e) Class 5..$ 20

7. (1) Fee payable for a search of official records is...$ 20

(2) The fee payable for a duplicate certificate of inspection, a duplicate certificate of qualification, a duplicate welding per-

formance qualification card or a photocopy or certified copy of a report or other document is...$ 20

(3) The fee payable for a copy of a design that has been registered is the actual cost of the copy.

Province of Nova Scotia

SUMMARY OF QUALIFICATIONS:

1. Exam Location, Address and Phone Number

Department of Labour
Stationary Engineers Section
P.O. Box 697
Halifax, Nova Scotia, Canada B3J 2T8
902/424-7521

2. Exam Days and Times

Exams are prescheduled fifteen times a year at different locations. Minimum of 15 exam dates for public. Also have special settings for community college level - they will go to community college for special groups - sometimes do this thirty times a year.

3. Examination Cost

$10.00 per examination paper.

4. Certificate of Qualification and Renewal

From Appendix A
  Renewal - $25.00

  Stationary Engineer-First Class-

  8 examination papers for a total fee of $80.00

  Stationary Engineer-Second Class-

  6 examination papers for a total fee of $60.00

  Stationary Engineer-Third Class-

4 examination papers for a total fee of $40.00

Stationary Engineer-Fourth Class-

2 examination papers for a total fee of $20.00

5. Type of Test

All Written

6. Grades of Certificates

Stationary Engineer-First Class

Stationary Engineer-Second Class

Stationary Engineer-Third Class

Stationary Engineer-Fourth Class

7. Experience and Education Requirements

See Act and Regulations-Nova Scotia

8. Canadian Citizenship Required?

No

9. Local Residency Required?

No

10. Recognition of licenses from other locales?

Have standard program throughout Canada except Ontario and Quebec they will accept reciprocal of certificate for same class of certificate.

STATIONARY ENGINEERS ACTS AND REGULATIONS for the Province of Nova Scotia

Statutes of Nova Scotia

Chapter 7 Regulations for Canadian Provinces and Territories

Acts of 1980

Chapter 18

CERTIFICATE OF QUALIFICATION AS STATIONARY ENGINEER

5.(1) On the recommendation of the Board and on payment of the fees prescribed by the regulations, the Minister may issue a certificate of qualification as a Stationary Engineer to any person:

(a) who has passed the examination prescribed by the board; or

(b) who, in the opinion of the Board, is the holder of a certificate of an equivalent rating.

Certificate of Plant Registration

(2) The Minister, on the recommendation of the Board, may issue to the owner of any kind of plant to which this Act applies a certificate of registration.

Cancellation or suspension of Certificate

(3) Any certificate may be revoked, cancelled or suspended by the Minister at any time for any violation of this Act or of the regulations.

Duty to Furnish Information

(4) Every person who is the owner of any plant to which this Act applies shall furnish to the Board on the form to be supplied by the Board such information as is required by this Act or by the regulations. 1980, c18, s.5.

APPEAL

6. Any person aggrieved by a decision of the Board may appeal therefrom to the Minister and the decision of the Minister shall be final. 1980. c.18, s.6.

DISPLAY OF CERTIFICATE OF QUALIFICATION

7.(1) The certificate of qualification shall at all times be exposed to view in the plant in which the holder thereof is employed, except the certificate shall be carried upon the person of the operator of a hoisting plant.

Display Of Certificate Of Plant Registration

(2) The certificate of plant registration shall at all times be exposed to view.

Failure To Display Is Prime Facie Evidence

(3) Failure to comply with subsections (1) and (2) shall prime facie be evidence of the lack of qualification or registration under this Act. 1980, c 18, s.7.

ONLY STATIONARY ENGINEER TO PERFORM DUTIES

9.(1) No person other than a Stationary Engineer shall perform or be employed by any person to perform the work or duties of a Stationary Engineer.

Only Work Authorized by Certificate May be Performed

(2) No stationary engineer may perform work or duties within the meaning of this Act which are not authorized by the scope of his certificate of qualification.

Effect of Cancellation Or Suspension of Certificate

(3) No person whose certificate of qualification has been cancelled or suspended may after cancellation or during suspension perform the duties of a Stationary Engineer.

Chapter 7   Regulations for Canadian Provinces and Territories

Restriction On Employer

(4) No person may employ a Stationary Engineer at a class or kind of work within the meaning of this Act other than the authorized by the certificate of qualification of the Stationary Engineer.

Where Insufficient Notice Of Absence

(5) If a Stationary Engineer is absent from his duties without having given his employer at least seven days notice of his intended absence; the duties of the stationary engineer may be performed by any person under competent supervision for a period not exceeding fourteen days.

Operation Of Plant In Absence Of Stationary Engineer

(6) Where a Stationary Engineer is absent from any plant due to sickness or while on holidays a Stationary Engineer holding a certificate not more than one class lower than the certificate required of the Stationary Engineer who is absent may during the absence operate any plant for not more than thirty days per year or such greater number of days per year as may be authorized by the Board in writing.

Performance Of High Class Duties

(7) With the approval of the Board and in special circumstances the holder of a certificate may perform, in accordance with conditions prescribed by the Board and for a period specified by the Board, the duties of a person holding a certificate of a higher class. 1980, c 18, s.9.

PART IV - STATIONARY ENGINEERS

14. Stationary engineers shall be classified as follows:

(a) Stationary Engineer - First Class;

(b) Stationary Engineer - Second Class;

(c) Stationary Engineer - Third Class;

(d) Stationary Engineer - Fourth Class;

(e) Stationary Engineer - (Refrigerator Plant)-First Class;

(f) Stationary Engineer - (Refrigerator Plant)-Second Class;

(g) Stationary Engineer - (Compressor Plant)-First Class;

(h) Stationary Engineer - (Compressor Plant)-Second Class;

(i) Stationary Engineer - (Hoisting Plant)-Mobile Crane;

(j) Stationary Engineer - (Hoisting Plant)-Tower Crane;

(k) Stationary Engineer - (Hoisting Plant)-Overhead Traveling Crane.

PART VI - EXAMINATIONS

20. Candidates for examination shall be not less than eighteen years of age and possess the experience and education requirements provided for in Sections 31 to 34, inclusive, of these regulations.

21. A candidate for examination shall:

(a) present himself for his examination at the time, date and place of which he has been notified;

(b) not leave the examination room without the permission of the person directing the examination of candidates;

(c) not communicate with any other candidate during the examination;

(d) not have any books or papers other than those approved by the Board for reference at the examination; and

Chapter 7   Regulations for Canadian Provinces and Territories

(e) obtain an average mark of 60 percent on all examinations with no single mark below 50 percent at the Stationary Engineer Fourth and Third Class classification levels, and obtain an average mark of 70 percent on all examinations with no single mark below 60 percent at the Stationary Engineer Second and First Class classification levels.

22. A candidate for examination may ask the person directing such examination to explain any examination questions which the candidate does not understand.

23. A candidate for examination shall write an examination of a minimum of two examination papers at any one sitting, except in the case of First and Second Class certificates of qualification where a single paper may be written, and in the case of the re-writing of any failed examination.

24. Any candidate who fails to pass his examination shall not be eligible for re-examination until a period of not less than sixty days has elapsed from his examination and a request in writing has been made on behalf of, or by the candidate to the Board for the re-examination.

25. Where a candidate has written an examination for a certificate of qualification of any class and has not passed that examination, but his answers have satisfied the Board that he possesses the qualifications required for a certificate of qualification of a lower class, he shall be deemed to have passed the examination for the lower class of certificate of qualification.

26. Any candidate who fails to appear for examination at the time, date and place he was advised of shall:

(a) forfeit his fee unless at least seven days prior to the time set for the examination he has notified the Board that he will not be attending the examination; or

(b) not be required to forfeit his fee in the event of his being suddenly taken ill; and

(c) be required to provide proof of illness under the conditions of Section 26(b) of these regulations; and

(d) be eligible for further examination upon application in writing to the Board.

27. Examination shall be:

(a) held at such times, on such dates and at such places as may be designated from time to time by the Board and conducted in the presence of the Board, the Chairman of the Board, a Board Member, or an examiner or invigilator (legal test monitor) appointed by the Minister; and

(b) by means of questions approved by the Board.

28. An apprentice registered under the Apprenticeship and Trademen's Qualification Act shall not be eligible for any examination under this Act until approval in writing has been received from the Director of Apprenticeship and Trademen's Qualifications for the apprentice to be examined.

29. The approval in writing mentioned in Section 28 shall accompany any application for examination submitted by an apprentice to the Board.

30. The Board shall cancel a certificate of qualification that is issued under this Act to an apprentice who writes and passes an examination but fails to meet the requirements of Section 28.

PART VII - EDUCATIONAL QUALIFICATIONS

Chapter 7   Regulations for Canadian Provinces and Territories

31.(1) Applicants for certificates of qualification as stationary engineers shall possess the following education qualifications:

(a) for any first class certificate of qualification, the completion of Grade 11 of a Nova Scotia high school or its equivalent;

(b) for any certificate of qualification other than a first class certificate of qualification, completion of Grade X of a Nova Scotia high school or its equivalent.

32. Upon an application for a qualification the Board may recognize and approve in place of some or all of the educational qualifications required under these regulations:

(a) service or training in Her Majesty's Canadian Armed Forces that is pertinent to the application; or

(b) Successful completion of courses in an apprenticeship, technical or trade school recognized by the Board; or

(c) time and experience in the operation, construction, repair, running in, setting up or testing of a plant pertinent to the application; or

(d) the completion in whole or in part of any correspondence or formal course of study in stationary engineering that is recognized by the Board.

PART VIII - PRACTICAL EXPERIENCE QUALIFICATIONS

33. In lieu of an applicant's practical experience in a boiler plant-high pressure, the Board may substitute a measure of the total experience of an applicant in a boiler plant-low pressure.

34. Applicants for certificates of qualification as Stationary Engineers shall possess the following practical experience on regular shift for the following certificates of qualification:

(a) **Stationary Engineer-First Class, applicants shall have:**

not less than sixty months' experience and since the issue of a second class certificate of qualification have;

- (i) acted as the Chief Stationary Engineer of a second class plant for a period of not less than twelve months, or
- (ii) acted as the Shift Stationary Engineer of a first class plant for a period of not less than twelve months, or
- (iii) acted as the Assistant Shift Stationary Engineer of a first class plant for a period of not less than twelve months, or
- (iv) one-half of the experience required by subclauses (i), (ii) or (iii) and in addition have been employed for a period of not less than eighteen months in a supervisory capacity in the design, construction, installation or repair or equipment to which the Act applies, or
- (v) successfully completed a course in stationary engineering leading to a certificate of qualification as a stationary engineer first class, as approved by the Board, and have been granted by the Board twelve months' credit in lieu of the operating experience required by subclauses (i), (ii) or (iii);

(b) **Stationary Engineer-Second Class, applicants shall have:**

not less than forty-eight months' experience and since the issue of a Third Class certificate of qualification have;

- (i) acted as the Chief Stationary Engineer of a third class plant for a period of not less than twenty-four months, or

Chapter 7   Regulations for Canadian Provinces and Territories

- (ii) acted as the Shift Stationary Engineer of a second class plant for a period of not less than twenty-four months, or
- (iii) acted as the Assistant Shift Stationary Engineer of a second class plant for a period of not less than twenty-four months, or
- (iv) one-half of the experience required by subclauses (i), (ii) or (iii) and in addition have been employed for a period of not less than thirty months in the design, construction, installation or repair of equipment to which the Act applies, or
- (v) successfully completed a course in stationary engineering leading to a certificate of qualification as a Stationary Engineer Second Class, as approved by the Board, and have been granted by the Board nine months' credit in lieu of the operating experience required by subclauses (i), (ii) or (iii);

(c) Stationary Engineer-Third Class, applicants shall have:

not less than twenty-four months' experience in a high pressure plant and since the issue of a Fourth Class certificate of qualification have;

- (i) acted as the Chief Stationary Engineer of a fourth class plant for a period of not less than twelve months, or
- (ii) acted as the Shift Stationary Engineer of a third class plant for a period of not less than twelve months, or

- (iii) acted as the Assistant Shift Stationary Engineer in a third class plant for a period of not less than twelve months, or

- (iv) one-half of the experience required by subclauses (i), (ii) or (iii) and in addition have been employed for a period of not less than eighteen months in the design, construction, installation or repair of equipment to which the Act applies, or

- (v) successfully completed a course in stationary engineering leading to a certificate of qualification as a Stationary Engineer Third Class, as approved by the Board, and have been granted by the Board six months' credit in lieu of the operating experience required by subclauses (i), (ii) or (iii);

(d) **Stationary Engineer-Fourth Class, applicants shall have:**

- (i) not less than twelve months' experience in a boiler plant-high pressure over fifteen boiler horsepower, or

- (ii) been employed for a period of not less than twelve months in a low pressure heating plant of twenty-five boiler horse-power or over, or

- (iii) successfully completed a vocational course in stationary engineering satisfactory to the Board, or

- (iv) not less than six months' experience of the kind specified in subclauses (i) or (ii) and in addition have been employed for a period of not less than twelve months in the design, construction, installation or repairs of equipment to which the Act applies, or

- (v) successfully completed a course in stationary engineering leading to a certificate of qualification as a Stationary Engineer Fourth Class, as approved by the Board, and have been granted by the Board six months' credit in lieu of the operating experience required by subclauses (i) or (ii);

(e) Stationary Engineer-(Refrigerator Plant)

First Class, applicants shall have not less than twenty-four months' experience in a refrigerator plant of which not less than twelve months shall be in a refrigerator plant over two hundred motive horse-power;

(f) Stationary Engineer-(Refrigerator Plant)

Second Class, applicants shall have not less than twelve months' experience in a refrigerator plant over twenty-five motive horse-power;

(g) Stationary Engineer-(Compressor Plant)

First Class, applicants shall have not less than twenty-four months' experience in a compressor plant of which not less than nine months shall be in a compressor plant over two hundred motive horse-power;

(h) Stationary Engineer-(Compressor Plant)

Second Class, applicants shall have not less than twelve months' experience in a compressor plant over fifteen motive horse-power;

(i) Stationary Engineer-(Hoisting Plant)

Mobile Crane, applicants shall have not less than twenty-four months' experience under supervision in mobile hoisting plant after having performed a practical test satisfactory to the Board;

(j) Stationary Engineer-(Hoisting Plant)

Tower Crane, applicants shall have not less than twelve months' experience under supervision in a tower crane hoisting plant after having performed a practical test satisfactory to the Board;

(k) Stationary Engineer-(Hoisting Plant)

Overhead Travelling Crane, applicants shall have not less than twelve months' experience under supervision in an overhead hoisting plant or have practical experience satisfactory to the Board.

35 (1) An applicant for a certificate of qualification of any class shall hold a certificate of qualification one class lower, if any, than the class for which he applies or an equivalent certificate or qualification recognized by the Board.

(2) Upon application for a certificate of qualification the Board may recognize and approve in place of some or all of the experience required under these regulations;

(a) service or training in Her Majesty's Canadian Armed Forces that is pertinent to the application; or

(b) successful completion of courses in an apprenticeship, technical or trade school recognized by the Board; or

(c) time and experience in the operation, construction, repair, running in, setting up or testing of a plant pertinent to the application; or

(d) the completion in whole or in part of any correspondence or formal course of study in stationary engineering that is recognized by the Board.

(e) relevant experience as a registered Professional Engineer in Nova Scotia.

PART IX RESPONSIBILITIES OF STATIONARY ENGINEER

36. The holder of a Stationary Engineer-First Class certificate of qualification may act as the chief stationary engineer and the holder of a Stationary Engineer-Second Class certificate of qualification may act as the shift stationary engineer in any of the following plants having unlimited capacity and to which the Act applies:

(a) boiler plant-high pressure;

(b) boiler plant-low pressure;

(c) compressor plant;

(d) refrigerator plant.

37. The holder of a Stationary Engineer-Second Class certificate of qualification may act as the chief stationary engineer and the holder of a Stationary Engineer-Third Class certificate of qualification may act as the shift stationary engineer in any plant to which the Act applies and having a capacity not exceeding the following limits:

(a) boiler plant-high pressure-1000 BHP;

(b) boiler plant-low pressure-unlimited BHP;

(c) compressor plant-unlimited MHP;

(d) refrigerator plant-unlimited MHP.

38. The holder of a Stationary Engineer-Third Class certificate of qualification may act as the chief stationary engineer and the holder of a Stationary Engineer-Fourth class certificate of qualification may act as the shift stationary engineer in any plant to which the Act applies and having a capacity not exceeding the following limits:

(a) boiler plant-high pressure-500 BHP;

(b) boiler plant-low pressure-1000 BHP;

(c) compressor plant-600 MHP;

(d) refrigerator plant-600 MHP.

39. The holder of a Stationary Engineer-Fourth class certificate of qualification may act as the chief stationary engineer in any plant to which the Act applies and having a capacity not exceeding the following limits:

(a) boiler plant-high pressure-100 BHP;

(b) boiler plant-low pressure-200 BHP;

(c) Compressor plant-200 MHP;

(d) refrigerator plant-200-MHP.

40. Notwithstanding sections 36(d), 37(d), 38(d) and 39(d) of these regulations, the holder of a Stationary Engineer certificate mentioned therein shall not act as a Chief Stationary Engineer or Shift Stationary Engineer in any refrigerator plant to which the Act applies unless the holder of the Stationary Engineer certificate also is the holder of a Standardized Certificate as defined in these regulations.

41. The holder of a Stationary Engineer-(Refrigerator Plant)-First Class certificate of qualification may act as the chief stationary engineer in any refrigerator plant having unlimited capacity and to which the Act applies.

42. The holder of a Stationary Engineer-(Refrigerator Plant) Second Class certificate of qualification may act as the chief

stationary engineer in any refrigerator plant not exceeding a capacity of 600 motive horse-power and to which the Act applies.

43. The holder of a Stationary Engineer-(Compressor Plant)-First Class certificate of qualification may act as the chief stationary engineer in any compressor plant to which the Act applies.

44. The holder of a Stationary Engineer-(Compressor Plant)-Second Class certificate of qualification may act as the chief stationary engineer in any compressor plant not exceeding a capacity of 600 motive horse-power and to which the Act applies.

45. The holder of a Stationary Engineer-(Hoisting Plant)-Mobile Crane certificate of qualification may act as the chief stationary engineer in any mobile hoisting plant to which the Act applies.

46. The holder of a Stationary Engineer-(Hoisting Plant)-Tower Crane certificate of qualification may act as the chief stationary engineer in any tower hoisting plant to which the Act applies.

47. The holder of a Stationary Engineer-(Hoisting Plant)-Overhead Traveling Crane certificate of qualification may act as the chief stationary engineer in any overhead hoisting plant to which the Act applies.

PART X - CERTIFICATES

48. (1) On application from an employer the Board may issue a certificate of qualification to an employee of the employer with long term service in a registered plant of the employer.

(2) Any certificate of qualification issued pursuant to subsection (1) shall:

- (i) be restricted to the registered plant of the employer,
- (ii) be non-transferable, and
- (iii) not exceed the level of a certificate of qualification, second class.

(3) (a) every certificate of qualification issued after the effective date of these regulations shall remain in force for a period of 4 years from the date of issue, and renewal shall be during the month of original issue.

(b) Where the renewal fees have not been paid for more than 4 consecutive years an application for a certificate of qualification shall be processed as if the applicant had not previously been the holder of a certificate of qualification.

49. The fees payable for all examination under the Act and these regulations shall be as set out in Appendix "A" to these regulations.

50. The fees payable under Section 49 do not apply to an apprentice registered under the Apprenticeship and Trademen's Qualification Act who is enrolled in a course of study under that Act.

51. Where a person is the holder of a certificate of qualification as a Hoisting Engineer issued under the Stationary Engineer Act when these regulations come into force, the board may issue to the person one or more certificates of qualification as Stationary Engineer-(Hoisting Plant)-Mobile Crane, Stationary Engineer-(Hoisting Plant)-Tower Crane or Stationary Engineer-(Hoisting Plant)-Overhead Travelling Crane without fee and without examination where the Board is satisfied that the person is qualified to work in any or all of the hoisting plant classifications.

Chapter 7   Regulations for Canadian Provinces and Territories

APPENDIX A

TO REGULATIONS MADE BY THE MINISTER OF LABOUR RESPECTING FEES PAYABLE MADE PURSUANT TO SUBSECTION (1) OF SECTION 4 OF THE STATIONARY ENGINEERS ACT.

1. The fees payable by a candidate for a certificate of qualification shall be $10.00 per examination paper.

2. The papers to be written and the fees payable by a candidate for each certificate are as follows:

(a) **Stationary Engineer-First Class**

8 examination papers for a total fee of $80.00

(b) **Stationary Engineer-Second Class**

6 examination papers for a total fee of $60.00

(c) **Stationary Engineer-Third Class**

4 examination papers for a total fee of $40.00

(d) **Stationary Engineer-Fourth Class**

2 examination papers for a total fee of $20.00

(e) **Stationary Engineer-Refrigerator Plant First Class**

2 examination papers for a total fee of $20.00

(f) **Stationary Engineer-Refrigerator Plant Second Class**

2 examination papers for a total fee of $20.00

(g) **Stationary Engineer-Compressor Plant First Class**

2 examination papers for a total fee of $20.00

(h) **Stationary Engineer-Compressor Plant Second Class**

2 examination papers for a total fee of $20.00

(i) Stationary Engineer-Hoisting Plant Mobile Crane

2 examination papers for a total fee of $20.00

(j) Stationary Engineer-Hoisting Plant Tower Crane

2 examination papers for a total fee of $20.00

(k) Stationary Engineer-Hoisting Plant Overhead Traveling Crane

2 examination papers for a total fee of $20.00

3. Candidates re-writing any failed examination shall be required to pay a fee of $10.00 per paper.

4. The fee for a replacement certificate shall be $10.00 upon proof of loss or damage.

5. The fee for a replacement identification card shall be $2.00.

6. The fee for a renewal of a certificate of qualification shall be $25.00.

Chapter 7   Regulations for Canadian Provinces and Territories

Province of Ontario
SUMMARY OF QUALIFICATIONS:

1. Exam Location, Address and Phone Number

Toronto Central Apprenticeship Office
625 Church St.
1st Floor
Toronto, Ontario, Canada M4Y2E8
416/326-5800
FAX - 416/326-5799

2. Exam Days and Times

Monday through Friday 8 a.m. and 12 noon - no appointment is necessary

3. Examination Cost

No cost

4. Certificate of Qualification Cost and Renewal

1st issue of certificate............$60.00

To upgrade your certificate....$46.00

Renewal (for 2 years).............$60.00

5. Type of Test

All multiple choice

6. Grades of Certificates

First Class Stationary Engineer

Second Class Stationary Engineer

Third Class Stationary Engineer

Fourth Class Stationary Engineer

7. Experience and Education Requirements

Regulations for Canadian Provinces and Territories   Chapter 7

See Regulation 4 - 7.

8. Canadian Citizenship Required?

No

9. Local Residency Required?

Yes

10. Recognition of licenses from other locales?

From other provinces and territories must take certificate to the Ministry of Consumer and Commercial Relations.

Accept interprovincial from Quebec - Example: 3rd Class could receive 4th Class without exam or could take 3rd Class exam and receive 3rd Class certificate with pass mark. Residents of other provinces and territories could receive a certificate 1 Class lower with pass mark on exam.

ACTS AND REGULATIONS for Province of Ontario

Revised Statutes of Ontario, 1980

Chapter 363

as amended by 1982, Chapter 42 and 1988, Chapter 10

Classes of Operating Engineers

15.(1) Operating Engineers shall be classified as follows:

1. Stationary Engineer (fourth, third, second or first class)

Classes of Operators

(2) Operators shall be classified as follows:

1. Compressor operator.

2. Refrigeration operator (B or A class)

Chapter 7 Regulations for Canadian Provinces and Territories

Stationary Engineer's (4th Class) what qualified to do:

16.(1) A person holding a Stationary Engineer's (fourth class) certificate of qualification is qualified,

   (a) to act as Chief Operating Engineer in charge of,

- (i) any stationary power plant of not more than 50 Therm-hours where the Therm-hour rating of refrigeration compressors is not more than 2.544 and the Therm-hour rating of compressors, including any refrigeration compressors, is not more than 5.088,

- (ii) any low pressure stationary plant of not more than 134 Therm-hours,

- (iii) any steam-powered plant of not more than 7.632 Therm-hours,

- (iv) any refrigeration plant of not more than 5.088 Therm-hours,

- (v) any compressor plant of not more than 10.176 Therm-hours,

- (vi) any plant referred to in subclause (ii) or (iii) whose total Therm-hour rating includes the Therm-hour rating of refrigeration compressors of not more than 3.816 Therm-hours or the Therm-hour rating of compressors, including any refrigeration compressors, of not more than 7.632 Therm-hours;

   (b) to act as Shift Engineer in,

- (i) any stationary power plant of not more than 134 Therm-hours where the Therm-hour rating of refrigeration compressors is not more than 5.088 and the Therm-hour rating of compressors, including any refrigeration compressors, is not more than 10.176,

Regulations for Canadian Provinces and Territories   Chapter 7

- (ii) any low-pressure stationary plant of not more than 400 Therm- hours,
- (iii) any steam-powered plant,
- (iv) any refrigeration plant of not more than 20.352 Therm-hours,
- (v) any compressor plant,
- (vi) any plant referred to in subclause (ii) or (iii) whose total Therm-hour rating includes the Therm-hour rating of refrigeration compressors of not more than 15.264 or the Therm-hour rating of compressors, including any refrigeration compressors, of not more than 30.528 Therm-hours;

(c) to act as Assistant Shift Engineer in,

- (i) any stationary power plant of not more than 400 Therm-hours,
- (ii) any low-pressure stationary plant, steam-powered plant, refrigeration plant or compressor plant.

Idem, Stationary Engineers (3rd Class)

(2) A person holding a Stationary Engineer's (third class) certificate of qualification is qualified,

(a) to act as Chief Operating Engineer in charge of,

- (i) any stationary power plant of not more than 134 Therm-hours where the Therm-hour rating of refrigeration compressors is not more than 5.088 and the Therm-hour rating of compressors, including any refrigeration compressors, is not more than 10.176,
- (ii) any low-pressure stationary plant of not more than 400 Therm-hours,

Chapter 7   Regulations for Canadian Provinces and Territories

- (iii) any steam-powered plant,
- (iv) any refrigeration plant of not more than 20.352 Therm-hours,
- (v) any compressor plant,
- (vi) any plant referred to in subclause (ii) or (iii) whose total Therm-hour rating includes the Therm-hour rating of refrigeration compressors of not more than 15.264 Therm-hours or the Therm-hour rating of compressors, including any refrigeration compressors, of not more than 30.528 Therm-hours;

(b) to act as Shift Engineer in,

- (i) any stationary power plant of not more than 400 Therm-hours that includes the Therm-hour rating of refrigeration compressors of not more than 15.264 Therm-hours or the Therm-hour rating of compressors, including any refrigeration compressors, of not more than 30.528,
- (ii) any low-pressure stationary plant, steam-powered plant, compressors or refrigeration plant;

(c) to act as Assistant Shift Engineer in any plant.

Idem, Stationary Engineers (2nd Class)

(3) A person holding a Stationary Engineer's (second class) certificate of qualification is qualified,

(a) to act as Chief Operating Engineer in charge of,

- (i) a stationary power plant of not more than 400 Therm-hours that includes the Therm-hour rating of refrigeration compressors of not more than 15.264 Therm-hours or the Therm-hour rating of compressors, including any refrigeration compressors, of not more than 30.528 Therm-hours,

- (ii) any low-pressure stationary plant, steam-powered plant, compressor or refrigeration plant;

(b) to act as Shift Engineer in any plant.

Idem, Stationary Engineers (1st Class)

(4) A person holding a Stationary Engineer's (first class) certificate of qualification is qualified to act as Chief Operating Engineer in charge of any plant.

Idem, Compressor Operators

(5) A person holding a compressor operator's certificate of qualification is qualified,

to act as a Chief or Shift Operator in any compressor plant whose prime mover is not a steam engine or steam turbine.

Idem, Refrigeration Operators (class B)

(6) A person holding a Refrigeration Operator's (class B) certificate of qualification is qualified,

(a) to act as Chief Operator in a refrigeration plant of not more than 20.352 Therm-hours or in any compressor plant whose prime mover is not a steam engine or steam turbine;

(b) to act as a Shift Operator in any refrigeration or compressor plant whose prime mover is not a steam engine or steam turbine.

Idem, Refrigeration Operators (class A)

(7) A person holding a Refrigeration Operator's (class A) certificate of qualification is qualified,

to act as Chief or Shift Operator in any compressor or refrigeration plant whose prime mover is not a steam engine or steam turbine.

Chapter 7 Regulations for Canadian Provinces and Territories

(8), (9) Repealed: 1982, c. 42, s. 9 (1).

Idem, Stationary Engineers

(10) A person holding a certificate of qualification of any class of Stationary Engineer is qualified to operate a portable compressor plant, a temporary heating plant or a portable boiler used in connection with any portable machinery or a device for melting ice or snow.

Idem, holders of provisional certificates

(11) A person holding a provisional certificate of qualification under section 23 is qualified to perform the same work and duties as an Operating Engineer or operator holding a corresponding certificate of qualification.

Trainees

17. A person who is obtaining qualifying experience for his first certificate of qualification may not perform work in connection with the actual operation of a plant except under the personal direction and supervision of an operating engineer or operator.

Shift Operators for compressors in stationary plants

18. Where a low-pressure stationary plant or stationary power plant has a compressor or a refrigeration compressor, the user of the plant may employ one or more compressor operators or one or more refrigeration operators, as the case may be, as Shift Operator or Shift Operators for the compressor.

Absence due to sickness or holidays

19. Where an operating engineer or operator is absent from his plant due to sickness or while on holidays, an operating engineer or operator holding a certificate not more than one class lower than the certificate of the operating engineer or operator who is absent may, during the absence, operate the plant for

not more than thirty days per year or such greater number of days per year as the chief officer may authorize in writing in any particular case.

Temporary absences

20. While a plant is in operation, an operating engineer or an operator qualified to be in charge of the plant shall be present in its boiler room, compressor room or engine room, as the case may be, or, where the plant is not enclosed, he shall be present in its immediate vicinity,

(a) unless an operating engineer or an operator holding a certificate of qualification that is not more than one class lower is present during his absence; or

(b) unless his absence is authorized by the regulation, and unless, in either case, he is satisfied at the time of his leaving the plant that it is operating safely.

Increase in Therm-hour rating

21. Where a plant has been operated by an operating engineer or operator in compliance with this Act and the regulations and the Therm-hour rating of the plant is increased so that the operating engineer or operator, as the case may be, is no longer qualified to operate the plant and he has operated the plant continuously for three consecutive years immediately before the increase, he may continue to operate the plant for such period and under such terms and conditions as the regulations prescribe.

Certificate of qualification

22.(1) The chief officer shall issue, in accordance with the regulations, a certificate of qualification to any person who,

(a) shows proof satisfactory to the chief officer of having acquired the qualifying experience required by the regulations;

(b) passes the examination conducted by the chief officer, or furnishes evidence that he has successfully completed a course of training that the Minister has approved for the purposes upon the advice of the board of review; and

(c) pays the fee prescribed by the regulations.

(2) Every certificate of qualification remains in force as prescribed by the regulations.

23.(1) The chief officer shall, upon payment of the fee prescribed by the regulations, issue a certificate of qualification to every person who applies therefor and holds a subsisting certificate issued by another province or territory of Canada that qualifies the person to perform the work or duties of an operating engineer or operator in such province or territory.

(2) The certificate of qualification issued under subsection (1) shall be of a class that authorizes the holder of the certificate to perform the work and duties that, in the opinion of the chief officer, the holder is qualified to perform in Ontario having regard to the qualifications prescribed by the regulations for applicants for certificates of qualification.

Cancellation or suspension of certificate of qualification

24. Subject to section 25, the chief officer may cancel or suspend a certificate of qualification if the operating engineer or operator,

(a) is habitually intemperate in his use of alcoholic beverages or is addicted to the use of drugs;

(b) operates a plant when his ability to do so is impaired by alcohol or a drug;

(c) is declared to be mentally incompetent or becomes physically incapable of safely performing his duties;

(d) is incompetent or negligent in the discharge of his duties as an operating engineer or operator;

(e) has obtained his certificate through misrepresentation or fraud;

(f) maliciously destroys his employer's property;

(g) allows another person to operate under his certificate;

(h) attempts to obtain a certificate by false means for another person;

(i) fails to give the notice required by section 31;

(j) leaves the employ of his employer without having given his employer at least seven days notice in writing of his intention to leave;

(k) furnishes information for the use of the chief officer respecting an applicant for a certificate without knowing that the information is true; or

(l) contravenes any of the provisions of this Act or the regulations.

Notice of proposal to suspend, etc., certificate

25(1) Where the chief officer proposes to refuse to renew or proposes to suspend or cancel a certificate of qualification, he shall serve notice of its proposal, together with written reason therefor, on the holder of the certificate.

Hearing

(2) A notice under subsection (1) shall inform the holder of the certificate that he is entitled to a hearing by a judge if he ap-

plies therefor to a judge of the county or district court for the county or district in which he resides, within fifteen days after the notice under subsection (1) is served on him, and he may so apply for such a hearing.

Powers of chief officer where no hearing

(3) Where a holder of a certificate does not apply to a judge for a heating in accordance with subsection (2), the chief officer may carry out the proposal stated in his notice under subsection (1).

Powers of chief officer where hearing

(4) Where a holder of a certificate applies to a judge for a hearing in accordance with subsection (2), the judge shall appoint a time for and hold the hearing and, on the application of the chief officer at the hearing, may by order direct the chief officer to carry out his proposal or refrain from carrying out his proposal and to take such action as the judge considers the chief officer ought to take in accordance with this Act and the regulations, and for such purposes the judge may substitute his opinion for that of the chief officer.

Service of notice by chief officer

(5) The chief officer may serve notice under subsection (1) personally or by registered mail addressed to the holder of the certificate at his address last known to the chief officer and, where notice is served by registered mail, the notice shall be deemed to have been served on the third day after the day of mailing unless the person to whom notice is being given establishes to the judge to whom he applies for a hearing that he did not, acting in good faith, through absence, accident, illness or other cause beyond his control receive the notice or order until a later date.

Extension of time for application

(6) A judge to whom application is made by a holder of a certificate for a hearing under this section may extend the time for making the application, either before or after expiration of the time fixed therein, where he is satisfied that there are prima facie grounds for granting relief to the holder of the certificate pursuant to a hearing and that there are reasonable grounds for applying for the extension, and may give such directions as he considers proper consequent upon the extension.

Continuation of certificate pending renewal

(7) Where, within the time prescribed therefor or, if no time is prescribed, prior to the expiry of his certificate, a holder of a certificate has applied for renewal of his certificate and paid the prescribed fee, his certificate shall be deemed to continue,

(a) until the renewal is granted; or

(b) where he is served with notice that the chief officer proposes to refuse to grant the renewal, until the time for applying for a hearing by a judge has expired and, where a hearing is applied for, until the judge has made his decision.

Parties

26.(1) The chief officer, the holder of the certificate who has applied for the hearing and such other persons as are specified by the judge are parties to the proceedings before a judge under section 25.

Notice of hearing

(2) Notice of a hearing under section 25 shall afford to the holder of the certificate a reasonable opportunity to show or to achieve compliance before the hearing with all lawful requirements for the retention of the certificate.

Examination of documentary evidence

Chapter 7   Regulations for Canadian Provinces and Territories

(3) A holder of a certificate who is a party to proceedings under section 25 shall be afforded an opportunity to examine before the hearing any written or documentary evidence that will be produced or any report the contents of which will be given in evidence at the hearing.

Recording of evidence

(4) The oral evidence taken before the judge at a hearing shall be recorded and, if so required, copies of a transcript thereof shall be furnished upon the same terms as in the Supreme Court.

Findings of fact

(5) The findings of fact of a judge pursuant to a hearing shall be based exclusively on evidence admissible or matters that may be noticed under section 15 and 16 of the Statutory Powers Procedure Act.

Appeal from decision of judge to court

27.(1) Any party to proceedings before a judge under section 25 may appeal from the decision or order of the judge to the Divisional Court in accordance with the rules of court.

Records to be filed in court

(2) Where notice of an appeal is served under this section, the judge shall forthwith file in the Supreme Court the record of the proceedings before him in which the decision or order was made, which together with the transcript of the evidence before the judge if it is not part of the record of the judge, shall constitute the record in the appeal.

Minister entitled to be heard

(3) The Minister is entitled to be heard, by counsel or otherwise, upon the argument of an appeal under this section.

Regulations for Canadian Provinces and Territories   Chapter 7

Powers of court on appeal

(4) The Divisional Court may, on the appeal, affirm the decision of the judge appealed from or may rescind it and make such new decision as the court considers proper under this Act and the regulations and may order the chief officer to do any act or thing he is authorized to do under this Act and as the court considers proper and for such purpose the court may substitute its opinion for that of the chief officer or of the judge, or the court may refer the matter back to the judge for rehearing, in whole or in part, in accordance with such directions as the court considers proper.

Appeal from decision of chief officer

28.(1) Any person who deems himself aggrieved by a decision of the chief officer under this Act or the regulations may, within ten days after the decision comes to his attention, appeal to a judge of the county or district court for the county or district in which the plant, boiler or other subject-matter to which the decision relates is located, by notice in writing sent by prepaid mail to the chief officer and the judge.

Powers of judge on appeal

(2) Where a person has appealed to a judge under subsection (1), the judge shall appoint a time for a hearing and shall hear the appeal and may affirm, rescind or vary the decision of the chief officer and may direct the chief officer to take any action that he is authorized to take under this Act or the regulations and as the judge considers proper, and for such purpose the judge may substitute his opinion for that of the chief officer.

(3) Subsection 25 (6) applies with necessary modifications to an appeal under this section.

Parties

(4) The chief officer, the appellant and such other persons as the judge may specify are parties to an appeal under this section.

Decision of judge final

(5) A decision of a judge under this section is final.

Effect of decision pending disposal of appeal

29. The bringing of an appeal under section 27 or 28 does not affect the operation of the decision appealed from pending disposition of the appeal.

Posting of certificates

30. Every operating engineer or operator shall display conspicuously his certificate of qualification in the engine room, compressor room or boiler room of the plant in which the operating engineer or operator works.

Duty to notify of absence

31. Every operating engineer or operator who,

(a) knows that he will be absent from his duties; or

(b) is unable to commence or continue his duties,

shall immediately make every reasonable effort in the circumstances to so notify his chief operating engineer or chief operator or shift engineer or shift operator, or, if none, his employer.

Prohibitions, operation by other than Operating Engineer or Operator

32.(1) No person other than an operating engineer who holds a certificate of qualification shall perform the work and duties of an operating engineer, and no person other than an operating engineer or operator who holds a certificate of qualification shall perform the work and duties of an operator.

Employment of unqualified persons prohibited

(2) No person shall employ,

(a) any person who is not an operating engineer to perform the work and duties of an operating engineer or operator, or any person who is not an operator to perform the work and duties of an operator; or

(b) any operating engineer or operator to operate a plant that he is not qualified under this Act to operate.

Work prohibited, unless qualified therefor

(3) No operating engineer or operator shall perform any work or duties of an operating engineer or operator that he is not qualified under this Act to perform.

Operation of plants

33. No person shall use or operate a plant or cause a plant to be used or operated except in accordance with this Act and the regulations.

False Statements

34. No person shall knowingly make a false statement or entry in an application, log book or document required by this Act or the regulations to be submitted or kept or knowingly furnish information under this Act or the regulations that is false, or knowingly make use of any such false statement, entry or information.

Offences

35.(1) Every person who contravenes or fails to comply with any of the provisions of this Act or the regulations, fails to comply with an order of an inspector or hinders or obstructs

Chapter 7   Regulations for Canadian Provinces and Territories

any person in the performance of his duties under this Act or the regulations, is guilty of an offence and on conviction is liable to a fine of not more than $10,000 or to imprisonment for a term of not more than twelve months, or to both.

Continuing Offence

(2) Where the circumstances constituting an offense against this Act continue from day to day and a prosecution has been commenced in respect to the offense, the offense shall be deemed to have been repeated on each day the circumstances continue.

REGULATION 740 UNDER THE OPERATING ENGINEERS ACT

QUALIFICATIONS OF APPLICANTS FOR CERTIFICATES

4. (1) An applicant for a certificate of qualification as a Stationary Engineer (Fourth Class) shall,

(a) furnish evidence that he has operated a plant in Ontario for one year under a provisional certificate of qualification as a Stationary Engineer (Fourth Class); or

(b) furnish evidence of,

- (i) accreditation in the subjects that comprise the training profile for this class of certificate and,

- (ii) previous training and experience as prescribed by subsection (2),

(2) For the purposes of subclause (1 (b) (ii), an applicant shall,

(a) furnish evidence from his employer stating that he has had three months qualifying experience in a stationary power plant or a low pressure stationary plant;

(b) furnish evidence that he holds a certificate of competency as a Third Class Engineer (steam or with steam endorsement) issued under the authority of the Canada Shipping Act, as amended or re-enacted from time to time; or

(c) furnish evidence that he has at least twelve months operating experience on boilers, engines and auxiliaries of naval or merchant ships equipped with boilers.

5. (1) An applicant for a certificate of qualification as a Stationary Engineer (Third Class) shall,

(a) furnish evidence that he has operated a plant in Ontario for one year under a provisional certificate of qualification as a Stationary Engineer (Third Class); or

(b) furnish evidence of,

- (i) accreditation in the subjects that comprise the training profile for this class of certificate, and

- (ii) previous training and experience as prescribed by subsection (2).

(2) For the purposes of subclause (1) (b) (ii), an applicant shall,

(a) hold a certificate of qualification as a Stationary Engineer (Fourth Class) and shall furnish evidence from his employer stating that he has had, in addition to the qualifying experience prescribed by the regulations for a certificate of qualification as a Stationary Engineer (Fourth Class), a further twelve months qualifying experience in a stationary power plant or a low pressure stationary plant;

Chapter 7 Regulations for Canadian Provinces and Territories

(b) hold a certificate of competency as a Second Class Engineer (steam or steam endorsement),

- (i) issued under the Merchant Shipping Act (Imperial), as amended or reenacted from time to time, or
- (ii) issued under the Canada Shipping Act, as amended or re-enacted from time to time;

(c) furnish evidence that he has had two years experience on boilers, engines and auxiliaries of naval or merchant ships equipped with boilers, or

(d) furnish evidence of,

- (i) a minimum of nine months practical experience and nine months academic experience in a course of training that has been approved by the Minister under clause 22 (1) (b) of the Act and at a community college that is equipped with a registered plant used for teaching purposes and that requires the services of a Stationary Engineer holding a Third Class or higher certificate of qualification, and
- (ii) six months practical operating experience in a registered industrial plant that requires the services of a Stationary Engineer holding a Second Class or higher certificate of qualification.

6.(1) An applicant for a certificate of qualification as a Stationary Engineer (Second Class) shall,

(a) furnish evidence that he has operated a plant in Ontario for one year under a provisional certificate of qualification, as a Stationary Engineer (Second Class); or

(b) furnish evidence of,

- (i) accreditation in the subjects that comprise the training profile for this class of certificate; and
- (ii) previous training and experience as prescribed by subsection (2).

(2) For the purpose of subclause (1) (b) (ii), an applicant shall,

(a) hold a certificate of qualification as a Stationary Engineer (Third Class) and shall furnish evidence from his employer stating that he has had, in addition to the qualifying experience prescribed by the regulations for a certificate of qualification as a Stationary Engineer (Third Class), a further eighteen months qualifying experience in a stationary power plant having a Therm-hour rating exceeding 134;

(b) hold a certificate of competency as a First Class Engineer (steam or with steam endorsement),

- (i) issued under the authority of the Merchant Shipping Act (Imperial), as amended or re-enacted from time to time, or
- (ii) issued under the authority of the Canada Shipping Act, as amended or re-enacted from time to time;

(c) hold a certificate of competency as a Second Class Engineer (steam or with steam endorsement),

- (i) issued under the authority of the Merchant Shipping Act (Imperial), as amended or re-enacted from time to time, or
- (ii) issued under the authority of the Canada Shipping Act, as amended or re-enacted from time to time, and has at least one year of qualifying experience in a

Chapter 7   Regulations for Canadian Provinces and Territories

stationary power plant having a Therm-hour rating exceeding 134; or

(d) furnish evidence that he,

- (i) holds an engineering degree conferred by a university in Canada or by a university outside Canada, where the course of study leading to the degree is, in the opinion of the Board, equivalent to that leading to an engineering degree in a university in Canada, and

- (ii) has had at least twenty-four months of qualifying experience that is approved by the Board for the purpose of this section in a stationary power plant having a Therm-hour rating exceeding 12,000 that is generating electricity for Ontario Hydro.

7.(1) An applicant for a certificate of qualification as a Stationary Engineer (First Class) shall,

(a) furnish evidence of accreditation in the subjects which comprise the training profile for this class of certificate; and

(b) furnish evidence of previous training and experience as prescribed by subsection (2),

(2) An applicant shall,

(a) hold a certificate of qualification as a Stationary Engineer (Second Class) and shall furnish evidence from his employer stating that he has had, in addition to the qualifying experience prescribed by the regulations for a certificate of qualification as a Stationary Engineer (Second Class), a further thirty months qualifying experience in a stationary power plant and shall have spent not less than,

- (i) six months of the thirty months in a stationary power plant having a Therm-hour rating in excess of 134, and

- (ii) twenty-four of the thirty months in a stationary power plant having a Therm-hour rating in excess of 300;

(b) have the qualifications and experience prescribed in clauses 6 (2) (b) and (c), and not less than twelve months of qualifying experience in a stationary powerplant with a Therm-hour rating exceeding 300 ; or

(c) furnish evidence that he,

- (i) holds an engineering degree conferred by a university in Canada or by a university outside Canada where the course of study leading to the degree is, in the opinion of the Board, equivalent to that leading to an engineering degree in a university in Canada, and

- (ii) has had at least thirty-six months of qualifying experience that is approved by the Board for the purpose of this section in a stationary power plant with a Therm-hour rating exceeding 12,000 that is generating electricity for Ontario Hydro.

RE-EXAMINATIONS

14.(1) Subject to subsection (2), an applicant for a certificate of qualification who has failed to pass the examination required by the Board may, on payment of the fee prescribed in the Schedule, be re-examined upon presenting himself for his re-examination at a place, date and time appointed by the Board.

Chapter 7 Regulations for Canadian Provinces and Territories

(2) A re-examination under subsection (1) shall not be held within sixty days following the previous examination.

ISSUE OF CERTIFICATES

15.(1) No certificate of qualification or provisional certificate of qualification shall be issued until the applicant has,

(a) delivered to the Board his application in the prescribed form; and

(b) fulfilled the requirements referred to in section 22 or 23 of the Act, as the case may be.

(2) A certificate of qualification issued for the first time expires on the second birthday next following of the holder of the certificate.

(3) A renewal of a certificate of qualification, upon payment of the prescribed fee, shall be valid for a period of two years.

(4) Revoked.

(5) Where a certificate of qualification has not been renewed within one year of the date of its expiry the certificate shall not be reinstated until the fee prescribed in the Schedule is paid, and the Board is satisfied that the applicant is capable of performing the duties of an operating engineer or operator.

(6) A person holding a subsisting certificate of qualification shall notify the Board in writing within fifteen days of a change of his address.

(7) Where a person proves to the satisfaction of the Board that,

(a) his certificate of qualification has been lost or destroyed; or

(b) his name has been changed,

the Board shall, on payment of the fee prescribed in the Schedule, issue to him a duplicate certificate of qualification, or in the case of a change in name, reissue his certificate of qualification.

EVIDENCE OF AGE

16.(1) Where evidence of age is required under this Regulation, an applicant for a certificate of qualification or a provisional certificate of qualification shall furnish his birth certificate.

(2) Where the Board is satisfied that it is not practicable for the applicant to furnish his birth certificate, the Board may accept,

(a) one item of Class A evidence of birth; or

(b) two items of Class B evidence of birth,

as prescribed in section 8 or 10 of Regulation 942 of Revised Regulations of Ontario, 1980.

OPERATION BY AUTHORIZED PERSONS

17. A user shall ensure that his plant is operated by a person who holds a certificate of qualification that authorizes him to operate that plant.

CHIEF OPERATING ENGINEERS AND CHIEF OPERATORS

18. In addition to any powers and duties prescribed by the Act, a Chief Operating Engineer or a Chief Operator, as the case may be,

(a) shall take all measures necessary to maintain the plant in a safe operating condition and shall notify the user of the measures taken;

(b) shall maintain discipline among the persons employed in the plant who are under his control or supervision;

(c) shall direct and supervise Shift Engineers or Shift Operators, as the case may be, in their work and duties for the safe operation of the plant;

(d) shall ensure that an accurate record of matters that may affect the safety of that plant is made and maintained at all times as required by section 21; and

(e) shall supervise and be responsible for the repair and maintenance of the plant where the plant is a boiler or compressor plant; and

(f) shall, subject to section 24, supervise or perform operational and maintenance work on the premises on which the plant is situated.

SHIFT ENGINEERS AND SHIFT OPERATORS

19. In addition to the powers and duties prescribed by the Act, a Shift Engineer or Shift Operator, as the case may be, shall,

(a) under the direction and supervision of the Chief Operating Engineer or Chief Operator, as the case may be,

- (i) be responsible for the safe operation of the plant,
- (ii) supervise other employees on his shift who are under his control,

- (iii) subject to section 24, perform maintenance and operational work on the premises on which the plant is situated;

(b) maintain a close watch on the condition and repair of all equipment in the plant and report to the Chief Operating Engineer or Chief Operator, as the case may be, any condition that may impair the safety of the plant;

(c) take such measures as are necessary to prevent any immediate danger; and

(d) ensure that an accurate record of matters that may affect the safety for the plant is made and maintained at all times during the shift period as required by section 21.

20. Revoked.

21. (1) The user shall provide for use in his plant a a log book in a form approved by the Chief Officer.

(2) The person in charge of a shift in a plant shall record in the log book in respect of his shift,

(a) the date, the number or designation of the shift and his name;

(b) any change from normal operating procedure, and the time of such change;

(c) any special instructions that may have been given to achieve the change referred to in clause (b) and the name of the person who gave the instructions;

(d) any unusual or abnormal condition observed in the plant, and the time thereof;

Chapter 7 Regulations for Canadian Provinces and Territories

(e) repairs to any part of the plant and the time such repairs were commenced and, if completed on his shift, the time thereof;

(f) the time of commencing and terminating his shift; and

(g) the date and time of any test of a safety device carried out during his shift.

(3) No person shall deface, damage, destroy or, without permission of the user, remove the log book from the plant.

(4) The user shall ensure that the log book is kept accessible in the plant for at least one year after the last entry therein and shall produce the log book for examination upon the request of an inspector.

Regulations for Canadian Provinces and Territories   Chapter 7

Province of Prince Edward Island
SUMMARY OF QUALIFICATIONS:

1. Exam Location, Address and Phone Number

Exam Location
Marine & Fisheries Training Center
Holland College
Water Street E.
Summerside, Prince Edward Island, Canada
902/888-6489

Office Location
Department of Labour
P.O. Box 2000
Charlottetown, Prince Edward Island, Canada CIA 7N8
902/368-5564

2. Exam Days and Times

2nd Friday of every month from 9 a.m. to 12:30 p.m.

There are no examinations given June through August.

3. Examination Cost

Power Engineer First Class.........$30.00

Power Engineer Second Class.....$22.00

Power Engineer Third Class.......$15.00

Power Engineer Fourth Class......$ 8.00

4. License Cost and Renewal

Power Engineer First Class.........$40.00

Power Engineer Second Class.....$30.00

Power Engineer Third Class........$20.00

Chapter 7 Regulations for Canadian Provinces and Territories

Power Engineer Fourth Class......$10.00

5. Type of Test

All Written

If candidate does not pass the written test they may be considered for an oral exam by the Board - See Regulations - (8).

Essay and Multiple Choice for 4th Class Power Engineer

All Essay for 3rd, 2nd and 1st Class Power Engineers

6. Grades of Licenses

Power Engineer First Class

Power Engineer Second Class

Power Engineer Third Class

Power Engineer Fourth Class

7. Experience and Education Requirements

See Regulations for Prince Edward Island

8. Canadian Citizenship Required?

No

9. Local Residency Required?

No

10. Recognition of licenses from other locales?

Reciprocal from other provinces, up to the Board, see Acts (9).

ACTS AND REGULATIONS for Prince Edward Island

(including any amendments to December 31, 1990)

Laws of Prince Edward Island

Power Engineers Act

R.S.P.E.I. 1988, cap p-15

EXAMINATIONS

EXAMINATION PAPERS

7. (1) The examination papers for all classes of licenses shall be those that have been accepted by the Committee for the Standardization of Power Engineers Examinations in Canada as meeting national standards, and are in use in the standardization program.

APPROVAL OF BOARD

(2) Examination papers for refrigeration licenses shall be approved by the Board.

CANDIDATE REQUIREMENTS

(3) A candidate shall, at least 15 days before the date fixed for examination, submit

(a) an application in form approved by the Minister;

(b) the appropriate fee set out in the Schedule; and

(c) copies of testimonials or other evidence respecting the candidate's experience.

EXPERIENCE QUALIFICATION

(4) The qualifications of a candidate relating to his experience in the installation, operation, and repairing of boilers, pressure vessels, pressure piping, and related equipment, may be proved by testimonials signed by the employer or Chief Engineer of the plant in which he was employed or by statutory

Chapter 7 Regulations for Canadian Provinces and Territories

declarations made by responsible persons who have personal knowledge of the facts that are to be established.

EDUCATIONAL QUALIFICATION

(5) Educational qualifications shall be vouched for by documents issued by the institution in which the candidate received his training.

ORAL EXAMINATIONS - TRANSITIONAL PROVISIONS

CANDIDATE WITHOUT LICENSE

8. (1) The provisions of this section apply to candidates who do not hold licenses issued after written examination in accordance with standards established by the Committee for the Standardization of Power Engineers Examinations in Canada but in the opinion of the board have sufficient knowledge and experience to be considered for the issue of a license by way of oral examination.

FIRST CLASS LICENSE

(2) No person may be a candidate for a First Class license unless

he has not less than twelve years experience in the province in the operation and maintenance of boilers and pressure vessels as of June 30, 1973, and

(a) is presently employed in a heating plant or power plant having a rating greater than 800 therm hours;

(b) has had four years experience in the operation and maintenance of a heating plant or power plant having a rating greater than 800 therm hours; and

(c) has had two years experience serving as chief engineer or shift operator in a heating plant or power plant having a rating greater than 800 therm hours.

SECOND CLASS LICENSE

(3) No person may be a candidate for a Second Class license unless he had not less than ten years experience in the province in the operation and maintenance of boilers and pressure vessels as of June 30, 1973, and

(a) is presently employed in a heating plant or power plant having a rating greater than 400 therm hours;

(b) has had three years experience in the operation and maintenance of a heating plant or power plant having a rating greater than 400 therm hours; and

(c) has had two years experience serving as chief engineer or second class operator in a heating plant or power plant having a rating greater than 400 therm hours.

THIRD CLASS LICENSE

(4) No person may be a candidate for a Third Class license unless

he had not less than seven years experience in the province in the operation and maintenance of boilers and pressure vessels as of June 30, 1973, and

(a) is presently employed in a heating plant or power plant having a rating greater than 200 therm hours;

(b) has had two years experience in the operation and maintenance of a heating plant or power plant having a rating greater than 200 therm hours; and

Chapter 7 Regulations for Canadian Provinces and Territories

(c) has had one years experience serving as chief engineer or third class operator in a heating plant or power plant having a rating greater than 200 therm hours.

FOURTH CLASS LICENSE

(5) No person may be a candidate for a Fourth Class license unless he had not less than four years experience in the province in the operation and maintenance of boilers and pressure vessels as of June 30, 1973, and

(a) is presently employed in a heating plant or power plant having a rating greater than 15 therm hours (power plant) or 50 therm hours (heating plant);

(b) has had one years experience in the operation and maintenance of a heating plant or power plant having a rating great than 15 therm hours (power plant) or 50 therm hours (heating plant); and

(c) has had one years experience serving as chief engineer or fourth class operator in a heating plant or power plant having a boiler rating greater than 15 therm hours (power plant) or 50 therm hours (heating plant).

REFRIGERATION CLASS A LICENSE

(6) No person may be a candidate for a Refrigeration Class A license unless he had not less than four years experience in the province in the operation and maintenance of refrigeration plants as of June 30, 1973, and

(a) is presently employed in a refrigeration plant having a rating greater than 20 therm hours;

(b) has had two years experience in the operation and maintenance of a refrigeration plant having a rating greater than 20 therm hours; and

(c) has had one years experience acting as chief engineer or shift operator in a refrigeration plant having a rating greater than 20 therm hours.

REFRIGERATION CLASS B LICENSE

(7) No person may be a candidate for a Refrigeration Class B license unless he has not less than three years experience in the province in the operation and maintenance of refrigeration plant as of June 30, 1973, and

(a) is presently employed in a refrigeration plant having a rating greater than 2.544 therm hours;

(b) has had two years experience in the operation and maintenance of a refrigeration plant having a rating greater than 2.544 therm hours; and

(c) has had one years experience serving as chief engineer or shift operator in a refrigeration plant having a rating greater than 2.544 therm hours.

ORAL EXAMINATION

(8) Where a person meets the requirements of this section the board shall direct the chief inspector to arrange for and conduct an oral examination of the candidate and in the conduct of the oral examination the chief inspector may, if he deems it advisable, and after consultation with the board, obtain the services of persons knowledgeable in the field of power engineering to assist him in the examination of any candidate for a license under this section.(EC22/80)

ISSUE OF LICENSES TO PERSONS QUALIFIED IN ANOTHER PROVINCE

LICENSE TO QUALIFIED PERSONS IN ANOTHER PROVINCE

Chapter 7   Regulations for Canadian Provinces and Territories

9. (1) A person who has obtained a Power Engineer's license by successfully passing the standardized examination in any other Canadian province shall be issued a license under these regulations if:

(a) he completes and files with the board an application for transfer in a form approved by the Minister;

(b) he pays the appropriate fee set out in the Schedule; and

(c) the board obtains confirmation of the issue of the license from the issuing authority.

CONDITIONS

(2) A persons who holds a license issued by the appropriate authority in any jurisdiction which did not use a standardized examination to determine the competency of that person, may be granted a license under these regulations if:

(a) he completes and files with the board an application for transfer;

(b) he pays the required fee;

(c) the board obtains confirmation of the issue of the license from the issuing authority; and

(d) the board determines that the license held is equivalent to a license issued under this Act. (EC22/80)

POWER OF BOARD TO ISSUE DISPENSATION
CHIEF ENGINEER

10. If a person is employed as Chief Engineer in a plant, and during the course of his employment has been involved in a plant expansion with the result that a higher class of license is required for a chief engineer in that plant, the board may, on the recommendation of the Chief Inspector, grant him a dispen-

Regulations for Canadian Provinces and Territories   Chapter 7

sation from the requirements of these regulations that will enable him to continue to act as Chief Engineer in that plant. (EC22/80)

LICENSES

CLASSES OF LICENSES

11. (1) Power engineers licenses shall be classified as follows:

Power Engineer Fourth Class

Power Engineer Third Class

Power Engineer Second Class

Power Engineer First Class

Power Engineer Refrigeration B

Power Engineer Refrigeration A

FORM

(2) Licenses shall be issued in a form approved by the Minister.

DURATION

(3) Licenses shall expire on the thirty-first day of December in the year of issue but may be renewed on payment of the renewal fee set out in the Schedule.

COMPOSITION

(4) A license shall contain the following information:

(a) classification;

(b) whether the license is standardized or provincial;

(c) the positions that the holder of the license may be employed to fill;

(d) the date the license was first issued and the date of expiry.

(EC22/80)
QUALIFICATIONS FOR CANDIDACY
FOURTH CLASS

12.(1) Any person may be a candidate for a Fourth Class license who:

(a)

- (i) has not less than 12 months experience in installation, operation and repair of boilers, pressure vessels, pressure piping and related equipment, and
- (ii) has completed an upgrading course in power engineering fourth class as required by the board;

(b) has completed, at any recognized trade school or university, a full-time course in power engineering fourth class.

THIRD CLASS

(2) Any person may be a candidate for a Third Class license who is the holder of a valid Fourth Class license and has since the issue of that license:

(a)

- (i) for a period of one year operated as Chief Engineer or Shift Engineer in a heating plant or power plant, or
- (ii) has for a period of one year operated as assistant shift engineer in a heating plant or power plant ; and

(b) has completed an upgrading course in power engineering third class as required by the board.

SECOND CLASS

(3) Any person may be a candidate for a Second Class license who is the holder of a valid Third Class license and has, since the issue of that license, not less than 24 months experience in aggregate in the following capacities or any combination of them:

(a) Chief Engineer or Shift Engineer in a registered power plant having a rating greater than 100 therm hours;

(b) Shift Engineer or Assistant Engineer in a registered power plant having a rating great than 400 therm hours;

(c) has for a period of not less than 24 months operated as an Assistant Shift Engineer in a registered power plant having a rating greater than 800 therm hours.

FIRST CLASS

(4) Any person may be a candidate for a First Class license who is the holder of a valid Second Class license and has since the issue of that license not less than 24 months experience in aggregate in the following capacities or any combination of them:

(a) Chief Engineer or Shift Engineer in a registered power plant having a rating greater than 400 therm hours;

(b) Shift Engineer or Assistant Engineer in a registered power plant having a rating greater than 800 therm hours;

(c) has for a period of 12 months been employed as plant supervisor in a registered power plant having a rating greater than 400 therm hours.

Chapter 7 Regulations for Canadian Provinces and Territories

REFRIGERATION CLASS B

(5) Any person may be a candidate for a Refrigeration Class B license who:

(a) has not less than 12 months experience (at least 3 of which are in operation) in the installation, operation and repair of industrial refrigeration systems; or

(b) not less than 6 months experience in the operation of an industrial refrigeration plant having a therm hour rating greater than 2.544 therm hours.

REFRIGERATION CLASS A

(6) Any person may be a candidate for a Refrigeration Class A license who is the holder of a valid class B license and has, since the issue of that license, not less than 12 months experience in the following capacities or any combination of them:

(a) Chief Engineer or Shift Engineer in a registered refrigeration plant having a rating greater than 2.544 therm hours;

(b) Shift Engineer or Assistant Engineer in a registered refrigeration plant having a rating greater than 20 therm hours.

UNSUCCESSFUL CANDIDATE

(7) If a candidate fails an examination, 90 days shall elapse before he is eligible to rewrite that examination.

PASS MARKS

(8) Pass marks for all examinations shall be 65% (EC22/80)

RECOGNITION OF EQUIVALENT TRAINING AND EXPERIENCE

EQUIVALENT TRAINING

13. A person having special engineering training in a recognized university or having completed a course in power engi-

neering satisfactory to the board, or having experience in the construction or repair of boilers, may be granted such time in lieu of practical operating experience as the board deems fair and reasonable. (EC22/80)

CAPACITIES IN WHICH LICENSED ENGINEERS MAY BE EMPLOYED

FIRST CLASS EMPLOYMENT CAPACITY

14.(1) The holder of a valid First Class license may be employed as Chief Engineer or Shift Engineer in any registered plant.

SECOND CLASS

(2) The holder of a valid Second Class license may be employed as: (a) Chief Engineer of:

- (i) any registered heating plant.
- (ii) any registered power plant not exceeding 800 therm hours,
- (iii) any registered refrigeration plant;

(b) Shift Engineer of:

- (i) any registered heating plant,
- (ii) any registered power plant,
- (iii) any registered refrigeration plant.

THIRD CLASS

(3) The holder of a valid Third Class license may be employed as:

(a) Chief engineer of:

- (i) any registered heating plant,

Chapter 7   Regulations for Canadian Provinces and Territories

- (ii) a registered power plant not exceeding 400 therm hours,
- (iii) any registered refrigeration plant not exceeding 20 therm hours;

(b) Shift engineer of:

- (i) any registered heating plant,
- (ii) a registered power plant not exceeding 800 therm hours,
- (iii) any registered refrigeration plant;

(c) Assistant Engineer in any registered plant.

FOURTH CLASS

(4) The holder of a valid fourth class license may be employed as:

(a) Chief Engineer of:

- (i) a registered heating plant not exceeding 200 therm hours,
- (ii) a registered power plant not exceeding 100 therm hours;

(b) Shift Engineer of:

- (i) any registered heating plant
- (ii) a registered power plant not exceeding 400 therm hours,
- (iii) any registered refrigeration plant;

(c) assistant engineer in a registered plant not exceeding 800 therm hours.

REFRIGERATION CLASS A

(5) The holder of a valid Refrigeration Class A license may be employed as chief engineer or shift engineer of any registered refrigeration plant.

REFRIGERATION CLASS B

(6) The holder of a Refrigeration Class B license may be employed as:

(a) Chief Engineer of a registered refrigeration plant not exceeding 20 therm hours;

(b) Shift Engineer of any registered refrigeration plant.

(EC22/80)

DUTIES OF EMPLOYERS

CHIEF ENGINEER DESIGNATED

15. (1) In a registered plant where two or more power engineers are employed to operate the plant, the employer shall designate one of them as Chief Engineer of the plant.

LOG BOOK

(2) The employer shall provide a log book for use in his plant in a form approved by the Chief Inspector.

IDEM

(3) The employer or his designate shall note the entries made in the log book for each twenty-four hour period and shall sign or initial the log entries for each such period.

TOOLS AND EQUIPMENT

(4) The employer shall supply all the necessary tools, equipment, parts and supplies to enable power engineers to operate,

Chapter 7 Regulations for Canadian Provinces and Territories

maintain and repair all plant components as required by the employer.

STORAGE AREA

(5) The employer shall provide a suitable storage area or stock room for the retention of the tools, equipment, parts and supplies mentioned in subsection (4). (EC22/80)

DUTIES OF CHIEF ENGINEER
CHIEF ENGINEER

16. (1) The Chief Engineer shall be held accountable to the employer for the proper care and safe operation of the boilers, pressure vessels, and machinery under his charge.

CASUALTIES

(2) He shall report all accidents and casualties.

DEFECTS

(3) He shall report to the employer and to an inspector any defects that he may have discovered or that have been reported to him which could endanger the safety of the boilers or machinery.

DUTIES OF CHIEF ENGINEER

(4) The Chief Engineer shall:

(a) take all measures necessary to maintain the plant in a safe operating condition and notify the employer of the measures taken;

(b) direct and supervise shift supervisors or shift engineers, as the case may be, in their work and duties to ensure the safe operation of the plant;

(c) be responsible for the safe keeping of all tools, equipment, and supplies provided by the employer for the operation, maintenance, and repair of the plant;

(d) see that the engineer in charge of each shift records in the log book:

- (i) the date, number and designation of the shift and his name,
- (ii) that he has completed the applicable tasks and subtasks of Block D of the Analysis,
- (iii) any changes from normal operating procedures and the time of such change,
- (iv) any special instructions which may have been given to effect the change referred to in subclause (iii), and the name of the person who gave the instructions,
- (v) any unusual or abnormal conditions observed in the plant and the time thereof,
- (vi) repairs to any part of the plant and the time such repairs commenced and if completed on his shift, the time thereof, and (vii) the time of commencing and terminating his shift.(EC22/80)

SHIFT ENGINEER DUTIES

17. The Shift Engineer shall:

(a) under the direction of the Chief Engineer be responsible for:

- (i) safe operation of the plant, and
- (ii) supervision of other employees on his shift who are under his control;

Chapter 7   Regulations for Canadian Provinces and Territories

(b) maintain close watch on the condition and repair of all equipment in the plant and report to the chief engineer any condition that may impair the safety of the plant;

(c) take all measures that are necessary to prevent any immediate danger;

(d ensure that an accurate record of matters that may affect the safety of the plant is made and maintained at all times during the shift period;

(e) ensure that all maintenance and operational work performed on the plant is in accordance with safe operating procedures and accepted engineering practices.(EC22/80)

ASSISTANT SHIFT ENGINEER

18. The Assistant Shift Engineer shall be under the direction and supervision of the Chief Engineer or the shift engineer, as the case may be, and be responsible for:

(a) the safe operation of a particular section of the plant;

(b) ensuring that an accurate record of matters that may affect the safety of that section of the plant is made and maintained at all times during the shift period;

(c) the performance of such maintenance and operational work on the plant as may be directed by the Chief Engineer or the Shift Engineer.(EC22/80)

POWER ENGINEER'S DUTIES

19. The duties of Power Engineers include the tasks and subtasks outlined in the analysis applicable to the particular plant.(ec22/80)

OPERATIONAL REQUIREMENTS

SUPERVISION OF BOILER

20. (1) In any plant when the heat source of a boiler is created by the burning of a solid fuel, the boiler shall be under continuous supervision.

PROTECTIVE DEVICES

(2) A boiler that is not under continuous supervision shall be provided with protective devices satisfactory to the chief inspector or his designate which may include:

(a) a high pressure limiting device on a steam boiler or a high temperature limiting device on a hot water boiler, as the case may be;

(b) an independent low water cut out control which will shut off the fuel to the burner in the event of a low water condition;

(c) a prepurge and flame failure device that will automatically prevent the supply of fuel to the boiler when an abnormal condition occurs during the operation of the boiler;

(d) a high water level limiting device that controls the supply of feedwater to the boiler;

(e) an alarm system that is audible in any part of the premises on which the plant is situated and in which persons may be present.

IDEM

(3) The protective devices prescribed in subsection (2) must:

(a) be manually reset after shutdown; and

(b) maintain the warning until the abnormal condition has been corrected.

UNATTENDED PREMISES

Chapter 7 Regulations for Canadian Provinces and Territories

(4) A power plant or heating plant having a rating less than 100 therm hours may be left unattended and in operation for up to 8 hours if the premises are unoccupied and the plant is equipped with protective devices in accordance with this section.

IDEM

(5) A refrigeration plant having a rating less than 20 therm hours may be left unattended and in operation provided the premises are unoccupied and the plant is equipped with protective devices satisfactory to the chief inspector. (EC22/80; 241/83)

OFFENCES CONCERNING LOG BOOK

21. (1) No person shall deface, damage or destroy a log book.

REMOVAL OF LOG BOOK

(2) No person shall remove the log book from a plant without the permission of the employer.

ACCESS TO LOG BOOK

(3) The employer shall ensure the log book is kept accessible in the plant for at least one year after the last entry therein and shall produce the log book upon the request of an inspector.(EC22/80)

SCHEDULE

TABLE OF FEES

1. On application for examination for a Power Engineer's license:

First Class...$30.00

Second Class...$22.00

Third Class...$15.00

Fourth Class...$ 8.00

Refrigeration A...$15.00

Refrigeration B...$ 8.00

2. On application for transfer of a license issued in another province...$ 5.00

3. On application for renewal of license............$ 5.00

4. On application for registration of plant:

First Class...$40.00

Second Class..$30.00

Third Class...$20.00

Fourth Class...$10.00

Refrigeration A...$20.00

Refrigeration B...$10.00

5. On application for re-registration of a plant..$ 5.00

POWER ENGINEERS ACT for Prince Edward Island

LICENSE, ISSUE OF

6. Subject to the regulations, the board, on payment of the prescribed fee, may issue a license of the appropriate class to an applicant who:

(a) is of the required age;

Chapter 7 Regulations for Canadian Provinces and Territories

(b) completes the application form and files it with the board; and

(c) passes the examination, or holds a license issued by the proper authority in any other jurisdiction which the board deems equivalent to a license issued under this Act. 1977, c29, s.6.

LICENSE, SUSPENSION OR CANCELLATION OF

7. The board may cancel or suspend the license of a Power Engineer who:

(a) obtains his license through misrepresentation or fraud;

(b) becomes physically or mentally incapacitated.

(c) is incompetent or negligent in the discharge of his duties;

(d) allows some other person to have use of his license;

(e) during working hours while his plant is in operation, engages in any labour or pursuit not connected with the operation of the plant;

(f) operates or has charge of a plant other than a plant of such description or class as he is authorized by his license to operate or have charge of;

(g) is convicted of an offence under this Act or the regulations;

(h) leaves his plant unattended at any time it is in operation without first ensuring that his replacement is on shift; or

(i) fails to make any entry in the log book which he is required to make under the regulations,1977,c.29,2.7.

LICENSE, REQUIREMENT FOR

8. (1) No person shall operate a plant unless he holds a valid license authorizing him to do so.

LICENSE, SCOPE OF

(2) No person who is the holder of a license shall operate a plant:

(a) of larger therm hour rating than he is authorized to operate by his license;

(b) after his license has expired or been cancelled; or

(c) while his license is suspended. 1977, c.29, s.8.

ATTENDANCE WHILE PLANT IN OPERATION

9. Except in so far as the regulations may otherwise provide, at all times when a plant is in operation a power engineer who holds a valid license of a class appropriate for the operation of the plant shall be in attendance. 1977, c.29, s.9.

EMPLOYMENT OF NON-LICENSED OPERATOR

10. No employer shall employ any person for the purpose of operating or having charge of a plant unless that person is the holder of a license authorizing him to operate or have charge of the plant. 1977, c.29, s.10.

DESIGNATION OF PERSON HAVING CHARGE OF PLANT

11. Where two or more Power Engineers are employed in the operation of a plant, the employer shall designate one of them as having charge of the plant. 1977,c.29,s.11.

ACTIVITIES NOT CONNECTED WITH OPERATION

12. No employer shall permit a power engineer employed by him to engage during working hours while the plant is in op-

Chapter 7 Regulations for Canadian Provinces and Territories

eration in any labour or pursuit not connected with the operation of the plant. 1977,c.29,s.12.

INTERIM OPERATOR, AUTHORIZATION OF

13. When a power engineer is temporarily absent from a plant, the chief inspector may, in writing, authorize a power engineer who holds a license not more than one class lower than that of the absent engineer to act in his stead for a period not exceeding thirty days. 1977,c.29,s.13.

LICENSE, DISPLAY OF

14. Each employer shall cause the licenses of all Power Engineers employed in the plant to be displayed in a conspicuous place at the plant. 1977,c.29,s.14.

ENFORCEMENT

15. An inspector may for the purpose of enforcing this Act and the regulations:

(a) at all reasonable times enter and examine any building or premises if he has reason to believe that any boiler is being installed or operated therein;

(b) make such examination and inquiries as he may consider necessary to determine whether there is or has been any contravention of this Act or the regulations;

(c) make a decision or issue an instruction requiring an employer or a power engineer to comply with this Act or the regulations.1977,c.29,s.15.

AGENT

16. (1) Any person who considers himself aggrieved by a decision or action of an inspector or the board may, within thirty days thereof, appeal in writing to the Minister and the Minis-

ter, after giving notice of the hearing to all parties appearing to him to have an interest in the matter, shall hear the appeal and may approve, disapprove or vary the decision or action appealed against.

DECISION FINAL

(2) The decision of the Minister pursuant to subsection (1) is final and no appeal lies therefrom. 1977,c.29,s.16.

OFFENCES

17. (1) Every person who contravenes any of the provisions of this Act or the regulations or hinders or obstructs an inspector in the performance of his duties is guilty of an offence and on summary conviction is liable to a fine of not less than $50 and not more than $500.

CONTINUING OFFENCE

(2) Where the circumstances constituting an offence continue from day to day and an information has been laid in respect of the offense, the offence shall be deemed to be repeated on each day the circumstances continue. 1977,c.29,s.17.

REGULATIONS

18. The Lieutenant Governor in Council may make regulation for the better administration and enforcement of this Act and without limiting the generality thereof:

(a) concerning the duties of employers;

(b) concerning the duties of power engineers;

(c) specifying the classes of licenses and their form, duration and content;

Chapter 7 Regulations for Canadian Provinces and Territories

(d) prescribing the qualifications required of applicants for licenses and the evidence to be furnished by applicants in respect of previous training and experience;

(e) prescribing the fees for the issue and renewal of licenses;

(f) providing for the reciprocal recognition of power engineers licenses issued in other provinces;

(g) prescribing forms and providing for their use;

(h) prescribing procedures and bylaws in respect of the board;

(i) concerning the registration of plants and the class of license required to be held by a power engineer to operate a plant of a particular description;

(j) providing for the measurement of the capacity of boilers by reference to their therm hour rating;

(k) governing the isolation of boilers and the interconnection of boilers and plant systems;

(l) regulating the operating of a plant in the absence of power engineers;

(m) requiring the keeping of log books of plant operations. 1977, c.29,s.18.

PAYMENT OF FEES, ETC.

19. All fees and fines collected under this Act and the regulations shall be paid into the Consolidated Fund. 1977,c.29,s.19.

Regulations for Canadian Provinces and Territories   Chapter 7

Province of Quebec
SUMMARY OF QUALIFICATIONS

1. Exam Location, Address and Phone Number
Societe Quebecoise De Developpement De La Main Doeuvre
201 Cremazie Blvd.
1st Floor
Montreal, Quebec, Canada H2M 1L5
514/873-6531

2. Exam Days and Times

1st Monday every month at 8:30 a.m. except if it's a legal holiday then Monday of next week.

3. Examination Cost

$53.00 for all certificates

4. Certificate of Competency Cost and Renewal

$53.00 for two years

5. Type of Test

All written - multiple choice

(Ontario has same test as Quebec)

6. Grades of Certificates

Fourth Class Stationary Engineer

Third Class Stationary Engineer

Second Class Stationary Engineer

First Class Stationary Engineer

7. Canadian Citizenship Required?

Regulations do not mention this. Need Social Security #

8. Local Residency Required

Chapter 7   Regulations for Canadian Provinces and Territories

Regulations do not mention this. Generally, no.

9. Recognition of licenses from other locales

Accept interprovincial from Ontario - Example: 3rd Class could receive 4th Class without exam or could take 3rd Class exam and receive 3rd Class certificate with pass mark. Residents of other provinces and territories could receive a certificate 1 class lower with pass on exam.

Province of Saskatchewan

SUMMARY OF QUALIFICATIONS:

1. Exam Location, Address and Phone Number

Office Location

Saskatchewan Environment & Public Safety
1870 Albert Street
Regina, Saskatchewan,
Canada S4P 3V7
306/787-4522

Exam Locations:

Saskatoon and Regina 4 times a year in March, June, September and December

2. Exam Days and Times

Varies

3. Examination Cost

From Table 11 (Section 8) Examination Fees

First Class Engineer, Part A $80.00

First Class Engineer, Part B $80.00

Second Class Engineer, Part A $60.00

Second Class Engineer, Part B $60.00

Third Class Engineer, Part A $40.00

Third Class Engineer, Part B..............$40.00

Fourth Class Engineer.......................$40.00

Fifth Class Engineer..........................$40.00

Engineer's Special (Provisional).........$25.00

Fireman..$ 9.00

4. Certificate of Competency Cost and Renewal

First Class Engineer.........................$80.00

Second Class Engineer....................$40.00

Third Class Engineer.......................$40.00

Fourth Class Engineer.....................$40.00

5. Type of Test

5th Class - Multiple Choice

4th Class thru 1st Class - Multiple Choice and Essay

6. Grades of Certificates

First Class Engineer

Second Class Engineer

Third Class Engineer

Fourth Class Engineer

Fifth Class Engineer

Fireman

7. Experience and Education Requirements

See Below

8. Canadian Citizenship Required?

Chapter 7 Regulations for Canadian Provinces and Territories

No

9. Local Residency Required?

No

10. Recognition of licenses from other locales?

See Act #32

ACTS AND REGULATIONS for Province of Saskatchewan

THE BOILER AND PRESSURE VESSEL SAFETY AMENDMENT ACT, 1987

CHAPTER B-5

An Act respecting Boilers and Pressure Vessels and Steam Refrigeration and Compressed Gas Plants.

ENGINEER'S AND FIREMAN'S CERTIFICATES

Classes of Certificates

25. There shall be the following classes of final certificates:

(a) First Class Engineer's Certificate;

(b) Second Class Engineer's Certificate;

(c) Third Class Engineer's Certificate;

(d) Fourth Class Engineer's Certificate;

(e) Fifth Class Engineer's Certificate;

(f) Refrigeration Engineer's Certificate;

(g) Refrigeration Plant Operator's Certificate;

and the following classes of temporary certificates:

(h) Engineer's Special Certificate;

(i) Fireman's Certificate.

Issue of Certificates

26(1) Upon the recommendation of the Chief Inspector, the department shall issue a final or temporary certificate of the appropriate class to every person who has demonstrated his competence and who has complied with the regulations.

Duration of Certificates

27(1) A final certificate shall be valid for as long as it continues to be registered periodically in accordance with the regulations.

(2) A temporary certificate shall be valid for such period as may be prescribed by the regulations.

Scope of Certificates

28 (1) A valid First Class Engineer's Certificate entitles the holder:

to operate as Chief Engineer a boiler, steam plant or refrigeration plant of any capacity.

(2) A valid Second Class Engineer's certificate entitles the holder:

(a) to operate as Chief Engineer a boiler or steam plant having a capacity of not more than 10,000 kilowatts and a refrigeration plant of any capacity; and

(b) to assist in the operation of a boiler or steam plant of any capacity.

(3) A valid Third Class Engineer's Certificate entitles the holder:

(a) to operate as Chief Engineer:

- (i) a boiler or steamplant having a capacity of not more than 5,000 kilowatts; and
- (ii) a refrigeration plant having a capacity of not more than 500 tonnes; and

(b) to assist in the operation of:

- (i) a boiler or steam plant having a capacity of not more than 10,000 kilowatts; and
- (ii) a refrigeration plant of any capacity.

(4) a valid Fourth Class Engineer's Certificate entitles the holder:

(a) to operate as Chief Engineer:

- (i) a boiler or steam plant having a capacity of not more than 1,000 kilowatts; and
- (ii) a refrigeration plant having a capacity of not more than 200 tonnes; and

(b) to assist in the operation of:

- (i) a boiler or steam plant having a capacity of not more than 5,000 kilowatts; and
- (ii) a refrigeration plant having a capacity of not more than 500 tonnes.

(5) A valid Fifth Class Engineer's Certificate entitles the holder:

(a) to operate as Chief Engineer

- (i) a low pressure boiler or heating plant of any capacity;

- (ii) a high pressure boiler or steam plant having a capacity of not more than 500 kilowatts; and
- (iii) a refrigeration plant having a capacity of not more than 500 tonnes; and

(b) to assist in the operation of:
- (i) a boiler or steam plant having a capacity of not more than 1,500 kilowatts; and
- (ii) a refrigeration plant having a capacity of not more than 200 tonnes.

(6) a valid Engineer's Special Certificate entitles the holder:

to operate a boiler or steam plant of the capacity indicated in the certificate, provided that the capacity of the boiler or steam plant does not exceed 500 kilowatts.

(7) a valid Fireman's Certificate entitles the holder:

(a) to operate either or both of the following as indicated in the certificate:
- (i) a low pressure boiler or steam plant having a capacity of not more than 1,000 kilowatts;
- (ii) a high pressure boiler used for heating purposes only and having a capacity of not more than 300 kilowatts; and

(b) to act as assistant to the holder of a valid Fifth Class Engineer's Certificate.

(8) A valid refrigeration engineer's certificate entitles the holder to operate a refrigeration plant of any capacity.

Chapter 7   Regulations for Canadian Provinces and Territories

(9) A valid refrigeration plant operator's certificate entitles the holder to operate, but not to repair or overhaul, a refrigeration plant having a capacity of not more than 100 tonnes, provided that a refrigeration engineer or refrigeration repair company is on call

Permit in urgent cases

29(1) Where the owner of a boiler, steam plant or refrigeration plant shows to the satisfaction of the Chief Inspector that he is unable, because of some exceptional circumstance, to obtain the services of a person holding a certificate of the appropriate class, the Chief Inspector may in his discretion issue a permit to operate the boiler, steam plant or refrigeration plant as Chief Engineer, Shift Engineer or otherwise, as the case may require, to a person recommended by the owner, provided that that person holds a certificate that is not more than one class lower than the class of certificate required.

(2) A permit issued pursuant to subsection (1) shall be valid for a period of time, not exceeding 90 days, that is specified by the Chief Inspector.

Special certificates in certain cases

30. Where a candidate for a Second Class, Third Class, Fourth Class or Fifth Class Engineer's Certificate has twice failed in the examination for such certificate but has obtained such marks as may be specified in the regulations, the Chief Inspector, upon receiving a recommendation from the owner of a boiler or steam plant, may in his discretion issue a special certificate which shall entitle the candidate to operate that boiler or steam plant, but no other, in such capacity as is indicated in the certificate.

Foreign certificates

31. Upon receipt of an application accompanied by evidence of qualifications and identity satisfactory to the Chief Inspector, the Chief Inspector may, in his discretion, recommend the issue, subject to any conditions he may prescribe and upon payment of the prescribed fee, of a certificate of the class determined by the Chief Inspector to be appropriate to the holder of a certificate of qualification as an engineer from the Government of Canada or of any province of Canada or from any competent authority in any other jurisdiction.

Interprovincial certificates

32. The minister may arrange with the competent authority of any other province for the granting of interprovincial engineer's certificates upon such terms and conditions as may be agreed upon.

Certificate to be produced

33(1) Every holder of a certificate who is engaged in the operation of a boiler, steam plant or refrigeration plant shall produce evidence of such certificate and current registration upon the demand of an inspector.

(2) Failure to produce a certificate on demand as required by subsection (1) shall be prima facie evidence that the person concerned has no certificate.

Cancellation, suspension, or recall or certificates, etc.

34. Upon the recommendation of the Chief Inspector and upon due cause being shown, the minister may cancel, suspend or recall any certificate, permit, license or other authorization issued under this Act or the regulations, including any permit, license or other authorization issued to a person who directly or indirectly aids or abets any other person in the violation of any of the provisions of this Act or the regulations.

Chapter 7   Regulations for Canadian Provinces and Territories

THE BOILER AND PRESSURE VESSEL FEES AMENDMENT REGULATIONS, 1988

CHAPTER B-5 REG. 1

8(1) The fee payable for an examination, including the fee for the appropriate certificate, is the fee set out in Table 11 of the Appendix.

(2) Where an examinee receives a passing grade after his examination is remarked the fee payable for remarking is refundable.

Table 11

Section 8

Examination Fees

First Class Engineer, Part A	$80.00
First Class Engineer, Part B	$80.00
Second Class Engineer, Part A	$60.00
Second Class Engineer, Part B	$60.00
Third Class Engineer, Part A	$40.00
Third Class Engineer, Part B	$40.00
Fourth Class Engineer	$40.00
Fifth Class Engineer	$40.00
Refrigeration Engineer	$40.00
Engineer's Special (Provisional)	$25.00
Fireman	$ 9.00
Refrigeration Plant Operator	$ 9.00

Regulations for Canadian Provinces and Territories   Chapter 7

Part of an examination (per paper)..................$20.00

Re-marking an examination (per paper).........$80.00

9 (1) The fee payable for the periodic registration of a final certificate pursuant to section 27 of the Act, including the fee for the appropriate registration certificate, is:

(See Bulletin below)

August 3, 1992

BULLETIN

TO: Power Engineers and Fireman's Certificate Holders

FROM: Public Safety Division, Boiler and Pressure Vessel Safety Branch

RE: Notification of Certificate Registration Fee Increases

The Boiler and Pressure Vessel Fees Amendment Regulations, 1992, were passed through Order-in-Council 694/92 on July 21, 1992 and Gazetted on July 31, 1992.

The new schedule of fees related to the five (5) year registration of the various classes of power engineers and fireman's certificates will come into force effective September 1, 1992.

The fee changes are summarized as follows:

CERTIFICATE CLASS

1. First Class Engineer's Certificate............$100.00

2. Second, Third, Fourth or Fifth Class

Engineer's Certificate..................................$ 50.00

3. Refrigeration Engineer's Certificate.........$ 50.00

4. Refrigeration Operator's Certificate.........$ 36.00

Chapter 7 Regulations for Canadian Provinces and Territories

5. Engineer's Special Certificate, Oilfield...$100.00

6. Fireman's Certificate...............................$ 36.00

7. Engineer's Special Certificate,
Traction Engine..$ 36.00

    30 day permit...$ 40.00

    90 day permit...$120.00

A copy of the related sections of the Boiler and Pressure Vessel Fees Amendment Regulations is available from the department upon request.

Regulations for Canadian Provinces and Territories   Chapter 7

SUMMARY OF QUALIFICATIONS:

1. Exam Location, Address and Phone Number

Administration Building
P.O. Box 2703
2071 Second Ave.
Whitehorse, Yukon, Canada Y1A 2C6
403/667-5765

2. Exam Days and Times

Varies with candidates

Once a month or once every three months

3. Examination Cost

See Examination and Certificate of Competency Fees - Schedule A

4. Certificate of Competency Cost and Renewal

First Class Engineer.............$6.00

Second Class Engineer.........$6.00

Third Class Engineer............$6.00

Fourth Class Engineer..........$3.00

Fireman...............................$3.00

Building Operator.................$3.00

5. Type of Test

Fourth Class Engineer - Multiple choice and Essay

3rd, 2nd, and 1st Class Engineer - All Essay

6. Grades of Certificates

First Class Engineer

Second Class Engineer

Chapter 7 Regulations for Canadian Provinces and Territories

Second Class Engineer

Third Class Engineer

Fourth Class Engineer

Fireman

Building Operator

7. Experience and Education Requirements

See Regulations for Yukon below

8. Canadian Citizenship Required?

Generally, yes, up to Chief Inspector

9. Local Residency Required?

Generally, yes, up to Chief Inspector.

10. Recognition of licenses from other locales?

See Regulation 20 and 21

Up to Chief Inspector.

REGULATIONS for Yukon Territories

These regulations are made under the "BOILERS AND PRESSURE VESSELS ORDINANCE."

These Regulations may be cited as the Engineers Regulations.

1. In these regulations:

(a) "Assistant Engineer" means a person who holds a certificate of competency permitting him to perform the functions of an assistant engineer;

(b) "Building Operator" means a person who holds a certificate of competency permitting him to perform the functions of a building operator;

(c) "Chief Engineer" means a person who holds a certificate of competency permitting him to perform the functions of a Chief Engineer;

(d) "Fireman" means a person who holds a Fireman's Certificate of competency who takes charge of a power plant not exceeding 500 KW or who takes charge of a shift in the operation of a power plant not exceeding 1000 kW;

(e) "Heating Surface" means any part of the surface of a boiler that is in contact with liquid under pressure on one side and the products of combustion on the other side;

(f) "Renewal Card" means the renewal card issued to the holder of a certificate of competency;

(g) "Shift Engineer" means a person who holds a certificate of competency permitting him to perform the functions of a shift engineer and who has charge of a shift in a power plant under the supervision of a Chief Engineer;

(h) "Special Oilwell Operator" means a person who holds a certificate of competency permitting him to perform the functions of a special oilwell operator.

GENERAL SUPERVISION REQUIRED FOR PLANTS

2. (1) Pursuant to the Ordinance, any person in charge of a power plant specified in this Subsection shall hold a certificate of competency of the type specified as follows:

(a) with respect to a power plant, it shall be under the general supervision of the holder of not less than a Fourth Class Engineer's Certificate of Competency;

(b) Notwithstanding subsection (a) the chief inspector may require an individual who has general supervision of a power plant to have a certificate of competency in excess of a Fourth Class Engineer;

(c) with respect to a power plant of over 20 kW but not exceeding an aggregate capacity of 250 kW, it shall be under the general supervision of the holder of not less than a Fireman's Certificate of Competency;

(d) with respect of a power plant of over 0.0425 cubic metres but not exceeding an aggregate capacity of 0.0850 m3, it shall be under the general supervision of the holder of not less than a Fireman's Certificate of Competency.

Supervision of heating plants

(2) Pursuant to Section 28, Subsection (1) of the Ordinance, no owner or person in charge of a heating plant exceeding 750 kW used primarily for the purpose of heating one or more buildings, shall operate it, or permit or cause it to be operated unless it is under the supervision of the holder of a valid certificate of competency issued pursuant to these regulations, the classification of which qualifies the holder to act as building operator of the heating plant.

Exemptions re Supervision of heating plants

(3) Pursuant to Section 28, Subsection (1) of the Ordinance, a heating plant not exceeding 750 kW is not required to be operated under the supervision of the holder of a certificate of competency.

Certificate of Competency

ISSUANCE OF CERTIFICATES OF COMPETENCY

3. (1) The certificate of competency specified in these regulations shall be issued by the chief inspector upon a person satisfying the requirements thereof.

ESTABLISHMENT OF BUILDING OPERATOR'S CERTIFICATE OF COMPETENCY

4. (1) Building Operator's Certificate of Competency are hereby established:

(a) Building Operator

(2) The holder of a Fourth Class Engineer's Certificate of Competency issued prior to October 31st 1980, may be issued a Building Operator's Certificate of Competency without examination, on payment of a fee in accordance with the Schedule, Examination and Certificate of Competency.

(3) The holder of a Fireman's Certificate of Competency issued prior to October 31st 1980, may be issued a Building Operator's Certificate of Competency without examination, on payment of a fee in accordance with the Schedule, Examination and Certificate of Competency.

AUTHORITY OF HOLDERS OF CERTIFICATE OF COMPETENCY FIRST CLASS ENGINEER

5. (1) A First Class Engineer's Certificate of Competency qualifies the holder to:

(a) take charge of the general care and operation of any power plant as chief engineer, and to supervise the engineers in that plant, or;

(b) take charge of a shift in any power plant as shift engineer.

SECOND CLASS ENGINEER

(2) A Second Class Engineer's Certificate of Competency qualifies the holder to:

(a) take charge of the general care and operation of a power plant not exceeding 10,000 kW as chief engineer and to supervise the engineers in that plant, or;

(b) take charge of the general care and operation of a power plant consisting of one or more coil type drumless boilers having an aggregate capacity not exceeding 15,000 kW when used for the sole purpose of underground thermal flooding in oil fields, as chief engineer and to supervise the engineers in that plant, or;

(c) take charge of a shift in any power plant as shift engineer.

THIRD CLASS ENGINEER

(3) A Third Class Engineer's Certificate of Competency qualifies the holder to:

(a) take charge of the general care and operation of a power plant not exceeding 5,000 kW as chief engineer and to supervise the engineers in that plant; or

(b) take charge of the general care and operation of a power plant consisting of one or more coil type drumless boilers having an aggregate capacity not exceeding 10,000 kW when used for the sole purpose of underground thermal flooding in oil fields, as chief engineer and to supervise the engineers in that plant; or

(c) take charge of a shift in a power plant not exceeding 10,000 kW as shift engineer; or

(d) take charge of a shift in a power plant consisting of one or more coil type drumless boilers having an aggregate capacity not exceeding 15,000 kW when used for the sole purpose of underground thermal flooding in oil fields, as shift engineer; or

(e) take charge of a section of any power plant as assistant engineer, under the supervision of the shift engineer in that plant.

FOURTH CLASS ENGINEER

(4) A Fourth Class Engineer's Certificate of Competency qualifies the holder to:

(a) take charge of the general care and operation of a power plant not exceeding 1000 kW as chief engineer and to supervise the engineers in that plant; or

(b) take charge of the general care and operation of a power plant consisting of one or more coil type drumless boilers having an aggregate capacity not exceeding 5000 kw when used for the sole purpose of underground thermal flooding in oil fields, as chief engineer; or

(c) take charge of a shift in a power plant not exceeding 5000 kW as shift engineer; or

(d) take charge of a shift in a power plant consisting of one or more coil type drumless boilers having an aggregate capacity not exceeding 10,000 kW when used for the sole purpose of underground thermal flooding in oil fields, as shift engineer; or

(e) take charge of a section of a power plant not exceeding 10,000 kW as assistant engineer, under the supervision of the shift engineer in that plant; or

(f) take charge of the general care and operation of a power plant not exceeding 5000 kW.

FIREMAN

(5) A Fireman's Certificate of Competency qualifies the holder to:

(a) take charge of the general care and operation of a power plant not exceeding 750 kW as fireman in charge and to supervise the firemen on shift in that plant; or

(b) take charge of a shift in a power plant not exceeding 1000 kW; or

SPECIAL FIREMAN

(c) special fireman certificate of competency qualifies the holder to take charge of the general care and operation of portable steam boilers not exceeding 750 kW.

(6) A Special Oilwell Operator's Certificate of Competency qualifies the holder to take charge of a power plant operating on an oil drilling site having an aggregate capacity not exceeding 1000 kW.

BUILDING OPERATOR

(7) A building Operator Certificate of Competency qualifies the holder to exercise general supervision of a heating plant not exceeding 3000 kW and take responsibility for its general care and operation.

(8) A Certificate of Competency of a grade higher than Fourth Class qualifies the holder to exercise general supervision of any heating plant and take responsibility for its general care and operation.

TEMPORARY CERTIFICATE OF COMPETENCY

6. (1) Where an employer applies for a temporary certificate of competency and certifies he is unable to obtain services of sufficient holders of certificates of competency, a certificate of competency may be issued on a temporary basis to:

(a) a shift engineer of a power plant to act as chief engineer of the same plant; or

(b) an assistant engineer or assistant shift engineer to act as shift engineer in the same plant.

(2) Where the chief inspector is not satisfied that the person referred to in subsection (1) should be issued a temporary certificate of competency, he may require that a person to take an examination as he considers necessary.

(3) A temporary certificate of competency issued by the chief inspector shall not be more than one grade higher than the certificate of competency held.

(4) An employer shall apply for a temporary certificate of competency on a form prescribed by the chief inspector which shall contain a declaration that the person in whose favor the application is being made is, to the best of his knowledge, capable of acting in the capacity for which the temporary certificate of competency is being requested.

(5) Where the chief engineer is sick or expects to be absent from the power plant for which he is responsible or a period exceeding ninety-six hours, the employer or chief engineer shall apply to the chief inspector for a temporary certificate of competency to be issued in accordance with subsection (3) (a).

(6) The duration of any temporary certificate of competency is at the discretion of the chief inspector, but in no case shall a temporary certificate of competency be issued for a period longer than six months.

(7) Where a temporary certificate of competency is issued, the chief inspector may impose such conditions on the holder of the certificate of competency as he considers necessary

Chapter 7   Regulations for Canadian Provinces and Territories

which may include a condition as to the person by whom the holder of the temporary certificate of competency is to be employed while he holds the temporary certificate of competency.

ANNUAL FEES

7. (1) Subject to subsection (4), the holder of a certificate of competency referred to in these regulations shall on or before April 1st each year pay an annual fee in respect of his certificate of competency.

(2) Where a person pays his annual fee pursuant to subsection (1) the chief inspector shall issue to him a renewal card in respect of the year for which the fee is paid.

(3) Where a person writes an examination in the months of January, February or March and obtains a certificate of competency as a result of that examination, no annual fee is payable by him until April 1st of the next calendar year.

(4) Where a temporary certificate of competency is issued pursuant to these regulations, a fee is payable in respect thereof in accordance with the Schedule.

SUSPENSION FOR FAILURE TO PAY ANNUAL FEE

8. Where the holder of a certificate of competency fails to pay his annual fee before April 1st in any year, he shall stand suspended until the fee in respect of that year is paid and any arrears of the fees are paid.

PROHIBITION FOR ARREARS OF ANNUAL FEE

9. No person who is the holder of a certificate of competency shall act in that capacity in any year in respect of which his annual fee remains unpaid.

QUALIFICATIONS AND EXAMINATIONS

QUALIFICATIONS FOR FIRST CLASS ENGINEERS CERTIFICATE OF COMPETENCY EXAMINATION

10. (1) To qualify to take a First Class Engineer's Certificate of Competency examination, a candidate shall:

(a) hold a Second Class Engineer's Certificate of Competency or equivalent; and

EXPERIENCE REQUIRED

(b) furnish evidence satisfactory to the chief inspector of employment for a period of

- (i) thirty months as chief engineer in a power plant having a capacity exceeding 5000 kW; or
- (ii) thirty months as chief engineer in a power plant consisting of coil type drumless boilers used for the sole purpose of underground thermal flooding having a capacity exceeding 10,000 kW; or
- (iii) thirty months as shift engineer in a power plant having a capacity exceeding 10,000 kW; or
- (iv) thirty months as shift engineer in a power plant consisting of coil type drumless boilers used for the sole purpose of underground thermal flooding having a capacity exceeding 15,000 kW; or
- (v) forty-five months as assistant shift engineer in a power plant having a capacity exceeding 10,000 kW; or
- (vi) thirty months as an inspector of boilers and pressure vessels under this Ordinance.

CREDIT IN LIEU OF PLANT OPERATING EXPERIENCE

Chapter 7   Regulations for Canadian Provinces and Territories

(2) Twelve months credit in lieu of power plant operating experience may be granted by the chief inspector on successful completion of a course in power engineering satisfactory to the chief inspector, leading towards a First Class Engineer's Certificate of Competency examination.

MINIMUM EDUCATIONAL REQUIREMENTS

(3) The minimum educational requirements to qualify for a First Class Engineer's Certificate of Competency examination are at least 50% standing in Physics (Grade 12), Mathematics (Grade 11) and English (Grade 11) or equivalent or a pass in Part "A" of a First Class Course in Power Engineering satisfactory to the chief inspector.

CANDIDATE'S ELIGIBILITY TO WRITE EXAMINATION

(4) The examination shall be divided into two parts, lettered A and B, and a candidate may:

(a) write any one or all papers for Part A at any scheduled examination after obtaining a Second Class Engineer's Certificate of Competency; or

(b) write any one or all papers for Part B at a time subsequent to successfully completing the papers for Part A, providing the qualifying experience has been obtained as specified in subsection (1); or

(c) write the papers for both parts at the same sitting providing the qualifying experience has been obtained as specified in subsection (1).

REFERENCE SYLLABUS

(5) The examination shall consist of questions relating to the subjects contained in the current reference syllabus as estab-

lished by the chief inspector for the First Class Engineer's Certificate of Competency examination.

PASS MARK

(6) To obtain a pass:

(a) when a candidate writes a single paper without having previously written any of the papers, the candidate must obtain 70% of the marks for the paper; or

(b) when a candidate writes more than one paper at the same sitting or when a candidate has previously received a pass in any paper or papers, the candidate must obtain an average of 70% of the total marks for the paper and not less than 60% of the marks for each paper.

QUALIFICATIONS FOR SECOND CLASS ENGINEER'S CERTIFICATE OF COMPETENCY EXAMINATION

11. (1) To qualify to take a Second Class Engineer's Certificate of Competency examination, a candidate shall:

(a) hold a Third Class Engineer's Certificate of Competency or equivalent; and

(b) furnish evidence satisfactory to the chief inspector of employment for a period of:

- (i) twenty-four months as chief engineer of a power plant having a capacity exceeding 1000 kW; or

- (ii) twenty-four months as chief engineer in a power plant consisting of coil type drumless boilers used for the sole purpose of underground thermal flooding having a capacity exceeding 5000 kW; or

Chapter 7   Regulations for Canadian Provinces and Territories

- (iii) twenty-four months as shift engineer in a power plant having a capacity exceeding 5000 kW; or

- (iv) twenty-four months as shift engineer in a power plant consisting of coil type drumless boilers used for the sole purpose of underground thermal flooding having a capacity exceeding 10,000 kW; or

- (v) thirty-six months as shift engineer in a power plant having a capacity exceeding 1000 kW; or

- (vi) twenty-four months as assistant engineer in a power plant having a capacity exceeding 10,000 kW.

CREDIT IN LIEU OF POWER PLANT OPERATING EXPERIENCE

(2) Nine months credit in lieu of power plant operation experience may be granted on successful completion of a course in power engineering satisfactory to the chief inspector leading towards a degree in mechanical engineering or equivalent from a university satisfactory to the chief inspector.

MINIMUM EDUCATIONAL REQUIREMENTS

(3) The minimum requirements to take a Second Class engineer's Certificate of Competency examination are at least a 50% standing in Science of Physics (Grade 11), Mathematics (Grade 11) and English (Grade 11),or equivalent, or a pass in Part "A" of a Second Class Course in power engineering, satisfactory to the chief inspector.

CANDIDATE'S ELIGIBILITY TO WRITE EXAMINATION

(4) The examination shall be divided into two parts, lettered A and B, and the candidate may write:

Regulations for Canadian Provinces and Territories   Chapter 7

(a) any one or all papers for Part A at any scheduled examination after obtaining a Third Class Engineer's Certificate of Competency; or

(b) any one or all papers for Part B at a time subsequent to successfully completing the papers for Part A, providing the qualifying experience has been obtained as specified in subsection (1); or

(c) all the papers for both parts at the same sitting, providing the qualifying experience has been obtained as specified in subsection (1).

REFERENCE SYLLABUS

(5) The examination shall consist of questions relating to the subjects contained in the reference syllabus as established by the chief inspector for the Second Class Engineer's Certificate of Competency examination.

PASS MARK

(6) To obtain a pass:

(a) when a candidate writes a single paper, without having previously written any of the papers, the candidate must obtain 70% of the marks for the paper; or

(b) when a candidate writes more than one paper at the same sitting, or when a candidate has previously received a pass in any paper or papers, the candidate must obtain an average of 70% of the total marks for the papers and not less than 60% of the marks for each paper.

QUALIFICATIONS FOR THIRD CLASS ENGINEER'S CERTIFICATE OF COMPETENCY EXAMINATION

12. (1) To qualify to take a Third Class Engineer's Certificate of Competency examination, a candidate shall furnish evi-

Chapter 7 Regulations for Canadian Provinces and Territories

dence satisfactory to the chief inspector of employment for a period of:

(a) twelve months as chief engineer in a power plant having a capacity exceeding 500 kW while holding a Fourth Class Engineer's Certificate of Competency; or

(b) twelve months as chief engineer in a power plant consisting of coil type drumless boilers used for the sole purpose of underground thermal flooding having a capacity exceeding 1000 kW while holding a Fourth Class Engineer's Certificate of Competency; or

(c) twelve months as shift engineer in a power plant having a capacity exceeding 1000 kW while holding a Fourth Class Engineer's certificate of Competency or equivalent; or

(d) twelve months as shift engineer in a power plant consisting of coil type drumless boilers used for the sole purpose of underground thermal flooding having a capacity exceeding 5000 kW while holding a Fourth Class Engineer's Certificate of Competency; or

(e) twelve months as assistant engineer in a power plant having a capacity exceeding 5000 kW while holding a Fourth Class Engineer's Certificate of Competency, or equivalent; or

(f) twenty-four months as building operator in a heating plant having a capacity exceeding 3000 kW while holding a Building Operator Certificate of Competency.

CREDIT IN LIEU OF POWER PLANT OPERATING EXPERIENCE

(2) Six months credit in lieu of power plant operating experience may be granted by the chief inspector on successful completion of a course in power engineering satisfactory to the

chief inspector, leading towards a Third Class Engineer's Certificate of Competency examination, a degree in mechanical engineering or equivalent from a university satisfactory to the chief inspector.

MINIMUM EDUCATIONAL REQUIREMENTS

(3) The minimum educational requirements to qualify to take a Third Class Engineer's Certificate of Competency examination are at least a 50% standing in Science or Physics (Grade 10), Mathematics (Grade 10) and English (Grade 10) or equivalent, or a pass in Part "A" of a Third Class Course in power engineering satisfactory to the chief inspector.

CANDIDATE'S ELIGIBILITY TO WRITE EXAMINATION

(4) The examination shall be divided into two parts, letter A and B and a candidate may write:

(a) any one or all papers for Part A at any scheduled examination after obtaining a Fourth Class Engineer's Certificate of Competency; or

(b) any one or all papers for Part B at a time subsequent to successfully completing the papers for Part A, providing the qualifying experience has been obtained as specified in subsection (1); or

(c) all the papers for both parts at the same sitting, providing the qualifying experience has been obtained as specified in subsection (1).

REFERENCE SYLLABUS

(5) The examination shall consist of questions relating to the subjects contained in the current reference syllabus as estab-

lished by the chief inspector for the Third Class Engineer's Certificate of Competency examination.

PASS MARK

(6) To obtain a pass:

(a) when a candidate writes a single paper, without having previously written any of the papers, the candidate must obtain 60% of the marks for the paper; or

(b) when a candidate writes more than one paper at the same sitting, or when a candidate has previously received a pass in any paper or papers, the candidate must obtain an average of 60% of the total marks for the papers and not less than 50% of the marks for each paper.

QUALIFICATIONS FOR FOURTH CLASS ENGINEER'S CERTIFICATE OF COMPETENCY EXAMINATION

13. (1) To qualify to take a Fourth Class Engineer's Certificate of Competency examination, a candidate shall furnish evidence satisfactory to the chief inspector:

(a) of employment for a period of twelve months as fireman of a power plant having a capacity exceeding 250 kW; or

(b) of being a holder of a degree in mechanical engineering or equivalent from a university satisfactory to the chief inspector; or

(c) of employment for a period of one-half that specified in clause (a) and in addition has been employed for a period of twelve months on a capacity satisfactory to the chief inspector on the design, construction, installation, repair, maintenance or operation of equipment to which the Ordinance applies; or

(d) of employment for a period of twelve months in a pressure plant in an operating capacity satisfactory to the chief inspector; or

(e) of having successfully completed a vocational course in power engineering satisfactory to the chief inspector; or

(f) of employment for a period of twelve months as a building operator in a heating plant exceeding 750 kW and is the holder of a certificate of Fourth Class grade in power engineering issued by an educational institution satisfactory to the chief inspector.

CREDIT IN LIEU OF POWER OR PRESSURE PLANT OPERATING EXPERIENCE.

(2) Six months credit in lieu of power or pressure plant operating experience as specified in subsection (1) (a) or (1) (d) may be granted by the chief inspector on successful completion of a course in power engineering satisfactory to the chief inspector, leading towards a Fourth Class Engineer's Certificate of Competency examination.

CREDIT FOR POWER ENGINEERING DAY COURSE

(3) A candidate who has successfully completed the first full term of a two year day course in power plant engineering from an educational institution satisfactory to the chief inspector is qualified to take a Fourth Class Engineer's Certificate of Competency examination.

CREDIT FOR HOLDER OF GAS TECHNOLOGY DIPLOMA

(4) A candidate who is the holder of a diploma in gas technology after completing a two year course from an educational institution satisfactory to the chief inspector is qualified to take

Chapter 7 Regulations for Canadian Provinces and Territories

a Fourth Class Engineer's Certificate of Competency examination.

CANDIDATE'S ELIGIBILITY TO WRITE EXAMINATION

(5) The examination shall be divided into two parts, lettered A and B and a candidate may write:

(a) Part A at any scheduled examination after obtaining a Fireman's Certificate of Competency or six months experience as specified in subsection (1) (a) or (1) (d); or

(b) Part B at a time subsequent to successfully completing Part A, providing the qualifying experience has been obtained as specified in subsection (1); or

(c) both parts at the same sitting, providing the qualifying experience has been obtained as specified in subsection (1).

REFERENCE SYLLABUS

(6) The examination shall consist of questions relating to the subjects contained in the current reference syllabus as established by the chief inspector for the Fourth Class Engineer's Certificate of Competency examination.

PASS MARK

(7) To obtain a pass:

(a) when a candidate writes a single paper, without having previously written any of the papers, the candidate must obtain 60% of the marks for the paper; or

(b) when a candidate writes more than one paper at the same sitting, or when a candidate has previously received a pass in any paper or papers, the candidate must obtain an average of

Regulations for Canadian Provinces and Territories   Chapter 7

60% of the total marks for the papers and not less than 50% of the marks for each paper.

QUALIFICATIONS FOR FIREMAN'S CERTIFICATE OF COMPETENCY EXAMINATION.

14. (1) To qualify to take a Fireman's Certificate of Competency examination a candidate shall furnish evidence satisfactory to the chief inspector.

(a) of having acted as fireman operating any boiler for a period of at least six months; or

(b) of having successfully completed a vocational course in boiler operation satisfactory to the chief inspector.

GRANT IN LIEU OF OPERATING EXPERIENCE

(2) Three months credit in lieu of operating experience as specified in subsection (1) may be granted by the chief inspector upon successful completion of a course in boiler operation satisfactory to the chief inspector, leading towards a Fireman's Certificate of Competency examination.

PASS MARK

(3) To qualify for a Fireman's Certificate of Competency, a candidate must receive 50% of the total marks allotted for the examination.

REFERENCE SYLLABUS

(4) The examination shall consist of questions relating to the subjects contained in the current reference syllabus as established by the chief inspector for the Fireman's Certificate of Competency examination.

QUALIFICATIONS FOR SPECIAL OILWELL OPERATOR'S CERTIFICATE OF COMPETENCY EXAMINATION

Chapter 7 Regulations for Canadian Provinces and Territories

15. (1) To qualify to take a Special Oilwell Operator's Certificate of Competency examination, a candidate shall furnish evidence satisfactory to the chief inspector that he has:

(a) obtained six months experience in a power plant on an oil drilling site; or

(b) successfully completed a vocational course in boiler operation satisfactory to the chief inspector.

PASS MARK

(2) To qualify for a Special Oilwell Operator's Certificate of Competency, a candidate must receive 50% of the total marks allotted for the examination.

REFERENCE SYLLABUS

(3) The examination will consist of questions relating to the subjects contained in the current reference syllabus as established by the chief inspector for the Special Oilwell Operator's Certificate of Competency examination.

QUALIFICATION FOR BUILDING OPERATOR CERTIFICATE OF COMPETENCY EXAMINATION EXPERIENCE REQUIRED

16 (1) To qualify to take a Building Operator Certificate of Competency examination, a candidate shall furnish proof of having been employed for a period of six months in the operation of any heating plant.

CREDIT IN LIEU OF HEATING PLANT OPERATING EXPERIENCE

(2) Three months credit in lieu of operating experience as specified in subsection (1) may be granted upon successful completion of a course satisfactory to the chief inspector, leading towards a Building Operator Certificate of Competency.

Regulations for Canadian Provinces and Territories   Chapter 7

PASS MARK

(3) To qualify for a Building Operator Certificate of Competency a candidate must receive 50% of the total marks allotted for the examination.

REFERENCE SYLLABUS

(4) The examination shall consist of questions relating to the subjects contained in the current reference syllabus as established by the chief inspector for the Building Operator Certificate of Competency examination.

CREDITS FOR TECHNICAL COURSES

17 (1) The chief inspector may provide a person credit for taking courses in power engineering or heating plant operation, that he considers satisfactory in lieu of practical experience but that credit shall only be permitted once for each class of engineer's or building operator's examination.

(2) A candidate shall not receive credit for a course in power engineering except upon completion of the specified minimum power plant operating experience as required for each class of engineer's examination.

ASSESSMENT OF EQUIVALENT EDUCATION

18. (1) The assessment of equivalent education which may be accepted in lieu of educational minimum requirements under these regulations shall be determined by the chief inspector.

CREDIT FOR ALTERNATIVE QUALIFICATIONS

19. (1) When a candidate has experience made up in part as chief engineer, shift engineer, assistant shift engineer, assistant engineer, building operator, or other experience, the chief inspector may evaluate that experience.

Chapter 7 Regulations for Canadian Provinces and Territories

(2) When a candidate can furnish proof of experience acceptable to the chief inspector with equipment to which the Ordinance applies, the chief inspector may allow such experience as credit in lieu of the specified qualification mentioned in section 10, 11 and 12.

(3) When a power plant is in operation for only part of a year and the engineer is retained for the non-operational period and is employed on plant maintenance the chief inspector may grant a credit of two-thirds of the maintenance time towards experience required for a higher level of examination.

(4) Credit for practical experience previously used in qualifying for an engineer's examination shall not be used again in qualifying for a higher level of examination.

(5) When a plant consists of a combination of a power plant and a heating plant, the boiler capacity of each plant shall be considered separately when assessing the experience required to qualify to take an engineer's or building operator's examination.

ISSUANCE OF EQUIVALENT CERTIFICATE OF COMPETENCY

(20). (1) Where the chief inspector considers that a person holds a certificate of competency from a jurisdiction outside Yukon equivalent to a certificate of competency specified in these regulations, the chief inspector may, upon application to him, issue an equivalent certificate of competency.

(2) Any certificate of competency issued pursuant to this section may be issued subject to such conditions as the chief inspector considers necessary.

(3) The chief inspector shall not issue a certificate of competency pursuant to this section until he is satisfied as to the appli-

cant's identity, experience and qualifications and for that purpose may require such evidence as he considers necessary.

(4) A candidate from a jurisdiction outside Yukon who has passed any paper of an engineer's examination in that jurisdiction may be given credit by the chief inspector for having passed that paper.

CANDIDATES FROM JURISDICTIONS OUTSIDE YUKON

21. (1) At the discretion of the chief inspector, a certificate of competency of a class determined appropriate by the chief inspector may be issued by the chief inspector to the holder of a valid certificate of competency as an engineer from the Government of Canada or any Province or Territory of Canada or any competent authority in any other jurisdiction who makes application therefor accompanied by evidence satisfactory to the chief inspector of the applicant's qualifications and identity.

APPLICATION AND CONDUCT OF EXAMINATIONS

PROCEDURE FOR SUBMITTING APPLICATIONS FOR EXAMINATIONS

22. (1) A candidate for examination shall apply on the form prescribed by the chief inspector at least twenty-one days before the date of the examination.

(2) Applications for examination shall be submitted for approval to the chief inspector.

REFERENCES

Chapter 7 Regulations for Canadian Provinces and Territories

(3) Originals of references or photocopies thereof vouching for the candidate's experience, ability and conduct shall accompany the application together with the fee specified in the Schedule hereto.

(4) Original documents shall be returned to the candidate after verification.

REFERENCES UNOBTAINABLE

(5) The qualifications of a candidate relating to plant operation, engineering experience, ability and general conduct may be proved by references signed by the owner or chief engineer of the plant where the candidate was employed, but if such references are not available, a written statement may be accepted if it is made by a person who has personal knowledge of the facts to be established.

STATUTORY DECLARATION

(6) Where a candidate for a Special Oilwell Operator, Fireman or Fourth Class Engineer's Certificate of Competency examination is unable to produce the statement referred to in subsection (5), a statutory declaration may be accepted if it is made by the candidate declaring that he has obtained the required operating experience to qualify him for the examination.

VERIFICATION OF TRAINING COURSES

(7) Educational qualifications shall be vouched for by documents issued by the institution from which the candidate received training.

TIME AND PLACE OF EXAMINATION

23. (1) A candidate for examination shall appear at such place and time as the chief inspector may direct.

(2) Examinations are under the direction of the chief inspector.

(3) Prior to the examination commencing, the existing certificate of competency and annual renewal card held by a candidate must be presented to the person conducting the examination.

AMANUENSIS OR ORAL EXAMINATION

24. (1) A candidate who is unable to write and who is qualified to take an examination may employ a person to write the examination:

(a) if the person selected is approved by the inspector designated to conduct the examination; and

(b) if the person selected signs a statement on the form prescribed by the chief inspector stating that he is not an engineer and does not have any knowledge of the construction or operation of boilers, pressure vessels, engines or other equipment to which the Ordinance applies and gives it to the person conducting the examination when the candidate takes the examination.

(2) the chief inspector may authorize an inspector to conduct an oral examination of any candidate who is unable to undertake a written examination, when a suitable amanuensis is not readily available.

(3) Notwithstanding subsections (1) and (2), in every case, candidates for examination respecting First, Second, Third Class Engineer's Certificate of Competency must complete a written examination and the use of an amanuensis is not permitted.

CAUSES FOR DISQUALIFICATION OF CANDIDATE

Chapter 7 Regulations for Canadian Provinces and Territories

25. (1) The inspector conducting any examination under these regulations may declare a candidate to have failed an examination if:

(a) formulae or other information not approved or authorized by the chief inspector have been added to or inserted into any published text of a book, table, regulation or code that is taken into the examination room;

(b) a candidate looks at or refers to any material not approved or authorized by the chief inspector during the examination;

(c) a candidate removes or attempts to remove any questions or part thereof from an examination room;

(d) a candidate copies from another candidate;

(e) a candidate communicates with another candidate in any manner during the examination.

(2) Any candidate who contravenes any provision of subsection (1) during an examination may be disqualified by the chief inspector from writing any further examination at a period not exceeding twelve months from the date of examination.

CANDIDATE TO PROVIDE NECESSARY EQUIPMENT

26. (1) Every candidate for examination shall provide pens, ink, pencils, drawing instruments and such other equipment as may be required and permitted for use during the examination.

MARKING OF PAPERS BY INSPECTOR

27. (1) Every paper in an examination specified in these regulations shall be marked by an inspector.

CANDIDATE FAILING TO PASS EXAMINATION

28. (1) A candidate failing to pass a paper in any part of an examination for any class of certificate of competency specified in these regulations on three consecutive attempts shall not be examined again for a period of six months from the date of examination.

(2) When the time interval between writing and rewriting a paper in any part of an examination for any class or certificate of competency specified in these regulations exceed five years, any previous papers which the candidate may have passed in that examination must be rewritten.

CALCULATION OF BOILER RATINGS

29. (1) Where calculations are made with respect to the application of this Ordinance or these regulations, boiler ratings shall be determined on the basis that:

(a) 0.5 square metre of heating surface equals 10 kilowatts; or

(b) where electric power is used as the heat source, the boiler rating shall be the maximum kilowatt capacity of the heating element; or

(c) where none of the above determinations are applicable, an hourly output of 36 megajoules is equivalent to 10 kilowatts.

(2) The heating surface of a boiler shall be determined by computing the area of the surface involved in square metres and where a computation is to be made of a curved surface, the surface having the greater radius shall be taken.

LOSS OF CERTIFICATE OF COMPETENCY OR RENEWAL CARD

30. (1) Where any certificate of competency or renewal card is lost or destroyed, a duplicate certificate of competency or renewal card may be issued upon evidence being furnished to the

satisfaction of the chief inspector, that the original certificate of competency or renewal card has been lost or destroyed.

(2) An application under subsection (1) must be accompanied by the fee for the issuance of a duplicate certificate of competency or renewal card as specified in the Schedule.

OBTAINING ANOTHER PERSON'S CERTIFICATE OF COMPETENCY

31. (1) Any persons acquiring any certificate of competency specified in these regulations other than the person whose name appears thereon shall send the certificate of competency to the chief inspector.

LOG BOOK TO BE MAINTAINED

32. (1) The chief engineer or building operator shall ensure that a log book is maintained to record any matter relating to the operation of the power plant, or heating plant, including a record of the testing and servicing of safety valves and other safety devices and controls.

33. The fees and charges as prescribed in these regulations are those set out in the schedule entitled schedule A that is annexed to and forms part of these regulations.

34. In these regulations, the metric system of calculation has legal effect.

SCHEDULE A - EXAMINATION AND CERTIFICATE OF COMPETENCY FEES

1. (1) The following fees are payable by any person applying to take an examination leading towards a certificate of competency;

Regulations for Canadian Provinces and Territories   Chapter 7

(a) First Class Engineer's Examination.............$40.00

Part A..$20.00

Part B..$20.00

Individual Paper...$ 5.00

(b) Second Class Engineer's Examination.........$30.00

Part A..$15.00

Part B..$15.00

Individual Paper...$ 5.00

(c) Third Class Engineer's Examination............$20.00

Part A..$10.00

Part B..$10.00

Individual Paper...$ 5.00

(d) Fourth Class Engineer's Examination.........$10.00

Part A..$ 5.00

Part B..$ 5.00

(e) Fireman's Examination...............................$ 5.00

(f) Special Oil Well Operator's Examination....$ 5.00

(g) Building Operator's Examination................$ 5.00

(2) The fee payable for any examination specified in subsection (1) shall be paid at the time of application.

(3) When an application referred to in subsection (2) is refused by the chief inspector, the examination fee shall be refunded.

Chapter 7   Regulations for Canadian Provinces and Territories

(4) An applicant for an examination who fails to appear at the designated examination time and place, forfeits his examination fee unless a reason satisfactory to the chief inspector, for failure to appear is submitted in writing to the chief inspector within seven day after the designated examination date.

(5) The fee payable for the re-marking of any examination paper shall be the same as the fee specified in subsection (1).

(6) Where a candidate obtains a pass in any paper after it has been re-marked the fee in subsection (5) shall be refunded to him.

2. (1) The fee payable for certificate of competency issued in lieu of a certificate of competency from another jurisdiction shall be the same as the examination fee specified in section 1 of this Schedule.

(2) Where a candidate receives credit as having passed any paper in any part of an engineer's examination as provided in section 18, subsection (3) (b) or section 21, subsection 4 of the Engineer's Regulations, the fees for such papers shall be the same as provided in subsection (1).

(3) In the case of an engineer's examination submitted by another jurisdiction for marking by an inspector, the fee payable shall be the same as specified in section 1 of this Schedule for the class of examination written.

3. (1) Specimen examination papers, approved for release by the chief inspector, may be purchased as follows:

(a) Special Oil Well Operator, Fireman, Fourth Class Engineer's and Building Operator's Certificate of Competency examination papers............................$2.00 each.

(b) First, Second and Third Class Engineer's Certificate of Competency examination papers....$3.00 each.

RENEWAL CARDS

4. (1) The annual fees payable in respect of a renewal card for an Engineer's, Fireman's, Special Oil Well or Building Operator's Certificate of Competency are;

(a) First Class Engineer's Certificate of Competency..$6.00

(b) Second Class Engineer's Certificate of Competency..$6.00

(c) Third Class Engineer's Certificate of Competency..$6.00

(d) Fourth Class Engineer's Certificate of Competency..$3.00

(e) Fireman's Certificate of Competency.....................$3.00

(f) Special Oil Well Operator's Certificate of Competency..$3.00

(g) Building Operator Certificate of Competency..$3.00

REINSTATEMENT OF CERTIFICATE OF COMPETENCY

5. (1) Where a person wishes to have his Engineer's, Special Oil Well Operator's, Fireman's or Building Operator's Certificate of Competency reinstated he shall pay any arrears of annual fees up to the amount of the examination fee payable under section 1 of this Schedule.

TEMPORARY CERTIFICATE OF COMPETENCY

Chapter 7　Regulations for Canadian Provinces and Territories

6. (1) Subject to subsection (2), the fees payable for the issue of a temporary Certificate of Competency are;

(a) temporary First Class Engineer's Certificate of Competency......$40.00

(b) temporary Second Class Engineer's Certificate of Competency......$30.00

(c) temporary Third Class Engineer's Certificate of Competency......$20.00

(d) temporary Fourth Class Engineer's Certificate of Competency......$10.00

(e) temporary Fireman's Certificate of Competency..$ 5.00

(f) temporary Special Oil Well Operator's Certificate of Competency......$ 5.00

(g) temporary Building Operator's Certificate of Competency......$ 5.00

(2) Where the chief inspector is satisfied that the temporary Certificate of Competency is required for holiday, emergency or sick relief, the fee payable for the issuance of the temporary Certificate of Competency shall be one-half the fee specified in subsection (1).

7. (1) The fees payable for any duplicate certificate of competency or duplicate renewal card are;

(a) Engineer's, Special Oil Well Operator's, Fireman's or Building Operator's Certificate of Competency......$ 5.00

(b) Engineer's, Special Oil Well Operator's, Fireman's or Building Operator's renewal card......$ 5.00

Chapter 8
Schools Teaching Power Engineering
LISTING OF COLLEGES, TRADE, AND VOCATIONAL SCHOOLS

Alaska

State of Alaska - Department of Education
Alaska Vocational Technical Center
P.O. Box 889
Seward, AK 99664
907/224-3322

University of Alaska Anchorage
College of Community and Continuing Education
707 A. Street, Suite 201
Anchorage AK. 99501
907/786-1525

University of Alaska Southeast
11120 Glacier Highway
Juneau, AK 99801
907/789-4458
907/465-8774

Arizona

Northland Pioneer College
1200 East Hermosa Drive
P.O. Box 610
Holbrook, AZ 86025
602/524-6111

Chapter 8 — Schools Teaching Power Engineering

California

Lassen College - Spot Program
P.O. Box 3000
Susanville, CA 96130
916/257-6181

Practical Schools
900 East Ball Rd.
Anaheim, CA 92805
714/535-6000

Illinois

Oakton Community College
1600 East Golf Rd.
Des Plaines, IL 60016-1268
708/635-1600

Indiana

Indiana Vocational Technical College
H.R.A.A.
501 S. Airport St.
Terre Haute, IN 47803
812/877/3616

Iowa

Des Moines Area Community College - Ankeny Campus
2006 Ankeny Blvd.
Ankeny, IA 50021
515/964-6241

Des Moines Public Schools - Adult Education
1800 Grand Ave.
Des Moines, IA 50307
515/281-5294

Western Iowa Tech Community College

4647 Stone Ave.
P.O. Box 265
Sioux City, IA 51102
712/274-6400

Louisiana
New Orleans Regional Vocational Technical Institute
980 Navarre Avenue
New Orleans, LA 70124
504/483-4666

Maine
Central Maine Technical College
1250 Turner Street
Auburn, ME 04210
207/784-2385

Maine Maritime Academy
Castine, ME 04420
207/326-4311

Maryland
Dundalk Community College
7200 Sollers Point Road
Baltimore, MD 21222-4692
410/282-6700

Massachusetts
The Foxboro Training Institute
The Foxboro Company
Dept. 0890/B51-2A
33 Commercial Street
Foxboro, MA 02035-2099
800/682-0022

Peterson School of Steam Engineering

Chapter 8 — Schools Teaching Power Engineering

25 Montvale Avenue
Woburn, MA 01801
617/938-5656

Michigan

Detroit Edison Training Center
38155 Cherry Hill Rd.
Westland, MI 48185
313/326-4573

Education Training Institute of Technology
16801 Wyoming
Detroit, MI 48221
313/345-1871

Ferndale/Oak Park Adult & Community Education
881 Pinecrest
Ferndale, MI 48220
313/542-2535

Ferris State College
901 S. State Street
Big Rapids, MI 49307-9989
616/592-2340

Grand Rapids Community College
143 Bostwick Northeast
Grand Rapids, MI 49503
616/771-4101

Henry Ford Community College
Searle Technical Building
13020 Osburn St.
Dearborn, MI 48126-3640
313/845-9609

International Union of Operating Engineers Local 547
24270 W. Seven Mile Road
Detroit, MI 48219

313/532-5345

Macomb County Community College, South Campus
14500 E. Twelve Mile Road
Warren, MI 48093-3896
313/445-7000

Michigan HVAC Vocational Training Center
1541 E. 8 Mile Rd.
Ferndale, MI 48220
313/544-1965

Northern Michigan University
3040 Cohodas
Marquette, MI 49855
906/227-2650

Oakland Community College
Auburn Hills Campus
2900 Featherstone
Auburn Heights, MI. 48304
313/340-6572

Ram Technical Center
8935 W. Eight Mile Road
Detroit, MI 48219
313/537-0505

Vocational Institute of Michigan
17421 Telegraph Rd., #201 North
Detroit, MI 48219
313/537-6120

Minnesota

Dunwoody Industrial Institute
818 Wayzata Blvd.
Minneapolis, MN 55403
612/374-5800

Metro Technical Colleges

Chapter 8 — Schools Teaching Power Engineering

1300 E. 145th St.
Rosemount, MN 55068
612/423-8412
612/423-2121 Metro
1-800-657-3555 Statewide

Alexandria Technical College
1601 Jefferson St.
Alexandria, MN 56308
612/762-0221
1-800-253-9884

Anoka Technical College
1355 W. Highway 10
Anoka, MN 55303
612/427-8365
1-800-247-5588

Brainerd Staples Technical College
300 Quince St.
Brainerd, MN 56401
218/828-5344

Dakota Co. Technical College
1300 E. 145th St.
Rosemount, MN 55068
612/423-8301
1-800-548-5502

Duluth Technical College
2101 Trinity Rd.
Duluth, MN 55811
218/722-2801
1-800-432-2884

Hennepin Technical College - South Campus
Eden Prairie Campus
9200 Flying Cloud Drive
Eden Prairie, MN 55347

612/944-2222
1-800-345-4655

Hennepin Technical College - North Campus
Brooklyn Park Campus
9000 Brooklyn Blvd.
Brooklyn Park, MN 55445
612/425-3800
1-800-345-4655

Hutchinson Technical College
2 Century Ave.
Hutchinson, MN 55350
612/587-3636
1-800-222-4424

Mankato Technical College
1920 Lee Blvd.
Mankato, MN 56001
507/625-3441
1-800-722-9359

Minneapolis Technical College
1415 Hennepin Ave. S.
Minneapolis, MN 55403
612/370-9400
1-800-247-0911

Minnesota Riverland Technical College
1225 SW 3rd St.
Fairbault, MN 55021
507/334-3965
1-800-642-9738

Minnesota-Riverland Technical College
1926 College View Rd. S.E.
Rochester, MN 55904
507/285-8631
1-800-247-1296

Northwest Technical College
905 Grant Ave. S.E.
Bemidji, MN 56601
218/751-4137

Northwest Technical College
900 Highway 34 E.
Detroit Lakes, MN 56501
218/847-1341
1-800-492-4836

Northwest Technical College-East Grand Forks
Highway #220 N
East Grand Forks, MN 56721
218/773-3443

Northwest Technical College - Moorhead Campus
1900 28th Ave. S.
Moorhead, MN 56560
218/236-6277

Northwest Technical College
Highway #1 E.
Thief River Falls, MN 56701
218/681-5424
1-800-222-2884

Northeast Metro Technical College
3300 Century Avenue North
White Bear Lake, MN 55110
612/770-2351
1-800-228-1978

Pine Technical College
1000 4th St.
Pine City, MN 55063
612/629-6764
Twin cities: 332-2118

Red Wing Technical College
308 Pioneer Rd.
Red Wing, MN 55066
612/388-8271
1-800-642-3344

Range Technical College-Eveleth Campus
Highway #53
Eveleth, MN 55734
218/744-3302

Range Technical College-Hibbing Campus
2900 E. Beltline
Hibbing, MN 55746
218/262-6185

Riverland Technical College
1900 8th Ave. NW
Austin, MN 55912
507/433-0600

St. Cloud Technical College
1540 Northway Dr.
St. Cloud, MN 56303
612/654-5000

St. Paul Technical College
235 Marshall Ave.
St. Paul, MN 55102
612/221-1300
1-800-227-6029

Southwestern Technical College
Highway #212 W.
Granite Falls, MN 56241
612/564-4511
1-800-247-6016

Willmar Technical College
P.O. Box 1097

Chapter 8   **Schools Teaching Power Engineering**

Willmar, MN 56201
612/235-5114

Winona Technical College
1250 Homer Rd.
Winona, MN 55987
507/454-4600
1-800-372-8164

Missouri

Hilliard N.S. A.V.T.I.
3434 Faraon
St. Joseph, MO 64506
816/232-5459

NUS Training Corporation
12855 Flushing Meadow Dr.
Suite 101
St. Louis, MO 63131
314/822-2040
800/878-2NUS

Ranken Technical College
4431 Finney Ave.
St. Louis, MO 63113
314/371-0236

New Jersey

Bergen County Technical Schools - Adult Education
200 Hackensack Avenue
Hackensack, NJ 07601
201/343-6000

Burlington County Institute of Technology
Westampton Campus
695 Woodlane Road
Mount Holly, NJ 08060
609/267-4226

Schools Teaching Power Engineering Chapter 8

Hudson County Vocational Technical School
8511 Tonnelle Avenue
North Bergen, NJ 07047
201/854-3500

Mercer County Vocational Technical Adult Evening School
1085 Old Trenton Road
Trenton, NJ 08690-1299
609/586-5146

Passaic County Vocational Technical School
45 Reinhardt Rd.
Wayne, NJ 07470
201/790-6000

Technical Institute of Camden County
343 Berlin Cross Keys Rd.
Sicklerville, NJ 08081
609/767-7002

New Mexico

San Juan College
4601 College Boulevard
Farmington, NM 87402
505/326-3311
1-800-232-6327

New York

Rochester City School District
Department of Adult and Continuing Instruction
Addison Technical
655 Colfax St.
$INew York;RochesterRochester, NY 14606
716/262-8327

Boces #1 Monroe
Adult and Community Education

Chapter 8 — Schools Teaching Power Engineering

41 O'Connor Rd.
Fairport, NY 14450
716/383-2250

Grace Community Education
P.O. Box 300
N. Greece, NY 14515
716/865-1010

Greece Central School District
Community Education
P.O. Box 300
N. Greece, NY 14515
716/621-1000

Senaca Vocational School
666 East Delavan Ave.
Buffalo, NY 14215
716/897-8170

North Dakota

Bismarck State College
1500 Edwards Avenue
Bismarck, ND 58501-1299
701/224-5400

Ohio

University of Cincinnati
Ohio College of Applied Sciences
2220 Victory Parkway
Cincinnati, OH 45206
513/556-6567
Mr. John K. Fisher
Instructor

Ohio Dept. of Education
Division of Vocational and Career Education
65 S. Front Street, Room 915

Schools Teaching Power Engineering   Chapter 8

Columbus, OH 43266-0308
614/466-2562

District 1 Toledo

Ehove Joint Vocational
316 W. Mason Rd.
Milan, OH 44846
419/627-9665

Northwest Technical College
Route 1, Box 246-A
Archbold, OH 43502
419/267-5511
Dr. J.E. Nagel
Dean

Owens Technical College
Toledo Campus
Box 10000
Toledo, OH 43699
419/666-6163
Mr. Rick Hoover
Coordinator

Owens Technical College
Findlay Campus
300 Davis st.
Findlay, OH 45840
419/423-6827

Penta County Vocational School
Adult Education
30095 Oregon Rd.
Perrysburg, OH 43551
419/666-1120
Mr. Robert N. Hanna
Instructor

Chapter 8   Schools Teaching Power Engineering

Bullard School of Steam Engineering
P.O. Box 11418
Toledo, OH 43611
419/729-4528
Mr. Richard L. Bullard
Director, Technical Instructor
Mr. Mark L. Reed
Technical Instructor
Mr. Robert J. Ferrel
Technical Instructor

Sandusky Adult Education
2130 Hayes Ave.
Sandusky, OH 44870
419/621-2731
Mr. Wes Hartsook
Director

Terra Technical College
2830 Napoleon Rd.
Fremont, OH 43420-9670
419/334-8400
Mr. Ron Rumschlag
Instructor

Washington Local Schools - Adult Education
5719 Clegg Dr.
Toledo, OH 43613
419/473-8439
Mr. Eugene Knauss
Director

District 2 Akron

Canton City Schools
Community Education
617 McKinley Ave. S.W.

Canton, OH 44707
216/438-2556
Mr. John W. Pieper
Director

Madison Adult Education
600 Esley Lane
Mansfield, OH 44905
419/589-6363

District 3 Cleveland

Cuyahoga Community College
2900 Community College Ave.
Cleveland, OH 44115-3196
216/987-4384
Dr. Fred Sutton
Coordinator, Engineering Technology
Mr. Larry Taylor
Instructor

Lorain County Community College
Continuing Education
1005 N. Abbe Rd.
Elyria, OH 44035
216/365-4191
216/233-7244 Ext. 7535 or 7543
Mr. M.L. Milovich
Instructor
Mr. G. Fote
Instructor

Max S. Hayes Vocational Evening School
4600 Detroit Ave.
Cleveland, OH 44102
216/651-2410

Parma Adult Continuing Education

Chapter 8   **Schools Teaching Power Engineering**

6726 Ridge Rd.
Parma, OH 44129
216/842-7102
Mr. Gary R. Soukup
Instructor

West Side Institute of Technology
9801 Walford Ave.
Cleveland, OH 44102
216/651-1656
Mr. Jim Ball
Instructor

District 4 Lima

Apollo Career Center
3325 Shawnee Rd.
Lima, OH 45806
419/999-3015
Mr. Jack Campbell
Instructor

Tri-Star Adult Education
585 E. Livingston st.
Celina, OH 45822
419/586-7060

Upper Valley Joint Vocational School
8811 Career Dr.
Piqua, OH 45356-9254
513/778-1980

Vantage Vocational School
Adult Education
818 N. Franklin St.
Van Wert, OH 45891
419/238-5411
Mrs. Dee Whitcraft

Schools Teaching Power Engineering — Chapter 8

Director
Mr. Thomas Brodbeck
Instructor

District 5 Columbus

Columbus State Community College
550 E. Spring St.
Columbus, OH 43215
614/227-2489
1-800-621-6407
Mr. Russell Lahut, B.S.M.E.,P.E.
Assistant Professor

Delaware J.V.S. - Adult Education
1610 SR 521
Delaware, OH 43015
614/363-1993 ext. 217
Mr. Jerome Donovan
Director
Mr. Herb Klingel
Instructor

Marion City Schools
Community Education Programs
910 East Church St.
Marion, OH 43302
614/387-3300
Mr. Edward L. Bell
Coordinator
Mr. William Gardner
Instructor

National Association of Power Engineers
Chapter #38 - Columbus
62 E. Weisheimer Rd.
Columbus, OH 43214

Chapter 8 Schools Teaching Power Engineering

614/268-7581

Ohio Point Adult Education
2280 SR 540
Bellefontaine, OH 43311
513/599-3010
Mr. Robert G. Mallow
Instructor

Southeast Career Center
3500 Alum Creek Dr.
Columbus, OH 43207
614/365-6660
Ms. Melodee Smith
Coordinator

District 6 Youngstown

Ashtabula County Joint Vocational School
1565 Route 167
Jefferson, OH 44047
216/576-6015
Mr. Frank M. Bishop
Instructor

Choffin Career Center
Adult Vocational Education
200 E. Wood St.
Youngstown, OH 44503
216/744-8700

Kent City Schools - Adult Education Program
1400 N. Mantua St.
Kent, OH 44240
216/673-9595

Mahoning County Joint Vocational
7300 North Palymra Rd.

Canfield, OH 44406
216/533-3923
Ms. Lynn Wright
Coordinator
Mr. Stephen A. Gerish, Instructor

District 7 Dayton

Clark State Community College
Continuing Education
P.O. Box 570
Springfield, OH 45501
513/325-0691

Montgomery County Joint Vocational School
Adult Education Division
6800 Hoke Rd.
Clayton, OH 45315-9740
513/837-7781
Mr. John E. Reichard
Director

District 8 Cincinnati

Miami University - Continuing Education
Middletown Campus
4200 E. University Blvd.
Middletown, OH 45042-3497
513/424-4444
Mr. John K. Fisher
Instructor

District 9 Portsmouth

Gallia-Jackson-Vinton JVSD
P.O. Box 157

Chapter 8 — Schools Teaching Power Engineering

Rio Grande, OH 45674
614/245-5334
Mr. Ponny Cisco
Adult Director

Lawrence County Joint Vocational School - Adult Ed.
Rt. 2, Box 262
Chesapeake, OH 45619
614/867-6641
Mr. Cermilus Melvin
Instructor

Ohio University
100 Factory St.
Athens, OH 45701-2979
614/593-1474

District 10 Cambridge

Belmont Technical College
120 Fox Shannon Place
St. Clairsville, OH 43950
614/695-9500
Mr. Ben Perzanowski
Training Coordinator
Mr. John R. McCabe
Instructor

Washington County Career Center
Post-Secondary Adult Vocational Education
 Route 2
Marietta, OH 45750
614/373-2766
Mr. Neal E. Eiber
Director
Mr. W.R. Waters
Instructor

Oklahoma

Metrotech - Foster Estes Campus
1900 Springlake Dr.
Oklahoma City, OK 73111
405/424-8324

Oklahoma State University - Technical Branch, Okmulgee
1801 East 4th Street
Okmulgee, OK 74447-3901
918/756-6211
1-800-722-4471

Pennsylvania

C.C.A.C. - Neville Technical Center
5800 Grand Ave.
, PA 15225
412/269-4944

International Correspondence Schools
Oak and Pawnee Streets
Scranton, PA 18515
1-800-233-0259

Rhode Island

Narragansett Electric Co. - Training Center
Providence Production
496 Eddy St.
Providence, R.I. 02903
401/455-3610 Ext. 288

Tennessee

State Technical Institute At Memphis
5983 Macon Cove
Memphis, TN 38134-7693
901/377-4111

Chapter 8   Schools Teaching Power Engineering

Texas

Houston Community College - Central Campus
1300 Holman
P.O. Box 7849
Houston, TX 77270-7849
713/868-0742

Washington

Bates Technical College
1101 South Yakima Ave.
Tacoma, WA 98405-4895
206/596-1588

Lake Washington Technical College
11605-132nd Ave NE
P.O. Box A
Kirkland, WA 98034-9987
206/828-5627

Renton Vocational Technical Institute
3000 N.E. 4Th street
Renton, WA 98055
206/235-2352

Spokane Community College - North
1810 Greene Street
Spokane, WA 99207-5399
509/533-7178

Wisconsin

Gateway Technical College - Kenosha Campus
3520 30th Ave.
Kenosha, WI 53144
414/656-6943

Gateway Technical College - Racine Campus
1001 South Main

Racine, WI 53403
414/631-7300

Lakeshore Technical College
1290 North Ave.
Cleveland, WI 53015-1414
414/458-4183

Milwaukee Area Technical College
Milwaukee Campus
700 W. State Street
Milwaukee, WI 53233-1443
414/278-6600

Milwaukee Area Technical College
North Campus Center
5555 West Highland Rd.
Mequon, WI 53092
414/242-6500

Milwaukee Area Technical College
South Campus Center
6665 South Howell Avenue
Oak Creek, WI 53154
414/762-2500

Milwaukee Area Technical College
West Campus Center
1216 South 71st Street
West Allis, WI 53214
414/476-3040

Northeast Wisconsin Technical College
Green Bay Campus
2740 W. Mason Street
P.O. Box 19042
Green Bay, WI 54307-9042
414/498-5600

Northeast Wisconsin Technical College

Marinette Campus
1601 University Drive
Marinette, WI 54143
715/735-9361

Northeast Wisconsin Technical College
Sturgeon Bay Campus
229 North 14th Avenue
Sturgeon Bay, WI 54235
715/743-2207
1-800-272-2740

University of Wisconsin-Madison
The College of Engineering
Department of Engineering Professional Development
432 North Lake Street
Madison, WI 53706
608/262-2061

Index

A

Alabama		72
Alaska		72 - 77, 788
	Anchorage	72, 788
	Fairbanks	72
	Juneau	788
	Kenai	72
	Seward	788
	Sitka	72
Alberta		520 - 564
	Calgary	520
	Edmonton	520
	Fort McMurrary	520
	Grande Prairie	520
	Lethbridge	520
	Medicine Hat	520
	Red Deer	520
	St. Paul	520
Arizona		77
	Holbrook	788
Arkansas		78 - 80
	Little Rock	40, 78

B

Boiler Examinations		
	how conducted	21
British Columbia		565 - 588
	Burnaby	71
	Fort St. John	565
	Kelowna	565
	Nanaimo	565

	Nelson	565
	Prince George	565
	Terrace	565
	Vancouver	565
	Victoria	565

C

California		47, 81, 789
	Anaheim	50, 789
	Los Angeles	40, 71, 81 - 88
	Newport Beach	50
	Oakland	40
	Paramount	51
	San Francisco	40, 70
	Stockton	40
	Susanville	789
	Vallejo	50
Canada		520
Colorado		89
	Colorado Springs	40
	Denver	40, 89 - 111
	Pueblo	108 - 109
Connecticut		41, 47, 112
	Meridan	17
	Bridgeport	112 - 120
	East Hampden	51
	Marlborough	51
	New Haven	17, 121 - 122
	Stamford	17

D

Delaware		127
	Wilmington	127 - 131

District of Columbia 70
    Washington D.C. 17

F

Florida 132
    Hillsboro County 133
    Jacksonville 41
    Miami 17
    Tampa 132 - 136

G

Georgia 47, 136
    Atlanta 17, 41, 53
    Covington 53
    Evans 53
    Macon 41
    Roswell 52
    Social Circle 52

H

Hawaii 136
History
    IUOE 68
    of N.A.P.E. 15
    of Power Engineering 5

I

Idaho 136
Illinois 47, 137, 789
    Argonne 53
    Chicago 17, 54, 71, 137 - 138
    Decatur 41
    Des Plaines 789

	Elgin	17, 139 - 144
	Evanston	145
	Great Lakes Naval Training Center 17	
	Gurnee	53
	Joliet	17
	Orland Park	53
	Peoria	146 - 149
Indiana		150, 789
	Evansville	17
	Fort Wayne	18
	Indianapolis	17, 41
	Kokomo	17
	Muncie	17
	Terre Haute	17, 150 -151, 789
International Union of Operating Engineers 68		
Iowa		47, 152, 789
	Ankeny	789
	Des Moines	18, 41, 152 - 161, 789
	Grimes	54
	Sioux City	162 - 172, 790
	Waukee	54

K

Kansas		47, 172
Kentucky		172

L

Louisiana		173, 790
	New Orleans	173 - 183, 790

M

Maine		42, 184 - 189, 790
	Auburn	790

	Augusta	184
	Castine	790
Manitoba		589 - 607
	Brandon	589
	Flinflon	589
	Tatpas	589
	Thompson	589
	Winnipeg	589
Maryland		41, 47, 190 - 194, 790
	Baltimore	42, 790
	Beltsville	55
	Bowie	55
	Glen Burnie	55
	Greenbelt	51
	Hyattsville	51
	Pasedena	18
	Pax River	18
	Prince Georges	18
	Towson	190
	Wheaton	18
Massachusetts		42, 47, 195 - 206, 790
	Boston	195
	Chicopee	56
	Fall River	18
	Foxboro	790
	Lowell	18
	Ludlow	55
	No. Dartmouth	56
	Springfield	18
	Whitinsville	55
	Woburn	791
	Worcester	18
Michigan		47, 207, 235, 791

	Ann Arbor	42
	Auburn Heights	235, 792
	Battle Creek	57
	Big Rapids	791
	Caledonia	56
	Cloverdale	56
	Dearborn	207 - 216, 234, 791
	Detroit	18, 21, 42, 71, 217 - 235, 791 - 792
	Ferndale	234, 791 - 792
	Flint	18
	Flushing	56
	Grand Rapids	236 - 238, 791
	Kalamazoo	18, 57
	Marquette	792
	Plainwell	57
	Saginaw	239 - 242
	Warren	234, 792
	Westland	235, 791
Minnesota		47, 243 - 251, 792
	Alexandria	793
	Anoka	793
	Apple Valley	58
	Austin	796
	Bemidji	795
	Brainerd	793
	Brooklyn Park	794
	Detroit Lakes	795
	Duluth	793
	East Grand Forks	795
	Eden Prarie	793
	Eveleth	796
	Fairbault	794
	Granite Falls	796

	Hibbing	796
	Hutchinson	794
	Mankato	794
	Minneapolis	18, 42, 57, 792, 794
	Moorhead	795
	Pine City	795
	Queen City (Rochester) 18	
	Red Wing	796
	Rochester	794
	Rosemount	793
	St. Cloud	18, 796
	St. Paul	70, 243, 796
	Thief River Falls 795	
	White Bear Lake 795	
	Willmar	797
	Winoa	797
	Winoma	18
Mississippi		252
Missouri		47, 252, 797
	Columbia	19
	Kansas City	19, 42, 58, 69, 252 - 266
	Kearney	58
	Raytown	58
	St. Joseph	267 - 276, 797
	St. Louis	42, 69, 277 - 288, 797
	St.Louis	19
Montana		289 - 300
	Billings	289
	Helena	289
	Missoula	289

N

N.I.U.L.P.E.		46
National Association of Power Engineers		17
Nebraska		47, 301
	Lincoln	19, 58
	Norfolk	19
	Omaha	19, 43, 301 - 313
	Syracuse	58
Nevada		47, 313
	Las Vegas	19, 43, 59, 71
New Brunswick		608 - 618
	Fredricton	608
New Hampshire		47, 313
	Keene	19, 60
	Manchester	19
	N. Swanzey	60
	W. Chesterfield	55
New Jersey		314 - 327, 797
	Bergen	19
	Central	19
	Emerson	60
	Hackensack	797
	Manahawkin	61
	Middlesex	43
	Monmouth	43
	Mount Holly	797
	Newark	43
	North Bergen	798
	Sicklerville	798
	Trenton	43, 314, 798
	W.Caldwell	70
	Wayne	798
New Mexico		328, 798
	Farmington	798

New York		47, 328, 344, 798
	Albany	43
	Bronx	60
	Buffalo	43, 328 - 341, 799
	Fairport	799
	Flower City (Rochester)	19
	Limerick	60
	Mt. Vernon	342
	N. Greece	799
	N.Greece	799
	Nassau County	19
	New York City	19, 43, 70, 343 - 344
	Niagara Falls	19, 345 - 346
	Phoenix	19
	Richmond Hill	70
	Rochester	347 - 360
	Southern Tier	19
	Westchester	19
	White Plains	361 - 362
Newfoundland		619 - 646
	Corner Brook	619
	Grand Falls	619
	Labrador City	619
	St. Johns	619
North Carolina		47, 362
	Charlotte	61
	Raleigh	43
North Dakota		43, 47, 362, 799
	Bismarck	799
Northwest Territories		
	Yellowknife	647
Nova Scotia		671 - 690
	Halifax	671

O

Ohio 47, 363 - 375, 799
- Akron 363
- Archbold 800
- Athens 807
- Beaver Creek 62
- Bellefontaine 805
- Cambridge 363
- Canfield 806
- Canton 802
- Celina 803
- Chesapeake 807
- Cincinnati 20, 43, 61 - 62, 363, 799, 806
- Clayton 806
- Cleveland 20, 43, 363, 802 - 803
- Columbus 20, 363, 800, 804 - 805
- Delaware 804
- Elyria 802
- Fairfield 62
- Findlay 800
- Fremont 801
- Hamilton 62
- Jefferson 805
- Kent 805
- Lima 44, 803
- Mansfield 802
- Marion 804
- Marrietta 807
- Middletown 806
- Milan 800
- Napoleon 63
- North Ridgeville 61

	Oregon	800
	Parma	803
	Perrysburg	800
	Piqua	803
	Portsmouth	44
	Rio Grande	807
	Sandusky	44, 801
	Springfield	806
	St. Clairsville	807
	Sylvania	63
	Toledo	44, 61 - 62, 363, 801
	Van Wert	803
	Youngstown	363, 805
Oklahoma		376
	Oklahoma City	376 - 380, 808
	Okmulgee	808
	Tulsa	381 - 390
Ontario		691 - 717
	Toronto	691
Oregon		391

P

Pay Scales		5, 40 - 45
Pennsylvania		47, 391, 808
	Erie	391 - 396
	Indiana	63
	Philadelphia	44, 397 - 401
	Pittsburgh	20, 44, 63, 402 - 409
	Scranton	44, 808
	Verona	63
	York	44
Power Engineers		
	Other names for	32

Prince Edward Island 718 - 743
    Charlottetown 718
    Summerside 718

Q

Quebec
    Montreal 744

R

Resumes 34
Rhode Island 47, 410, 808
    Coventry 64
    North Kingston 64
    Providence 20, 410 - 415, 808
    Warren 63

S

Saskatchewan 745 - 755
    Regina 745
    Saskatoon 745
South Carolina 47, 415
    Darlington 64
    Florence 20
    Georgetown 64
    Graniteville 52
    Upstate 20
South Dakota 415
Study Guides 86, 225, 227, 229, 283 - 285, 287
    357 - 359, 427, 430, 453,
    455 - 468, 470 - 475

Summary of Qualifications

Alaska	72
Arkansas	78
Bridgeport, Connecticut	112
Buffalo, New York	328
Chicago, Illinois	137
Dearborn, Michigan	207
Denver, Colorado	89
Des Moines, Iowa	152
Detroit, Michigan	217
Elgin, Illinois	139
Erie, Pennsylvania	391
Grand Rapids, Michigan	236
Hillsboro County, Florida	132
Houston, Texas	417
Kansas City, Missouri	252
Kenosha, Wisconsin	496
Los Angeles, California	81
Maine	184
Maryland	190
Massachusetts	195
Milwaukee, Wisconsin	503
Minnesota	243
Montana	289
Mt. Vernon, New York	342
New Haven, Connecticut	121
New Jersey	314
New Orleans, Louisiana	173
New York City, New York	343
Niagara Falls, New York	345
Northwest Territories	647
Ohio	363
Oklahoma City, Oklahoma	376

Omaha, Nebraska	301
Peoria, Illinois	146
Philadelphia, Pennsylvania	397
Pittsburgh, Pennsylvania	402
Providence, Rhode Island	410
Province of Alberta	520
Province of British Columbia	565
Province of Manitoba	589
Province of New Brunswick	608
Province of Newfoundland	619
Province of Nova Scotia	671
Province of Ontario	691
Province of Prince Edward Island	718
Province of Quebec	744
Province of Saskatchewan	745
Pueblo, Colorado	108
Racine, Wisconsin	512
Rochester, New York	347
Saginaw, Michigan	239
Salt Lake City, Utah	431
Seattle, Washington	439
Shelby County, Tennessee	416
Sioux City, Iowa	162
Spokane, Washington	476
St. Joseph, Missouri	267
St. Louis, Missouri	277
Tacoma, Washington	480
Tampa, Florida	132
Terre Haute, Indiana	150
Tulsa, Oklahoma	381
White Plains, New York	361
Wilmington, Delaware	127
Yukon Territories	756

T

Tennessee		416, 808
	Chattanooga	44
	Memphis	44, 416, 808
	Oak Ridge	71
	Shelby County	416
Texas		44, 417, 809
	Beaumont	44
	Dallas	44
	El Paso	44
	Fort Worth	44
	Houston	45, 417 - 430, 809
	San Antonio	45
	Wichita Falls	45

The National Institute for the Uniform Licensing of Power Engineers 46

U

Utah		431
	Salt Lake City	431 - 438

V

Vermont		45, 438
	Brattleboro	59
	Milton	56
	Richmond	56, 59
Virginia		41, 45, 47, 438
	Annandale	52
	Arlington	52
	Berryville	64
	Central	20
	Dale City	51

Fairfax 52
Ft. Belvoir 66
Manassas 64
Norfolk 45
North Virginia 20
Richmond 20, 45

W

Wage Rates 40 - 45
Washington 41, 47, 439, 809
    Cheney 65
    Kirkland 809
    Renton 809
    Richland 71
    Seattle 45, 439 - 475
    Spokane 65, 476 - 479, 809
    Tacoma 480 - 495, 809
West Virginia 495
    Moundsville 62
Wisconsin 47, 496, 809
    Burlington 512
    Chippewa Falls 66
    Cleveland 810
    Green Bay 20, 66, 810
    Kenosha 496 - 502, 512, 809
    Madison 20, 45, 811
    Manitowac 20
    Marinette 811
    Menomonie 66
    Mequon 810
    Milwaukee 45, 503 - 511, 810
    Neenah 65
    Oak Creek 810

	Racine	65, 512 - 519, 810
	Sheyboygan	20
	Stoughton	65
	Sturgeon Bay	811
	Union Grove	20
	Verona	66
	Walworth	512
	West Allis	66, 810
Wyoming		519

Y

Yukon Territories		756 - 786
	Whitehorse	756

Practical Boiler Operation
The Guide to License Examinations

This book brings together all the information needed to pass your High and Low pressure Boiler examinations. Explained in a straight forward manner without the outdated English common to Engineering texts. Richly illustrated with a total of over 300 photographs and drawings. Takes the reader from the basics of steam, through all the components of a steam plant, and then puts it all together with a photographic tour of a typical High Pressure Steam Plant. Explains the operation of boilers found in the Low and High Pressure Boiler Room. You Get:

- Fundamentals of Steam
- History of Boilers
- Fire Tube Boilers
- Water Tube Boilers
- Safety Valves
- Feedwater Regulators
- All types of valves used in the Boiler Room
- Steam Traps
- Simplex Pumps
- Duplex pumps
- Valve setting for pumps
- Fuels
- Boiling out a new Boiler
- Preparing a Boiler for Inspection
- Gauge glass replacement
- Emergency Procedures
- Oil Burners and fuel oil piping
- Natural Gas burners and piping
- Automatic Combustion Controls
- Underfeed Stokers
- Traveling and Chain Grate Stokers
- Inclined Overfeed Stokers
- Spreader Stokers
- Crushers
- ASME Codes
- Steam piping and valves required on boilers connected together in the Plant
- Refractory and firebrick

And of course you get hundreds of typical questions and answers from License examinations to help pass the Exams.

Available December 1993

From *PowerPlant Press*
P.O. Box 431219
Pontiac, MI 48343

$5.00 OFF COUPON

Practical Boiler Operation: The Guide to License Examinations

The normal price of 29.95 is reduced to only $24.95 when this coupon accompanies your check or money order.

Name _____

Address _____

City _____ State _____ Zip _____

Add $3.00 for shipping and handling